Bacteria
IN BIOLOGY, BIOTECHNOLOGY AND MEDICINE

FOURTH EDITION

Paul Singleton

JOHN WILEY & SONS

Chichester · New York · Weinheim · Brisbane · Singapore · Toronto

1st edition 1981 (reprinted 1985, 1987, 1989, 1991)
2nd edition 1992
3rd edition 1995 (reprinted 1995, 1996)
4th edition 1997

Japanese edition 1982
French editions 1984, 1994
German edition 1995

Other Wiley Editorial Offices

John Wiley & Sons, Inc., 605 Third Avenue,
New York, NY 10158–0012, USA

Weinheim. Brisbane. Singapore. Toronto

Library of Congress Cataloging-in-Publication Data

Singleton, Paul.
    Bacteria in biology, biotechnology, and medicine / Paul Singleton.
    – 4th ed.
        p. cm.
    Includes bibliographical references and index.
    ISBN 0–471–97534–6 (hbk) ISBN 0–471–97468–4 (pbk)
    1. Bacteria. I. Title.
QR75.S55 1997
589.9–dc20   95–7379
                                                        CIP

British Library Cataloguing in Publication Data

A catalogue record for this book is available from the British Library

ISBN 0 471 975346 (hardback) 0 471 974684 (paperback)

Typeset in 10/12pt Palatino by Vision Typesetting, Manchester, UK
Printed and bound in Great Britain by Biddles Ltd, Guildford and King's Lynn

This book is printed on acid-free paper responsibly manufactured from sustainable forestation,

Contents

Preface

This edition expands and updates topics from the 3rd edition, and covers many new areas in bacterial physiology, recombinant DNA technology/biotechnology and medicine. The style and format of the previous edition have been retained; the text, which assumes no prior knowledge of the bacteria, presents a sequential treatment of the subject to a level which gives the reader access to new ideas and developments in the current literature.

References to papers and reviews are included to help the reader acquire further detail and to extend his/her knowledge in particular direction(s). The references are given in square brackets, and a key to the abbreviated names of journals is given inside the front and back covers of the book. In this edition the literature base has been broadened by a further sixteen journals.

Several firms have contributed useful information/material: Perstorp Analytical LUMAC B.V. (Landgraaf, The Netherlands); QIAGEN GmbH (Hilden, Germany); Mast Diagnostics (Merseyside, UK); Stratagene (California, USA); Pharmacia (Uppsala, Sweden); Oxoid (Basingstoke, UK); and Molecular Probes Inc. (Oregon, USA). I am also indebted to the manufacturers of Scottish Pride dairy products (Glasgow, Scotland) for details of their current procedures for UHT milk processing and quality control.

I would like to acknowledge the invaluable help of the Medical Library, University of Bristol, and of the British Library (Science Reference Department).

Publication has been facilitated by the professional skills of Sally Betteridge and Lisa Tickner (Life and Medical Sciences), Lindsay Jackson (Production Editor) and other members of staff at the Chichester office of John Wiley & Sons, to whom I am most grateful.

Paul Singleton
Clannaborough Barton, Devon
January, 1997

1 The bacteria: an introduction

1.1 WHAT ARE BACTERIA?

Bacteria are minute organisms which occur almost everywhere. They sometimes reveal their presence – wounds 'go septic', milk 'sours', meat 'putrefies' – but usually we are unaware of them because their activities are less obvious and because they are so small. Indeed, the very existence of bacteria was unknown until the development of the microscope in the 17th century.

In most cases a bacterium is a single, autonomous cell. However, the bacterial cell has a *prokaryotic* organization and differs markedly from the *eukaryotic* cells of animals and plants; some of the differences are listed in Table 1.1 (prokaryotic features referred to in the table are described in later chapters). Prokaryotes and eukaryotes are believed to have 'split' from a common ancestor about 2–3.5 billion years ago [see e.g. Doolittle *et al.* (1996) Science *271* 470–477].

Bacteria are included in the category 'microorganisms'. The microorganisms include several distinct types of organism – algae, fungi, lichens, protozoa, viruses and subviral agents – as well as bacteria; hence, though all bacteria are microorganisms, not all microorganisms are bacteria.

1.1.1 A note on the use of the term 'bacteria'

The term 'bacteria' is still widely used in a general sense to mean 'the prokaryotic organisms' – and in this sense it includes *all* the prokaryotes. However, molecular studies have indicated that the prokaryotes can be divided into two fundamentally different groups (*domains*): the domain Bacteria (note 'B', not 'b') and the domain Archaea. [For a discussion see: Olsen, Woese & Overbeek (1994) JB *176* 1–6.] Because of this, the term 'bacteria' will often be used in books and journals to refer *specifically* to some or all members of the domain Bacteria, and the reader should therefore try to determine the sense in which the term is being used in any given context.

In this book, 'bacteria' refers to members of the domain Bacteria; the book deals almost exclusively with the organisms in this group.

The domain Archaea (see Fig. 1.1) contains a relatively small group of organisms which were previously classified as members of the kingdom Archaebacteria. Members of this group include species which tend to live in 'extreme' habitats characterized by high temperatures, high salinity etc.; information about these organisms can be found in reports from the First

Table 1.1   Eukaryotic and prokaryotic cells: some major differences

Eukaryotic cells	Prokaryotic cells
The chromosomes are enclosed within a sac-like, double-layered 'nuclear' membrane	There is no nuclear membrane: chromosomes are in direct contact with the cytoplasm
Chromosome structure is complex; the DNA is usually associated with proteins called histones	Chromosome structure is relatively simple
Cell division involves mitosis or meiosis	Mitosis and meiosis are not involved
The cell wall, when present, includes structural compounds such as cellulose or chitin, but never peptidoglycan	The cell wall, when present, usually contains peptidoglycan (though a peptidoglycan-like compound occurs in some members of the Archaea) but never cellulose or chitin structural components
Mitochondria are generally present; chloroplasts occur in photosynthetic cells	Mitochondria and chloroplasts are never present
Cells contain ribosomes of two types: a larger type in the cytoplasm and a smaller type in mitchondria and chloroplasts	Cells contain ribosomes of only one size
Flagella, when present, have a complex structure	Flagella, when present, have a relatively simple structure

International Congress on Extremophiles [see: FEMSMR (1996) *18* 89–288]. The Archaea also includes all the known methane-producing organisms. Aspects of archaeal biology are mentioned in various parts of the book – mainly in order to give some idea of the way in which these organisms differ from the 'true bacteria'.

1.2   WHY STUDY BACTERIA?

One important reason is the conquest of disease. Bacteria cause some major diseases as well as a number of minor ones; the prevention and control of these diseases depend largely on the efforts of medical, veterinary and agricultural bacteriologists. Pathogenic (i.e. disease-causing) bacteria are considered in Chapter 11.

Important though they are, the pathogenic bacteria are only a small proportion of the bacteria as a whole. Most bacteria do little or no harm, and many are positively useful to man. Some, for example, produce antibiotics which have revolutionized the treatment of disease, while others provide enzymes for 'biological' washing powders. Some are used as 'microbial

Prokaryotes

Bacteria	Archaea
The larger group	The smaller group
Includes all the medically important prokaryotes	No species is known to be medically important
Low proportion of species live in extreme environments	High proportion live in extreme environments
Some species can carry out photosynthesis	No species is photosynthetic (but see section 5.2.2)
No species forms methane	All methane-forming species are in this group

Fig. 1.1 The domains Bacteria and Archaea: some general features. Members of the Archaea differ from bacteria e.g. in their cell membrane (section 2.2.8.2), cell wall (section 2.2.9.3), flagella (section 2.2.14.1) and nucleic acids (section 16.2.2.2).

insecticides' – protecting crops from certain insect pests. Bacteria are even used to make biodegradable plastics ('Biopol' – section 13.7) and to leach out metals from certain low-grade or refractory ores (biomining – section 13.2).

Perhaps surprisingly, bacteria contribute a lot to the food industry (Chapter 12). We usually think of bacteria as a nuisance where food is concerned, causing 'spoilage' and 'food poisoning', but certain types of bacteria are actually employed in the production of food. For example, in the manufacture of butter, cheese and yoghurt, certain bacteria are used to convert 'milk sugar' (lactose) to lactic acid; the bacteria also form compounds which give these products their characteristic flavours. Xanthan gum, formed by certain strains of bacteria, is used as a gelling agent and thickener in the food industry. Vinegar is produced from alcohol (ethanol) by bacterial action, and bacteria also play a part in the manufacture of cocoa and coffee.

Some of these activities can be understood by studying the chemical reactions (metabolism) of bacterial cells (Chapters 5 and 6). Chapters 12 and 13 look at the activities of some 'useful' bacteria.

Bacteria (and their components) also play central roles in biotechnology – with applications in medicine, industry and agriculture. For example, one use of recombinant DNA technology is the production of agents such as streptokinase (used e.g. for the treatment of thrombosis) (section 8.5.10).

Not least, bacteria have essential roles in the natural cycles of matter (Chapter 10) – on which, ultimately, all life depends. In the soil, bacteria affect fertility and structure – agricultural potential – so that a better understanding of bacterial activity will permit better management of land and crops; in the future this will be vital to the survival of our ever-expanding population.

From this brief summary it should be clear that the more we learn about bacteria the more effectively we can minimize their harmful potential and exploit their useful activities.

1.3   CLASSIFYING AND NAMING BACTERIA

How is one type of bacterium distinguished from another, and how are bacteria classified? Bacteria may differ, for example, in their shape, size and structure, in their chemical activities, in the types of nutrients they need, in the form of energy they use, in the physical conditions under which they can grow, and in their reactions to certain dyes.

Features such as those listed above are widely used for classifying (and identifying) bacteria – such features being easily checked even in a modestly equipped laboratory. As in other areas of biology, bacteria are classified in a hierarchy of categories – e.g. families, genera, species; *species* which are sufficiently alike are placed in the same *genus*, and genera with a certain level of similarity are grouped into a *family*. A species may be subdivided into two or more *strains* – organisms which conform to the same species definition but which have minor differences. In general, members of (say) a bacterial family would have similar structure, would use the same form of energy, and would typically react in a similar way to certain dyes; the species in such a family may be grouped into genera on the basis of differences in chemical activities, nutrient requirements, conditions for growth and (to some extent) shape and size.

Although useful for everyday purposes, the kind of classification described above does not necessarily indicate *evolutionary* relationships among the bacteria. In the last decade or so, studies have been carried out on what are believed to be more fundamental characteristics of bacteria; such characteristics are believed to reveal major evolutionary pathways. This aspect of classification is referred to in later chapters.

As in the case of animals and plants, each species of bacterium is given a name in the form of a Latin binomial. A binomial consists of (i) the name of the genus to which a given organism belongs, followed by (ii) the 'specific epithet' which acts as a label for one particular species; for example, *Escherichia coli* gives the name of the genus (*Escherichia*) and the specific epithet (*coli*). By convention, a Latin binomial is printed in italics, or is underlined once if handwritten; the name of a genus always begins with a capital letter, but a specific epithet always begins with a lower-case letter.

The name of a species may be abbreviated by abbreviating the name of the genus – e.g. *Escherichia coli* may be written '*E. coli*'; however, this should be done only when the full name of the genus has been mentioned earlier so that the meaning of the abbreviation is clear. (Being a very common experimental organism, *E. coli* is mentioned frequently in this book; the name – and spelling – of the genus should be noted.)

The names of families and orders of bacteria are not printed in italics, but each has a capital initial letter. These names also have standardized endings, the name of a family always ending in '-aceae' (e.g Enterobacteriaceae) and the name of an order always ending in '-ales' (e.g. Actinomycetales).

The naming of bacteria is formally governed by various rules made by the International Committee on Systematic Bacteriology. The advantages of an internationally standardized system of naming are obvious, but the rules are not always adhered to in the literature – owing to a lack of awareness (of the rules, or of revised names) or to disagreement with published opinions.

2 The bacterial cell

2.1 SHAPES, SIZES AND ARRANGEMENTS OF BACTERIAL CELLS

2.1.1 Shape

Bacterial cells vary widely in shape, according to species. Rounded or 'spherical' cells – of any species – are called *cocci* (singular: *coccus*). Elongated, rod-shaped cells of any species are called *bacilli* (singular: *bacillus*), or simply *rods*. Cocci are not necessarily exactly spherical, and not all bacilli have exactly the same shape; for example, some cocci are more or less kidney-shaped, and some bacilli taper at each end (*fusiform* bacilli) or are curved (*vibrios*). Ovoid cells, intermediate in shape between cocci and bacilli, are called *coccobacilli* (singular: *coccobacillus*). There are also two types of spiral cell: those which are more or less rigid (*spirilla*, singular: *spirillum*; Plate 2.1, *bottom*, *left*),) and those which are flexible (*spirochaetes*, singular: *spirochaete*); although it is generally assumed that all spirochaetes are spiral (helical), it was recently reported that the spirochaete *Borrelia burgdorferi* (causal agent of Lyme disease) swims as a *flattened* waveform rather than as a circular helix (see Spirochaetales in the Appendix). Then there are the so-called 'square bacteria' (flat, square bacteria) and 'box-like bacteria' (variously-shaped, angular bacteria). Finally, there are the fungus-like *actinomycetes*: bacteria most of which grow in the form of fine threads called *hyphae* (singular: *hypha*); a group or mass of hyphae is referred to as *mycelium*. Bacteria of various shapes are shown in Fig. 2.1 and in Plate 2.1.

As seen in the caption of Fig. 2.1, some of the names used to describe shapes of bacterial cells are also used as names for bacterial genera. Care should be taken, for example, not to confuse 'bacillus' with '*Bacillus*' (note that the latter has a capital 'B' and is printed in italics); some bacilli belong to the genus *Bacillus*, others do not!

Although the cells of a given species of bacterium are usually more or less uniform in shape, in some species the shape of the cell typically varies from one cell to another – sometimes quite markedly; this phenomenon is called *pleomorphism*.

L-form cells are irregularly-shaped or spherical cells which are produced spontaneously by some species of bacteria, and which can be induced in other species e.g. by temperature shock and by other kinds of physicochemical stimulus; these cells were named after the Lister Institute of Preventive Medicine (London).

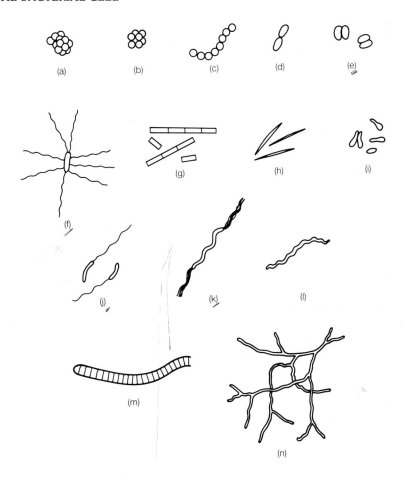

Fig. 2.1 Shapes and arrangements of some bacteria with named examples (not drawn to scale). (a) Uniform spherical cells (cocci) in irregular clusters: *Staphylococcus aureus*. (b) Cocci, in regular packets of eight cells: *Sarcina ventriculi*. (c) Cocci in chains: *Streptococcus pyogenes*. (d) Slightly elongated cocci in pairs (diplococci): *Streptococcus pneumoniae*. (e) Pairs of cocci (diplococci) in which each cell is flattened or slightly concave on the side next to its neighbour: *Neisseria gonorrhoeae*. (f) Rod-shaped cell (bacillus): *Escherichia coli*. The lines arising from the bacillus represent fine, hair-like appendages called *flagella* which are described later in the chapter. (g) Blunt-ended bacilli, singly and in chains: *Bacillus anthracis*. (h) Bacilli with tapered ends (fusiform bacilli): *Fusobacterium nucleatum*. (i) Irregularly-shaped (pleomorphic) cells: *Corynebacterium diphtheriae*. (j) Curved bacilli (vibrios), each with one flagellum: *Vibrio cholerae*. (k) A rigid spiral cell (spirillum) with a tuft of flagella at each end: *Spirillum volutans*. (l) A flexible spiral cell (spirochaete): *Treponema pallidum*. (m) One end of a filament (*trichome*) of a cyanobacterium: *Oscillatoria limnetica*. Trichomes are discussed in section 2.3. (n) Thin, branched filaments (hyphae): *Streptomyces albus*.

2.1.2   Size

Bacterial cells are usually measured in *micrometres*, μm (formerly called microns, μ); 1 μm = 0.001 mm. The smallest bacteria are about 0.2 μm (e.g. cells of *Chlamydia*); at the other end of the scale, some of the cells of *Spirochaeta* are about 250 μm in length, but the largest bacterium known − >600 μm − is *Epulopiscium fishelsoni*, which inhabits the gut of the surgeon fish (*Acanthurus nigrofuscus*) [Angert, Clements & Pace (1993) Nature 362 239–241]. However, these are extreme cases; in most species the maximum dimension of a cell lies within the range 1–10 μm. Note that the smallest bacteria are of the same order of size as the limit of resolution of a good light microscope, which is about 0.2 μm.

2.1.3   Arrangements of bacterial cells

Under the microscope, bacteria of a given species may be seen as separate (individual) cells or as cells in characteristic groupings. According to species, cells may occur in pairs, in irregular clusters, in chains or filaments, in regular *packets* of four, eight or more cells, or in *palisade* form − a number of elongated cells side-by-side in a row with adjacent cells touching. The species *Pelodictyon clathratiforme* is unusual in that it forms three-dimensional networks of cells. In a number of species the cells form *trichomes* (section 2.3.1). Some arrangements of cells are shown in Fig. 2.1. These different arrangements of cells do not result from the aggregation of previously single cells; they occur because (i) cells of the different species divide (reproduce) in different ways, and (ii) two or more cells may remain attached after the process of cell division.

In nature, certain bacteria occur in stable, mixed-species groups of cells called *consortia* [see e.g. Paerl & Pinckney (1996) Microbial Ecology 31 225–247].

2.2   THE BACTERIAL CELL: A CLOSER LOOK

Are all bacteria basically similar in structure? No: cells of different species may differ greatly both in their fine structure (ultrastructure) and chemical composition; for this reason there is no 'typical' bacterium. Figure 2.2 shows a 'generalized' bacterium in a very diagrammatic way; it is important to note that not all bacteria have all the features shown in the diagram, and that some bacteria have structures not shown in the diagram.

Figure 2.2 shows the cell's *chromosome* ('genetic blueprint') − a typically loop-like structure of DNA (Chapter 7) which is extensively folded to form a body called the *nucleoid*. Bathing the nucleoid is a complex fluid, the *cytoplasm*, which fills the interior of the cell. The cytoplasm contains *ribosomes*: minute bodies involved in the synthesis of proteins; sometimes there are also *storage granules* of reserve nutrients etc. The nucleoid, cytoplasm, ribosomes and storage granules all occur within the space bounded by a membranous

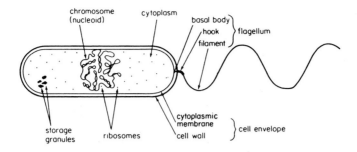

Fig. 2.2   Cross-section of a generalized bacterium (diagrammatic).

sac, the *cytoplasmic membrane* (also called cell membrane or plasma membrane). The outermost layer in Fig. 2.2 is the tough (mechanically strong) *cell wall*. The cytoplasmic membrane and cell wall are referred to, jointly, as the *cell envelope*. Between the cytoplasmic membrane and the cell wall is the so-called *periplasmic region*. The *flagellum* is a thin, hair-like proteinaceous appendage (involved in cell motility – i.e. movement) which is attached by specialized structures to the cell envelope; some cells have more than one flagellum, and some have many flagella. These and other features of the bacterial cell are considered in the following sections.

   The bacterial cell is sometimes said to be 'simple' because it lacks the specialized compartments (nucleus, mitochondrion etc.) found in eukaryotic cells. However, at the *molecular* level, bacteria use subtle and sophisticated strategies which involve a high degree of functional and structural organization. [Principles of functional and structural organization in the bacterial cell: Mayer (1993) FEMSMR *104* 327–346.] One intriguing aspect of bacterial organization is that particular structures (e.g. flagella) or enzymes etc. are often located at one pole (one 'end') of the cell; examples of this structural and/or functional specialization have been found in a diverse range of species, suggesting that this phenomenon of 'polarity' may be more general than previously supposed [Maddock, Alley & Shapiro (1993) JB *175* 7125–7129].

2.2.1   The nucleoid

In most bacteria the chromosome is essentially a loop of *deoxyribonucleic acid* (DNA) (described in Chapter 7); in some species, however, the chromosome is a linear molecule of DNA. The DNA is associated with certain types of protein, some of which are believed to be involved in the folding of the chromosome. Within the cell, the extensively folded DNA forms a dense body, the nucleoid, which is associated with the cytoplasmic membrane. In the bacterium *Escherichia coli* the chromosome is about 1.3 mm long; all of this DNA fits into a cell of only a few micrometres in length – with room to spare!

A bacterium may contain more than one copy of the chromosome, according to species and conditions of growth (Chapter 3); in *E. coli*, for example, cells undergoing rapid growth have more chromosomes per cell than do those undergoing slow growth. In some species each cell normally has many copies of the chromosome.

[The bacterial nucleoid revisited (review): Robinow & Kellenberger (1994) MR *58* 211–232.]

2.2.2 The cytoplasm

The cytoplasm is an aqueous (water-based) fluid containing ribosomes, nutrients, ions, enzymes, waste products and various molecules involved in synthesis, cell maintenance and energy metabolism; storage granules may be present under certain conditions. Bacterial cytoplasm lacks the equivalent of the endoplasmic reticulum in eukaryotic cells.

Exceptionally (in a few species), the cytoplasm may contain some unique and interesting items. In *Bacillus thuringiensis*, for example, it sometimes contains crystals that are poisonous for various insects; this species is used in agriculture for biological control (Chapter 13). Other bacteria, themselves parasites of protozoa, contain curious rolled-up ribbons of protein called 'R bodies'. In some aquatic bacteria, tiny particles of magnetite (Fe_3O_4) called

---

Plate 2.1 Some shapes, sizes and arrangements of bacteria. *Top left*. A typical view of *Helicobacter pylori* (×21 000), a helical (twisted) bacillus from the human gastrointestinal tract. This cell has three flagella (Fig 2.2, section 2.2.14), each flagellum being covered by an extension of the cell's outer membrane (section 2.2.9.2); the bulbous structures at/near the ends of the flagella appear to be regions where the membrane has 'ballooned out' – due perhaps to the techniques used in electron miscroscopy. Each flagellum is about 3 μm long. *Top right*. A multicellular, filamentous bacterium, *Simonsiella*, which is present as part of the mouth microflora in about 25% of humans. This section, stained with ruthenium red, shows the filament adhering to the surface of a buccal epithelial cell. The filament is about 3.5 μm in length. *Centre*. In a 'filled' (i.e. repaired) tooth: bacteria colonizing the small space between the filling material and the wall of the cavity. Each of the 'corn-cob' structures is composed of a mass of small cocci (each less than 1 μm in diameter) clustered around the end of a hypha – perhaps indicating a symbiotic relationship between two types of organism. *Bottom left*. Cells of *Aquaspirillum peregrinum*, a motile, nitrogen-fixing bacterium found in various freshwater habitats. The cell on the right is about 7.3 μm in (axial) length, 0.6 μm in thickness. *Bottom right*. A single cell of *Escherichia coli*: a straight, round-ended bacillus about 2 μm in length.

Photograph of *Helicobacter pylori* courtesy of Dr Alan Curry, PHLS, Withington Hospital, Manchester. *Simonsiella* courtesy of Dr Caroline Pankhurst, King's College, London, and Blackwell Scientific Publications Ltd. 'Corn-cob bacteria' courtesy of Prof. Dr Wolfgang Klimm, Medizinische Akademie 'Carl Gustav Carus', Dresden, Germany. *Aquaspirillum peregrinum* courtesy of Dr Hisanori Konishi, Yamaguchi University School of Medicine, Ube, Japan, and the Society for General Microbiology. *Escherichia coli* courtesy of Dr Markus B. Dürrenberger, University of Zürich, Switzerland.

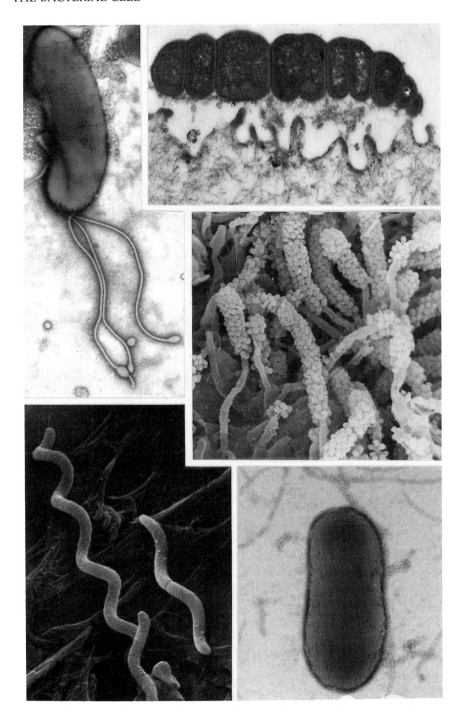

magnetosomes [review: Stolz (1993) JGM *139* 1663–1670] cause the cells to align in a magnetic field.

2.2.3   Ribosomes

Ribosomes are minute, rounded bodies, each about 0.02 μm, made of RNA (a polymer similar to DNA – Chapter 7) and protein. They are sites where proteins are synthesized (Chapter 7), and the cytoplasm contains a large number of them (Plate 8.1, *top*).

Bacterial ribosomes are smaller than those in the cytoplasm of eukaryotic cells, but they are similar (in size) to those in the chloroplasts of plants and algae. Ribosomes are usually described not by their diameters but by their rate of sedimentation in an ultracentrifuge; measurement is made in Svedberg units (S) – the higher the value the more rapid is the rate of sedimentation. Each bacterial ribosome (70S) consists of one 50S subunit and one 30S subunit (yes, 50S + 30S = 70S!); the parts of a ribosome are held together probably by hydrogen bonding and by ionic and hydrophobic interactions – magnesium ions generally being important in maintaining the structure.

Most of a ribosome (about 70% by mass) is RNA (ribosomal RNA, or rRNA); a ribosome's function seems to depend primarily on its rRNA. Like the ribosomes themselves, rRNA molecules are also measured in Svedberg units (S). The 30S ribosomal subunit contains 16S rRNA, while the 50S subunit contains 5S and 23S rRNA.

rRNA is a polymer containing four different types of subunit (nucleotide), and in a given rRNA molecule the nucleotides are arranged in a definite and meaningful sequence. In rRNA molecules the *sequence* of nucleotides appears to remain relatively unchanged over evolutionary periods of time. Different sequences in related organisms may therefore indicate evolutionary divergence between such organisms; for example, in different species of the same genus, 16S rRNA typically differs by at least 1.5%. In recent years, 16S rRNA has been used to classify bacteria into categories that are generally believed to reflect natural (evolutionary) relationships; for example, a difference in 16S rRNA was a major reason for distinguishing the domains Archaea and Bacteria.

Ribosomal proteins (r-proteins) account for about 30% (by mass) of a ribosome. In *E. coli*, the 30S subunit has 21 different types of r-protein, while the 50S subunit has over 30 different types.

2.2.4   Storage granules

Under appropriate conditions, many bacteria produce polymers which are stored as granules in the cytoplasm; these compounds include poly-β-hydroxybutyrate and polyphosphate.

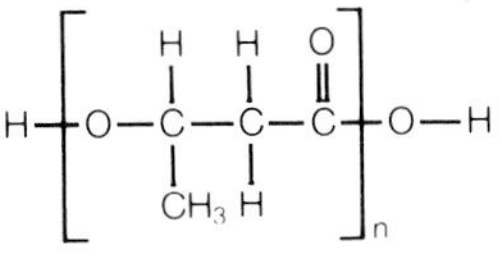

Fig. 2.3   A common storage compound in bacteria: poly-β-hydroxybutyrate.

2.2.4.1   Poly-β-hydroxybutyrate (PHB)

PHB is a linear polymer of β-hydroxybutyrate (Fig. 2.3). In some species, granules of PHB accumulate when decreasing supplies of nutrients (other than carbon) restrict the cell's rate of growth; in these cells the PHB acts as a reservoir of carbon and energy (Chapters 5 and 6) – to be used when other nutrients become more plentiful. In the soil bacterium *Azotobacter beijerinckii*, PHB accumulates (up to 80% of the cell's mass) when oxygen is scarce; in this species PHB can replace oxygen as a source of oxidizing power.

Enzymes involved in the synthesis and de-polymerization of PHB appear to occur at the surface of the PHB granules. Each granule is supposed to be bounded by a membrane, about 2–4.5 nm thick, but it has been pointed out that we do not have precise information about such a membrane [Mayer (1992) FEMSMR *103* 265–268]. PHB granules can be stained *in situ* (i.e. in the cell) by dyes such as Nile blue A.

PHB is the basis of a biodegradable plastic (Biopol: see section 13.7).

Other poly-β-hydroxyalkanoates (PHAs) occur in some bacteria; for example, granules of poly-β-hydroxyoctanoate form in *Pseudomonas oleovorans* when it is grown on *n*-octane. [Bacterial PHAs, including PHB and Biopol (multi-author symposium report): FEMSMR (1992) *103* 91–376.]

2.2.4.2   Polyphosphate (polymetaphosphate; volutin)

Polyphosphate (PO_3^{2-}-O-$[PO_3^-]_n$-PO_3^{2-}) granules occur in most types of bacteria. They are believed to act as reservoirs of phosphate and, in some cases, they appear to be involved in energy metabolism; polyphosphate may also serve to store or sequester cations.

When treated with certain dyes (e.g. polychrome methylene blue), polyphosphate granules develop a colour different to that of the dye used to stain them; this phenomenon is called *metachromacy*, and the granules are sometimes called *metachromatic granules*.

2.2.5   Gas vacuoles

Gas vacuoles (see Plate 2.2) occur only in certain (typically aquatic) prokaryotes; they are found e.g. in bloom-forming cyanobacteria (such as *Anabaena flos-aquae*) and in members of the Archaea (such as *Halobacterium* and *Methanosarcina*). Each vacuole consists of a cluster of tiny, elongated, hollow, gas-filled *vesicles*; each vesicle, which has a protein wall, is commonly about

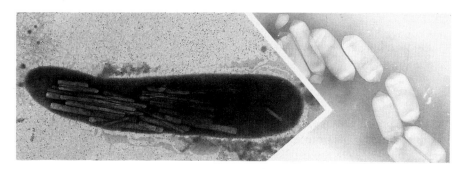

Plate 2.2   Gas vacuoles and vesicles (see section 2.2.5). *Left*. A single gas-vacuolated cell from the polar sea-ice. *Right*. Individual gas vesicles under higher magnification.

Photographs courtesy of Dr John J. Gosink, University of Washington, Seattle, USA.

60–250 nm in diameter. In some cases gas vacuoles are formed constitutively, in others they are inducible.

Gas vacuoles affect the buoyancy of free-floating cells; for cells in the aquatic environment, this will affect the intensity of received light – important in the ecology of e.g. photosynthetic organisms. Buoyancy can be regulated by at least two main mechanisms. New gas vesicles can be formed under low-intensity light, thereby increasing buoyancy. Alternatively, under high-intensity light, no new vesicles may be formed during cell growth and division – so that existing vesicles are 'diluted out' into new cells; this latter mechanism operates e.g. in the cyanobacterium *Oscillatoria agardhii*.

2.2.6   Carboxysomes

Carboxysomes are intracellular bodies, each about 100–500 nm in diameter, which are found in many *autotrophic* bacteria, i.e. bacteria which can use carbon dioxide for most or all of their carbon requirements (Chapter 6). Each carboxysome consists of a membranous sac or shell containing many copies of an enzyme (RuBisCO) involved in the 'fixation' of atmospheric carbon dioxide (Fig. 6.1).

2.2.7   Thylakoids

Thylakoids are flattened, intracellular membranous sacs which occur in most cyanobacteria; they usually occur close to, and parallel with, the cell envelope – but seem to be structurally distinct from the cytoplasmic membrane. Thylakoid membranes contain chlorophylls etc. and are the sites of photosynthesis (Chapter 5); in at least some cases they are also sites of respiratory activity (Chapter 5).

Structures similar to thylakoids (but called *chlorosomes* or *chlorobium vesicles*) are formed by bacteria of the suborder Chlorobiineae; their functions are similar to those of thylakoids.

2.2.8 Cytoplasmic membrane

The cytoplasmic membrane (CM) is a double layer of lipid molecules, about 7–8 nm thick, in which protein molecules are partly or wholly embedded – some proteins spanning the entire thickness of the membrane (Fig. 2.4); the arrangement of lipid molecules is such that the inner and outer faces of the membrane are hydrophilic ('water-loving') – stained darkly in Plate 2.3 (*top, left*) – while the interior is hydrophobic.

The lipids are mainly phospholipids. One phospholipid, phosphatidyl-glycerol (Fig. 2.5), seems to occur in most bacteria; the presence of other types of lipid depends mainly on whether a given bacterium belongs to one or other of two broad categories: the Gram-positive (G +ve) and Gram-negative (G −ve) bacteria – see section 2.2.9. Phosphatidylethanolamine (Fig. 2.5) is generally more common and abundant in G −ve bacteria. Phosphatidylcholine (*lecithin*) (Fig. 2.5) occurs in some G −ve bacteria but not in G +ve ones. Small amounts of glycolipids are common in bacterial CMs; sphingolipids are rare, and sterols occur in bacteria of the family Mycoplasmataceae. In *E. coli* the main lipid is phosphatidylethanolamine; phosphatidylglycerol and diphos-phatidylglycerol (DPG, *cardiolipin*) are minor components.

The membrane is not a rigid structure: the lipid molecules are actually in a fluid state; it 'hangs together' as a result of inter-molecular forces.

The CM proteins include various enzymes – e.g. the 'penicillin-binding proteins' (PBPs) involved in the synthesis of the cell envelope polymer *peptidoglycan* (sections 2.2.9.1 and 6.3.3.1); the PBPs may form part of a CM protein complex [Gittins, Phoenix & Pratt (1994) FEMSMR 13 1–12]. There are also components of *transport systems* (which translocate ions and molecules across the membrane), and components of *energy-converting systems* such as ATPases and electron transport chains (Chapter 5). In at least some bacteria there are 'sensory proteins' which detect changes in the cell's external environment (see e.g. sections 2.2.15.2 and 7.8.6).

The CM is not freely permeable to most molecules. Some small, uncharged or hydrophobic molecules – e.g. O_2, CO_2, NH_3 (but not NH_4^+) and water – can pass through more or less freely. Other molecules (including e.g. nutrients) and ions have to be transported across by special mechanisms, some of which require expenditure of energy by the cell; these mechanisms may allow the cell to accumulate a particular substance to a concentration far greater than that which occurs in the surrounding environment.

2.2.8.1 *Protoplasts*

If a cell loses its cell wall (Fig. 2.2) the resulting structure is called a *protoplast*. Although bounded only by the CM, a protoplast can survive (in the

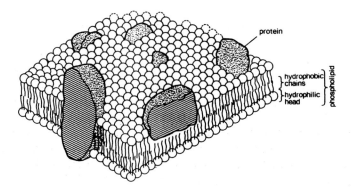

Fig. 2.4 Cytoplasmic membrane (diagrammatic): protein molecules in a fluid bilayer of phospholipid molecules – the so-called 'fluid mosaic model'. Both surfaces of the membrane are hydrophilic; the interior is hydrophobic.

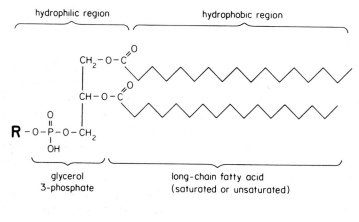

Fig. 2.5 Some of the diacyl glycerol phospholipids in bacterial cytoplasmic membranes. The bonds between glycerol 3-phosphate and the two long-chain fatty acid residues are ester bonds.

laboratory) and can carry out many of the processes of a normal living cell. However, if a protoplast be suspended in a medium more dilute than its cytoplasm, water will pass in through the CM (by osmosis) and the protoplast will swell and burst – an event known as *osmotic lysis*. In an intact bacterium, the delicate protoplast is usually saved from osmotic lysis by the mechanical strength of the cell wall (section 2.2.9).

2.2.8.2   *The cytoplasmic membrane in members of the Archaea*

In these organisms the CM contains lipids which do not occur in members of the Bacteria. Unlike the ester-linked glycerol–fatty acid bacterial lipids (Fig. 2.5), the archaeal lipids are characteristically ether-linked molecules which contain e.g. isoprenoid or hydro-isoprenoid components. Some archaeal and bacterial CM lipids are structurally analogous – e.g. the di-ether and di-ester lipids, both types of molecule having a single polar end. However, some archaeal lipids (e.g. tetra-*O*-di(biphytanyl) diglycerol) contain two ether-linked glycerol residues at *each* end of the molecule – which therefore has two polar ends; such molecules possibly span the width of the CM.

Given the extreme habitats typical of these organisms, it is likely that most or all will contain lipids (and proteins) whose characteristics reflect an adaptation to the environment.

2.2.9   Cell wall

In most bacteria a tough outer layer – the cell wall (Fig. 2.2) – protects the delicate protoplast (section 2.2.8.1) from mechanical damage and osmotic lysis; it also determines a cell's shape: an *isolated* protoplast is spherical, regardless of the shape of its original cell. Additionally, the cell wall acts as a 'molecular sieve' – a permeability barrier which excludes various molecules (including some antibiotics). However, the cell wall should not be thought of merely as an 'inert box' enclosing a living cell: it also plays an active role e.g. in regulating the transport of ions and molecules.

The cell walls of different species may differ greatly in thickness, structure and composition. However, there are only two major types of cell wall; whether a given cell has one or the other type of wall can generally be determined by the cell's reaction to certain dyes. Thus, when stained with crystal violet and iodine, cells with one type of wall retain the dye even when treated with solvents such as acetone or ethanol; cells with the other type of wall do not retain the dye (i.e. they become decolorized) under similar conditions. This important staining procedure was discovered empirically in the 1880s by the Danish scientist Christian Gram; the *Gram stain* is described in section 14.9.1. Bacteria which retain the dye (and which have one type of wall) are called *Gram-positive* bacteria; those which can be decolorized (and which have the other type of wall) are called *Gram-negative* bacteria. Some named examples of Gram-negative and Gram-positive bacteria are listed in Table 2.1. The cell walls of Gram-positive and Gram-negative bacteria are described below. (In practice, some bacteria do not give a clear-cut result in the Gram stain: see section 14.9.1 for the meaning of *Gram-type-positive* and *Gram-type-negative*.)

Table 2.1 Some genera of gram-negative and Gram-positive bacteria

Gram-negative	Gram-positive
Aeromonas	Bacillus
Bacteroides	Clostridium
Bordetella	Lactobacillus
Brucella	Listeria
Escherichia	Propionibacterium
Haemophilus	Staphylococcus
Pseudomonas	Streptococcus
Salmonella	Streptomyces
Thiobacillus	Streptoverticillium

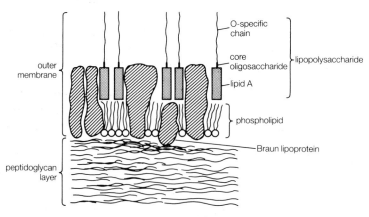

Fig. 2.6 The Gram-negative-type cell wall (diagrammatic). O-specific chains form the outermost part of the wall. The outer membrane is linked covalently to peptidoglycan through the Braun lipoproteins.

2.2.9.1   Gram-positive-type cell walls

The Gram-positive-type wall is relatively thick (about 30–100 nm) and it generally has a simple, uniform appearance under the electron microscope. Some 40–80% of the wall is made of a tough, complex polymer, *peptidoglycan* (also called *murein*). Essentially, peptidoglycan consists of linear hetero-polysaccharide chains that are cross-linked (by short peptides) to form a three-dimensional net-like structure (the *sacculus*) (Fig. 2.7) which envelops the protoplast.

In this type of wall the sacculus consists of *multilayered* peptidoglycan which, during growth, develops by the 'inside-to-outside' mechanism. In this process, new peptidoglycan is added to the inner (cytoplasmic) face of the wall; as growth continues, layers move outwards towards the cell surface, the oldest layers eventually being shed in fragments.

Covalently bound to peptidoglycan are compounds such as *teichoic acids*: typically, substituted polymers of glycerol phosphate or ribitol phosphate. In some bacteria (e.g. *Mycobacterium*) the wall contains lipids, while in others (e.g. strains of *Streptococcus*) it contains carbohydrates.

The composition of the wall can vary with growth conditions; for example, in *Bacillus*, the availability of phosphate affects the amount of cell wall teichoic acids.

2.2.9.2   Gram-negative-type cell walls

The Gram-negative-type wall (20–30 nm thick) has a distinctly layered appearance under the electron microscope. The inner layer (Fig. 2.6) – nearest the cytoplasmic membrane – is widely believed to consist of a 'periplasmic gel' of peptidoglycan, 15 nm thick, representing 1–10% of the dry weight of the cell; this view (based e.g. on electron microscopy) suggests that the peptidoglycan is multilayered. However, experiments on the turnover rate of peptidoglycan in growing cells of *Escherichia coli* suggest that the lateral cell wall may contain a *monolayer* of peptidoglycan [Park (1993) JB *175* 7–11]; multilayered peptidoglycan may occur at the poles. The chemical composition of the peptidoglycan in *E. coli* is shown in Fig. 2.7, and its mode of synthesis is outlined in section 6.3.3.1. Peptidoglycan is discussed further in sections 3.2.1. and 5.4.4.

The outer layer of the wall, the so-called *outer membrane* (Fig. 2.6, Plate 2.3, *top, left*), is essentially a protein-containing lipid bilayer – i.e. it resembles the cytoplasmic membrane. However, while the inward-facing lipids are phospholipids, the outward-facing lipids are macromolecules called *lipopolysaccharides* (LPS). Lipid A (Fig. 2.6) is a glycolipid. The core oligosaccharide (in e.g. *E. coli* and related bacteria) contains glucose and galactose residues and substituted residues of other sugars (including heptose phosphate). The O-specific chains, which form the outermost part of the cell wall, consist of linear or branched chains of oligosaccharide subunits; the chemical composition of the O-specific chain can vary from strain to strain, and this is exploited in the identification of particular strains by serology (section 16.1.5.1).

About half the mass of the outer membrane consists of proteins; in e.g. *E. coli* and related bacteria these include thousands of molecules of the *Braun lipoprotein* (Fig. 2.6) which are each linked covalently to the underlying peptidoglycan. There are also enzymes, proteins involved in specific uptake ('transport') mechanisms, and *porins*. A porin is one of three (sometimes two) similar protein molecules which, together, span the thickness of the membrane to form a water-filled 'pore'; such pores allow certain ions and molecules to pass through the outer membrane. Porins are linked via ionic bridges to the peptidoglycan. In *E. coli* and related bacteria, the outer membrane contains

(a)

(b)

several different types of porin which are designated e.g. OmpC, OmpF etc. (Omp = outer membrane protein); the relative proportions of these porins can vary according to the cell's environment (see e.g. section 7.8.6).

Components of the outer membrane are held together by ionic and other interactions. Adjacent core oligosaccharides appear to be linked by divalent cations, particularly Mg^{2+} and Ca^{2+}; experimental removal of these cations (by ion chelators) can disrupt the outer membrane. Lipid A is hydrophobically bound to the fatty acid residues of phospholipids. Some proteins appear to be linked to the core oligosaccharides.

The outer membrane is generally permeable to small ions and to small

---

Fig. 2.7 Peptidoglycan. The structures shown are typical of those in *Escherichia coli*; similar types of peptidoglycan are found in many other Gram-negative bacteria.

(a) The (three-dimensional) net-like peptidoglycan molecule: backbone chains of alternating residues of *N*-acetylglucosamine (G) and *N*-acetylmuramic acid (M) are held together by short peptides which link *N*-acetylmuramic acid residues. Each of the peptide bridges shown in the diagram links a given backbone chain to *one* other backbone chain. However, recent work indicates that there are also peptide bridges which link together three (or even four) backbone chains; the different types of peptide bridge in *E. coli* are described briefly in (b), below.

(b) Part of adjacent backbone chains showing the mode of peptide linkage between them. (The numbers in italic are used to refer to particular carbon atoms within the molecule.) Each *N*-acetylmuramic acid residue bears a short *stem peptide* – in this example the tetrapeptide L-alanine–D-glutamic acid–*meso*-diaminopimelic acid(*meso*-DAP)–D-alanine (shown in dotted boxes); in this kind of peptide bridge there is a direct, covalent link between the D-alanine of one stem peptide and the ε-amino group of *meso*-DAP in the other stem peptide. As this particular type of peptide bridge consists of two stem peptides, each of four amino acid residues, it has been referred to as a *tetra-tetra dimer* (or a tetra-tetra). Another type of peptide bridge, the *tetra-tri dimer* (tetra-tri), is formed from one tetrapeptide and one tripeptide. A *trimer* peptide bridge is one which links together *three* of the glycan backbone chains; this occurs when a stem peptide on a *third* backbone chain forms a covalent linkage with the free ε-amino group (see diagram) of a dimer bridge. An appreciation of these different types of peptide bridge is essential for understanding a newly proposed model for the replication of the peptidoglycan sacculus (section 3.2.1).

The enzyme *lysozyme* hydrolyses the *N*-acetylmuramyl-$(1 \rightarrow 4)$ linkages in the backbone chain (see diagram); such activity weakens the cell envelope.

In Gram-positive bacteria the peptidoglycan often differs from that shown above. For example, residues of *N*-glycollylmuramic acid occur in the backbone chains of *Mycobacterium*, and there are many differences in the types and positions of the cross-links. In the peptidoglycan of *Staphylococcus aureus* some of the muramic acid residues are not acetylated, and the stem peptides are linked by a penta- or hexa-glycine bridge; this bridge is susceptible to cleavage by the enzyme *lysostaphin* – an enzyme which can therefore lyse cells of this species.

Synthesis of peptidoglycan depends on the enzymic roles of the *penicillin-binding proteins* (PBPs) which occur in the cell envelope (sections 2.2.8 and 6.3.3.1); PBPs can be inactivated by penicillin and by related β-lactam antibiotics (section 15.4.1).

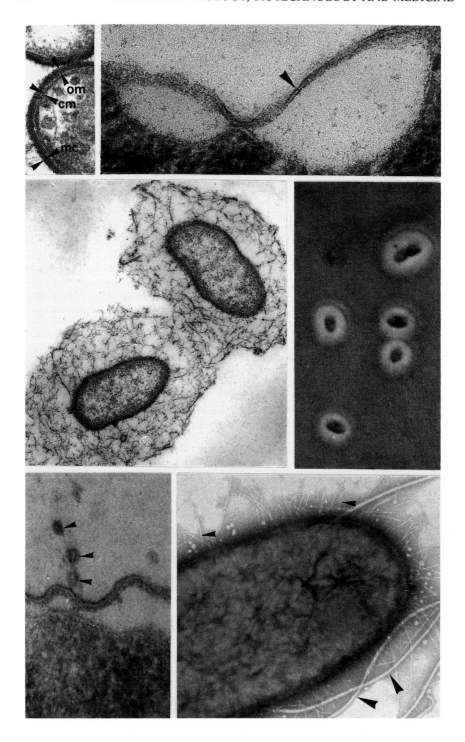

hydrophilic molecules – but much less permeable (or impermeable) to hydrophobic or amphipathic molecules.

2.2.9.3 Cell walls in members of the Archaea

In some of these organisms (e.g. species of *Halobacterium*, *Methanococcus* and *Thermoproteus*) the wall consists mainly or solely of a so-called 'S layer' (section 2.2.12) closely associated with the cytoplasmic membrane. In e.g. *Methanobacter* the wall contains *pseudomurein*: a peptidoglycan-like polymer in which the backbone chains contain N-acetyl-D-glucosamine (and/or N-acetyl-D-galactosamine, depending on species) and N-acetyl-L-talosaminuronic acid. Unlike peptidoglycan, pseudomurein is not cleaved by the enzyme *lysozyme*.

2.2.9.4 Layers external to the cell wall

In some bacteria there are one or more layers external to the wall; these layers include e.g. capsules, S layers and M proteins (see later).

Plate 2.3   The cell envelope and surface structures in some Gram-negative bacteria. *Top left.* The outer membrane (om) and cytoplasmic membrane (cm), and a microcapsule (mc), in a thin section of *Bacteroides fragilis* (×52 000). The microcapsule seen in this electronmicrograph would not be seen by light microscopy. *Top right.* The cell envelope in a thin section of a plasmolysed cell of *Escherichia coli* (×100 000). (Plasmolysis, in which water is withdrawn from the cell, causes the cytoplasmic membrane to shrink away from the outer membrane.) Here, gaps are seen between the outer membrane (arrowhead) and the cytoplasmic membrane below it. Near the centre of the photograph, the outer membrane is joined to the cytoplasmic membrane; this 'fused' region is called an adhesion site. *Centre left.* Two capsulated cells of *B. fragilis* as seen by electron microscopy (×28 000). The macrocapsules have been stained with ruthenium red. *Centre right.* Light microscopy of cells of *B. fragilis* from the same culture as those seen to the left (approx. ×1000). Here, the background has been stained darkly with eosin–carbolfuchsin ('negative staining') so that each capsule can be seen as a bright 'halo' surrounding its cell. The cell at the top is dividing. *Bottom left.* An F pilus (a specialized, hair-like appendage) arising from the cell envelope in *E. coli* (×150 000). The F pilus itself (less than 10 nm in diameter) is barely visible; however, in order to make it detectable, it has been 'labelled' with a particular bacteriophage (MS2) which binds specifically to the sides of these pili. Here, three MS2 bacteriophages (each about 25 nm in diameter) have bound, close together, along part of the length of the F pilus; each bacteriophage is indicated by an arrowhead. *Bottom right.* Part of an *E. coli* cell specially stained to show the large number of fimbriae (indicated by small arrowheads) (×54 000). The large arrowheads point to fragments of flagella.

Photographs of *B. fragilis* courtesy of Dr Sheila Patrick, Queen's University of Belfast. Adhesion site and F pilus in *E. coli* courtesy of Dr Manfred E. Bayer, Fox Chase Cancer Center, Philadelphia. *E. coli* fimbriae courtesy of Dr Anne Field, Public Health Laboratory Service, London.

2.2.9.5　Wall-less bacteria

Prokaryotic cells which lack cell walls are found among members of the Bacteria (e.g. *Mycoplasma*) and the Archaea (e.g. *Thermoplasma*).

2.2.10　Adhesion sites (in Gram-negative bacteria)

Adhesion sites are localized 'fusions' between the outer membrane and the cytoplasmic membrane (Plate 2.3: *top, right*). They can be seen under conditions of plasmolysis, and may give important clues about the cell's physiology [Woldringh (1994) MM *14* 597–607].

2.2.11　Capsules and slime layers

In some bacteria the outer surface of the cell wall is covered with a layer of material called a *capsule*. A *macrocapsule*, or 'true' capsule, is thick enough to be seen (in suitable preparations) under the ordinary light microscope (i.e. thicker than about 0.2 μm), while a *microcapsule* can be detected only by electron microscopy or e.g. serological techniques (Plate 2.3: *centre*, and *top left*). A *slime layer* is a watery secretion which adheres loosely to the cell wall; it commonly diffuses into the medium when a cell is growing in a liquid environment.

Capsules are composed mainly of water; the organic part is usually a homopolysaccharide (e.g. cellulose, dextran) or a heteropolysaccharide (e.g. alginate, colanic acid, hyaluronic acid), but in some strains of e.g. *Bacillus anthracis* the capsule is a homopolymer of D-glutamic acid. Species of *Xanthobacter* can form an α-polyglutamine capsule together with copious polysaccharide slime. Capsule-to-wall binding may be ionic and/or covalent.

Capsules have various functions. For example, they may (i) help to prevent desiccation; (ii) act as a permeability barrier to toxic metal ions; (iii) prevent infection by bacteriophages (Chapter 9); (iv) act as a nutrient reserve; (v) promote adhesion – important e.g. in those bacteria which form dental plaque; and (vi) help the cell to avoid phagocytosis. In pathogenic bacteria, capsule formation often correlates with pathogenicity (i.e. the ability to cause disease): in a given, normally capsulated pathogen, capsule-less strains are typically non-pathogenic. (See also section 11.5.1.)

Some secreted polysaccharides are used industrially. For example, the heteropolysaccharide *xanthan gum* (produced by strains of *Xanthomonas campestris*) is used in the food industry as a gelling agent, gel stabilizer, thickener, and inhibitor of crystallization.

2.2.12　S layers

Some cells have a so-called S layer – usually as the outermost layer of the cell; an S layer consists of a repeating pattern of protein or glycoprotein subunits arranged e.g. in squares or hexagons. In those bacteria which have an S layer, the S layer usually overlays the cell wall – e.g. in Gram-negative bacteria it covers the outer membrane. In some members of the Archaea the S layer *is* the cell wall, i.e. it overlays the cytoplasmic membrane. Double S layers occur e.g. in strains of *Aquaspirillum* and *Bacillus*.

2.2.13   M proteins

Molecules of M protein form a thin layer on the cell wall of the pathogen *Streptococcus pyogenes* – a Lancefield group A streptococcus (see *Streptococcus* in the Appendix) which can cause e.g. 'sore throat', scarlet fever and necrotizing fasciitis; this layer of M protein makes the cell less susceptible to phagocytosis, and hence contributes to the organism's virulence. The M protein differs slightly in different strains of *S. pyogenes,* and this enables strains to be classified into about 60 groups (*Griffith's serogroups*) – all the strains in a given group having a similar M protein; this is one example of *typing* (section 16.1.5) involving serological tests (section 16.1.5.1).

Another virulence factor of *S. pyogenes* is the hyaluronic acid capsule (section 11.5.1); the contributions of this capsule and the M protein to the virulence of *S. pyogenes* have recently been assessed [Moses *et al.* (1997) INFIM 65 64–71].

2.2.14   Flagella, fimbriae and pili

In many bacteria there are fine, hair-like proteinaceous filaments extending from the cell surface; these filaments can be divided into three main types: flagella, fimbriae and pili.

2.2.14.1   Flagella

Flagella (singular: *flagellum*) enable a cell to swim through a liquid medium, i.e. they are involved in cell motility (section 2.2.15). Depending on species, a cell may have a single flagellum (*monotrichous* arrangement); one flagellum at each end (*amphitrichous* arrangement); a tuft of flagella at one or both ends (monopolar or bipolar *lophotrichous* arrangement); or flagella which arise at various points on the cell surface (*peritrichous* arrangement) (see Fig. 2.1).

Each flagellum consists of a *filament*, a *hook* and a *basal body* (Fig. 2.2). The filament is helical, $5–20 \times 0.02$ μm, and is composed of eleven protein fibrils arranged like the strands of a rope; a fine channel (about 70 Å in diameter) runs through the axis of the filament. (In some species of e.g. *Pseudomonas* and *Vibrio*, and in *Helicobacter pylori*, the filament is sheathed, i.e. covered by an extension of the outer membrane – see e.g. Plate 2.1 *top, left*.) The hook is a hollow, flexible, proteinaceous structure (Fig. 2.8). The basal body consists of several co-axial ring-shaped components on a central, hollow rod-shaped core – and (in some accounts), also includes the recently discovered 'switch'

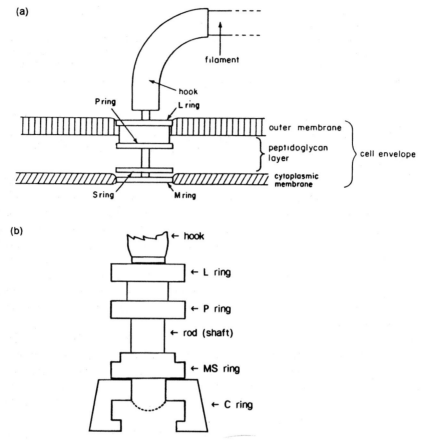

Fig. 2.8 The origin of a flagellum (see section 2.2.14.1) in the cell envelope of a Gram-negative enteric bacterium (e.g. *E. coli*, *Salmonella typhimurium*) (diagrammatic, not drawn to scale).

(a) An earlier model of the attachment of a flagellum to the cell envelope.

(b) A current view based on e.g. [Kihara *et al.* (1996) JB *178* 4582–4589; Katayama *et al.* (1996) JMB *255* 458–475]. In this model, the L and P rings are still associated with the outer membrane and peptidoglycan layer, respectively (compare (a), above). Though the S and M rings appear distinct under the electron microscope, it is now known that there is actually a single structure, the MS ring, composed of multiple copies of the subunit protein FliF (see Table 7.3); the MS ring is located in the cytoplasmic membrane.

The C ring (= 'switch') is bound to the MS ring and is located in the cytoplasm. It consists of three types of protein: FliG, FliM and FliN (Table 7.3); FliG forms the proximal part of the C ring, i.e. that part in contact with the MS ring. Control of flagellar rotation in chemotaxis (section 2.2.15.2) appears to involve interaction between FliM and the effector molecule CheY~P (see Fig. 7.13).

Encircling the MS/C rings in the cytoplasmic membrane are the MotA and MotB proteins (not shown); these are probably stationary – MotB, for example, having a known link with peptidoglycan. Torque generation (leading to flagellar rotation) is believed to result from interaction between MotA/MotB and the MS/C complex.

The structure which includes the L, P, MS and C rings, and the shaft, may be referred to as the *basal body* or *flagellar motor*.

Katayama *et al.* (cited above), using a quick-freeze procedure, have located a new structure within the C ring (dotted line), and have suggested that this structure possibly corresponds to the putative 'export apparatus' involved in flagellar assembly.

(C ring) (Fig. 2.8b); precise details – e.g. the number of rings seen under the electron microscope – differ according to species. Other essential parts of the flagellar apparatus occur in the cytoplasmic membrane, surrounding the basal body (see legend, Fig. 2.8b).

Flagella can be seen by light microscopy only after special staining; a simple flagellar stain is described by Heimbrook *et al.* [JCM (1989) 27 2612–2615].

The flagellum *rotates on its axis*. (Rotation has been demonstrated by anchoring the free end of the flagellum to a glass surface and observing rotation of the cell body.) The current view of flagellar action (in e.g. *E. coli* and *Salmonella typhimurium*) is that the MS ring, with at least part of the switch, rotates – thereby causing rotation of the shaft and, hence, filament; the L and P rings are believed to be stationary and to form a 'bush' for the rotating shaft.

Flagellar rotation needs energy in the form of an ion gradient across the cytoplasmic membrane (Chapter 5).

Assembly of the flagellum. Initially, the MS ring is inserted into the cytoplasmic membrane. Subsequently, flagellar components pass through the channel in the MS ring in strict order – first, subunits for the rod (shaft), then for the hook, and finally for the filament; thus, the protein (*flagellin*) subunits for the (elongating) filament pass through the filament's axial channel and are added sequentially to the growing tip.

Unlike other flagellar proteins, components of the L and P rings have signal sequences (section 7.6) and seem to be transported across the cytoplasmic membrane via the general secretory pathway (section 5.4.3) rather than via the MS ring.

Clearly, an open channel from the cytoplasm (where components are synthesized) to the exterior could not be tolerated by the cell; hence a so-called *export apparatus* must perform a 'gatekeeper' function by permitting only appropriate proteins to pass through the MS ring. The 'export apparatus' and the C ring are believed to be assembled with, or soon after, the MS ring.

Genetic control of flagellar assembly is outlined in section 7.8.2.7.

Flagella in spirochaetes. Spirochaetes are Gram-negative bacteria which have a unique structure. One (or usually many) flagella originate from *each* end of the cell and extend – between the peptidoglycan layer and outer membrane – towards the centre of the cell (where, in some species, flagella from both ends overlap). The number of these so-called *periplasmic flagella*, per cell, varies with species. For example, in *Leptonema illini* a single flagellum arises at each end but there is no overlap at the centre of the cell; in *Borrelia burgdorferi* the

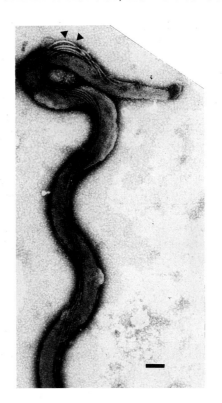

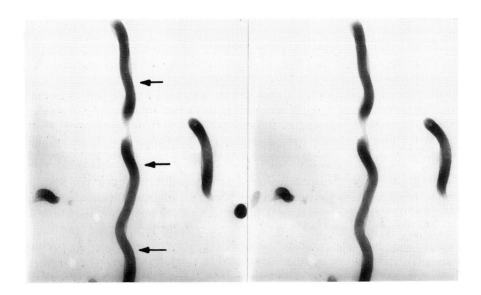

seven or so flagella from each end do overlap. The periplasmic flagella appear to *rotate* in a way similar to that of other flagella [Charon *et al.* (1992) JB *174* 832–840]. The periplasmic flagella of *Treponema pallidum* (causal agent of syphilis) are shown in Plate 2.4.

The flagellum in members of the Archaea. The archaeal flagellum differs markedly from that of bacteria in composition, structure and apparent mode of assembly. For example, the archaeal filament subunit protein, flagellin, is typically glycosylated, and the filament itself is typically much thinner than the bacterial filament. Moreover, archaeal flagellin contains a signal sequence (section 7.6), suggesting passage into the membrane, unlike bacterial flagellin (which passes through the hollow structures of the developing flagellum); this feature, and certain similarities between archaeal flagellins and type 4 fimbriae have suggested that the archaeal flagellum is actually assembled from the *base* (in contrast to the 'tip growth' in bacterial flagella) [Jarrell, Bayley & Kostyukova (1996) JB *178* 5057–5064].

2.2.14.2   Fimbriae

Fimbriae (singular: *fimbria*) are typically 2–10 nm in diameter and from 0.1 μm to several micrometres in length; they may occur all over the cell (as in Plate 2.3, *bottom right*) or may be localized. Each fimbria consists mainly of a linear sequence of identical protein subunits together with a small number of different (often specialized) subunits; subunits are commonly rich in non-polar amino acids, making the cell more hydrophobic. Fimbriae (chromosome- or plasmid-encoded – Chapter 7) occur mainly (and commonly) on Gram-negative bacteria – in which they arise from the outer membrane.

Fimbriae are sometimes (confusingly) called *pili* (but see section 2.2.14.3).

Plate 2.4   Spirochaetes. *Top.* Part of a cell of *Treponema pallidum* showing the periplasmic flagella (arrowheads). The scale bar is 0.1 μm.

Bottom. Stereoelectronmicrographs of part of a cell of *Borrelia burgdorferi* taken by high-voltage (1 MV) electron microscopy with a tilt angle of ~10° (scale: 13 mm = 1 μm). Arrows indicate regions where the flagellar bundle is on the 'observer-side' of the cell. The lower part of the cell is seen 'flat-on', i.e. a view which shows the *flattened* waveform maximally. An axial twist is located near the centre of the photograph; part of the cell which curved outwards (towards the observer) has been sliced off during preparation of the section – showing that the flagellar bundle runs underneath.

Photograph of *T. pallidum* reprinted, with permission, from *Medical Microbiology* 14th edn, Greenwood, D. *et al.* (eds), Churchill Livingstone, Edinburgh, 1992, p. 419, by courtesy of Dr Charles W. Penn, University of Birmingham, UK.

Photograph of *B. burgdorferi* reprinted from *Journal of Bacteriology* (1996) *178* 6539–6545 with permission from the American Society for Microbiology, and by courtesy of the authors: Dr S. Goldstein (University of Minnesota), Dr Karolyn Buttle (New York State Department of Health) and Professor N. W. Charon (West Virginia University).

Table 2.2 Fimbriae of *Escherichia coli*: some examples

Fimbria	Haemagglutination[1]	Comments
Type 1	MSHA	Found on many strains; agglutinates e.g. guinea-pig RBCs; chromosome-encoded; on-off switching
Type 4	MRHA	Found on various Gram-negative bacteria, e.g. species of *Neisseria*, *Proteus*, *Pseudomonas* and *Vibrio*, as well as *E. coli*; located at one pole of the cell; associated with 'twitching motility' and adhesion
CFA/I	MRHA	One of more than 10 types of adhesive fimbria found on strains of ETEC (Table 11.2) responsible for travellers' diarrhoea; agglutinates e.g. human, bovine RBCs; plasmid-encoded
P	–	On most strains of *E. coli* associated with human pyelonephritis or cystitis; mannose-resistant adherence to uroepithelium
S	MRHA	Found on many strains of *E. coli* which cause neonatal meningitis; binds to the sialyl galactoside receptors on, and agglutinates, human RBCs

[1] See section 2.2.14.2 for the meanings of MSHA and MRHA.

Many types of fimbria promote cell-to-cell adhesion. In a pathogen, this can assist the process of infection; for example, in *Neisseria* spp the fimbrial tips have specific protein *adhesins* which promote the binding of these pathogens to host cells [Rudel, Scheuerpflug & Meyer (1995) Nature 373 357–359]. In fact, for many pathogens (e.g. ETEC – Table 11.2) fimbriae are important *virulence factors* (section 11.5). In the laboratory, bacteria having certain types of adhesive fimbria can be detected by their ability to cause *haemagglutination*: the binding together of erythrocytes (red blood cells) into visible clumps. Haemagglutination caused by some types of fimbria does not occur in the presence of the sugar D-mannose; this is 'mannose-sensitive' haemagglutination (MSHA). Other types of fimbria cause 'mannose-resistant' haemagglutination (MRHA) – which is unaffected by mannose.

Some types of fimbria are listed in Table 2.2; more than one type can occur on a given cell. The synthesis of certain types (e.g. type 1 fimbriae of *E. coli*, and the fimbriae of *Haemophilus influenzae* type b) can be switched on or off by the cell – so that different cells may either have or lack such fimbriae. In e.g. *Neisseria gonorrhoeae*, a genetic mechanism periodically changes the composition of fimbrial subunits.

Assembly of fimbriae. Unlike flagella (section 2.2.14.1), most or all fimbriae appear to be assembled from the *base*; in at least some cases, assembly (in Gram-negative bacteria) seems to be as follows. Fimbrial subunits (which have a signal sequence – section 7.6) initially cross the cytoplasmic membrane.

A periplasmic *chaperone* protein then complexes each subunit, preventing it from forming abortive contact with other subunits, and transfers it to an *usher* protein in the outer membrane; the usher protein transfers each subunit to the base of the growing fimbria. Genes encoding the structure and assembly of fimbriae occur in *operons* (section 7.8.1); e.g. the operon for type 1 fimbriae includes genes *fimA* (major subunit), *fimC* (the chaperone) and *fimD* (the usher).

[Fimbriae in *E. coli* (review): Mol & Oudega (1996) FEMSMR *19* 25–52.]

2.2.14.3   Pili

Pili (singular: *pilus*) are elongated or hair-like proteinaceous structures which project from the cell surface; they are found specifically on those Gram-negative cells which have the ability to transfer DNA to other cells by *conjugation* (Chapter 8) – a process in which (in at least some cases) the pili themselves play an essential role. The genes encoding pili occur in genetic elements called *plasmids* (Chapter 7). Commonly, only one or a few pili occur on a given cell.

   The various types of pili differ e.g. in size and shape: for example, some are long, thin and flexible, while others are short, rigid and nail-like; the type of pilus correlates with the physical conditions under which conjugation can take place (Chapter 8). The best-studied flexible pilus, the F *pilus*, is one to several micrometres long, 8–9 nm in diameter; a 'labelled' F pilus is seen in Plate 2.3 (*bottom, left*). The F pilus, a tubular structure with an axial canal about 2.5 nm in diameter, is made of helically arranged protein subunits.

2.2.15   Motility and chemotaxis

Many bacteria are *motile*, i.e. they can actively move about in liquid media. Not only can they move, they can also move towards better sources of nutrients and away from harmful substances; such a directional response is called *chemotaxis*.

2.2.15.1   Motility

In most cases motility is due to the possession of one or more flagella (section 2.2.14.1). Flagellar motility involves energy-requiring *rotation* of the flagellum from the basal body.

   In peritrichously flagellate bacteria (e.g. *Salmonella*, *E. coli*), the flagella rotate independently of one another [Macnab & Han (1983) Cell *32* 109–117]. Each flagellum rotates counterclockwise (CCW) for most of the time (about 95%) and clockwise (CW) for the rest of the time; the timing of the switch from CCW to CW rotation depends on the given flagellum. When most of the flagella are rotating CCW they bunch together at one end of the cell so that the cell moves forward with the opposite end leading. Under uniform conditions such 'smooth swimming' is interrupted about once per second by *tumbling*: a brief random movement of the cell (lasting about 0.1 second) caused by a switch from CCW to CW rotation in some of the bunched flagella; when this switch occurs, the bundle of flagella is disrupted, causing the random movement. Smooth swimming is resumed when most of the flagella are again rotating CCW.

Alternate swimming and tumbling results in a three-dimensional *random walk*: the cell moves in a series of straight lines in randomly determined directions.

In general, monotrichously flagellated cells (e.g. those of *Pseudomonas aeruginosa*) can reach speeds of about 70 μm/s, compared with about 30 μm/s for peritrichously flagellated bacteria (though the peritrichously flagellated cells of *Thiovulum majus* may reach speeds >600 μm/s [Fenchel (1994) Microbiology *140* 3109–3116]). In monotrichously flagellated cells the flagellum rotates clockwise and counterclockwise for roughly equal periods of time; during change-over, randomization of direction (similar to that caused by tumbling in peritrichous cells) may be due to Brownian motion (see section 16.1.1.3).

Motility in spirochaetes is associated with rotation of their periplasmic flagella. In *Leptonema illini*, counterclockwise flagellar rotation causes the anterior (forward-pointing) end of the cell to assume a helical shape and the cell to rotate clockwise about its axis; the cell thus appears to move forward with a screw-like action. *Borrelia burgdorferi* swims as a flattened waveform, rather than as a helical form, although 'axial twists' along the length of the cell mean that different sections of the cell are in different planes; as in *L. illini*, the cell body appears to rotate clockwise about its axis. [Structure/motility of spirochaetes: Goldstein, Buttle & Charon (1996) JB *178* 6539–6545.]

Motility without flagella.   So-called *gliding motility* occurs in certain bacteria which lack flagella – e.g. species of *Beggiatoa* and *Oscillatoria*. It is a smooth form of motion which appears to occur only when cells are in contact with a solid surface; a gliding cell leaves a 'slime trail' but the mechanism of gliding is not known.

In thin films of water, cells of e.g. *Neisseria* and *Pseudomonas* with localized fimbriae exhibit *twitching motility* – apparently a passive movement involving extracellular physicochemical forces.

Serratia marcescens exhibits a flagellum-independent *spreading* in which surface tension may be involved [Matsuyama *et al.* (1995) JB *177* 987–991].

Actin-based motility is exhibited by certain pathogens *within* host cells (see e.g. section 11.3.3.2). By contrast, a recent report describes the movement of cells of enteropathogenic *E. coli* (EPEC – Table 11.2) on the *surfaces* of cultured eukaryotic cells by a mechanism involving polymerization of actin, underneath the EPEC cells, on the *inner* side of the eukaryotic cytoplasmic membrane [Sanger *et al.* (1996) CMC 34 279–287].

2.2.15.2   Chemotaxis

Chemotaxis means directional movement in response to a chemical concentration gradient. In a chemically uniform environment, flagellated cells typically adopt a 'random walk' (section 2.2.15.1). Suppose, however, that the concentration of nutrients increases in a certain direction; can a cell – which changes direction *randomly* – travel towards the higher concentration of nutrients? It can, simply by tumbling less frequently when moving towards and more frequently when moving away from the higher concentration. In this way, more time is spent going in the 'right' direction – so that the cell's overall (net) movement is towards the higher concentration of nutrients.

Substances which attract a cell are called *chemoattractants*, while those which cause a cell to move away are called *chemorepellents*; the general term *chemoeffector* is used to refer to both of these categories.

How does the cell 'sense' different concentrations, and how does it control its rate of tumbling? If a cell swims towards an increasing or decreasing concentration of chemoeffector it can detect the *change* in concentration e.g. by means of receptors in the cell envelope; depending on the direction of the gradient, a receptor may (in a given time) be either more or less likely to bind a molecule of the chemoeffector.

When a chemoattractant binds to, or leaves, a receptor, the effect is to inhibit or promote certain intracellular signals; these signals, which originate at the receptor complex in the cell envelope, are aimed at the flagellar motor(s) via a so-called *signal transduction pathway* (section 7.8.6; Fig. 7.13). For example, the release of a chemoattractant molecule by an MCP receptor enhances the signal for CW rotation (favouring tumbling); that is, tumbling occurs more often when the cell is moving into a decreasing concentration of the chemoattractant. On the other hand, movement towards a higher concentration of chemoattractant will increase the proportion of time spent in smooth swimming. Thus, concentration gradients in the environment produce a *biased random walk* by appropriately modifying the rate of tumbling when the cell is moving in a given direction. On moving into a uniform (high or low) concentration of chemoeffector the 'routine' rate of tumbling is resumed.

Chemotaxis independent of MCPs. For many carbohydrate chemoeffectors, the process of uptake across the cell envelope via the PTS transport system (section 5.4.2) provides a signal for chemotaxis.

Yet another system seems to operate in *Rhodobacter spheroides*. In this organism, chemotaxis appears to require *metabolism* of the chemoeffector – which is not necessary in MCP-mediated chemotaxis; this is suggested e.g. by the finding that the cell responds chemotactically to L-alanine (which is metabolized) but not to an analogue of L-alanine, 2-aminoisobutyrate, which is not metabolized [Poole, Smith & Armitage (1993) JB *175* 291–294].

Aerotaxis is a particular type of chemotaxis in which (motile) cells respond to a concentration gradient of dissolved oxygen; cells may move to a higher or lower concentration of oxygen according to that which is optimal for their growth (section 3.1.6). What is the mechanism of aerotaxis? In at least some cases, the signal for aerotactic movement may involve changes in the energy status of the cell's cytoplasmic membrane (section 5.3.5).

2.3   TRICHOMES AND COENOCYTIC BACTERIA

Earlier it was mentioned that most bacteria can live as single, autonomous cells. Thus, for example, each cell in a chain of streptococci leads an essentially independent life – except, of course, in that each cell shares a micro-environment with its neighbour(s). Some bacteria, however, normally exist in *trichomes* or as *coenocytic* organisms.

2.3.1   Trichomes

A trichome is a row of cells which have remained attached to one another following successive cell divisions; the cells are separated by *septa* (cross-walls, singular: *septum*), but, in at least some trichomes, adjacent cells communicate with one another via small pores (*microplasmodesmata*). (In a simple 'chain' of cells – as formed e.g. by some streptococci – such pores are not formed.) The positions of the septa may or may not be obvious (as constrictions) from the outside of the trichome. The cells of a trichome may or may not be covered by a common sheath. Trichomes are formed by many cyanobacteria and e.g. by species of *Beggiatoa*.

2.3.2   Coenocytic bacteria

The filamentous actinomycete *Streptomyces*, and some other bacteria, form tube-like hyphae which lack septa, the cytoplasm being continuous from one nucleoid to the next. Such a multinucleate organism is called a *coenocyte*.

3 Growth and reproduction

Growth in a bacterial cell involves a *co-ordinated* increase in the mass of its constituent parts; it is not simply an increase in total mass since this could be due, for example, to the accumulation of a storage compound within the cell.

Usually, growth leads to the division of a cell into two similar or identical cells. Thus, growth and reproduction are closely linked in bacteria, and the term 'growth' is generally used to cover both processes.

3.1 CONDITIONS FOR GROWTH

Bacteria grow only if their environment is suitable; if it's not optimal, growth may occur at a lower rate or not at all – or the bacteria may die, depending on species and conditions.

Essential requirements for growth include (i) a supply of suitable nutrients; (ii) a source of energy; (iii) water; (iv) an appropriate temperature; (v) an appropriate pH; (vi) appropriate levels (or the absence) of oxygen. Of course, none of these factors operates in isolation: an alteration in one may enhance or reduce the effects of another; for example, the highest temperature at which a bacterium can grow may well be lowered if the pH of the environment is made non-optimal.

3.1.1 Nutrients

Cells need nutrients as raw materials for growth, maintenance and division. As a group, the bacteria use an enormous range of compounds as nutrients; these include various sugars and other carbohydrates, amino acids, sterols, alcohols, hydrocarbons, methane, inorganic salts and carbon dioxide. However, no individual bacterium can use all of these compounds: it hasn't the range of enzymes to deal with them all, and (in any case) its cell envelope does not have uptake ('transport') systems for all of them. A given type of bacterium typically uses only a relatively limited range of compounds.

Whatever the organism, cells need sources of carbon, nitrogen, phosphorus, sulphur and other materials from which living matter is made. Some bacteria obtain all nutritional requirements from simple inorganic salts and substances such as carbon dioxide and ammonia; others need – to varying extents – more or less complex organic compounds derived from other organisms. Some important aspects of nutrition are discussed in Chapters 6 and 10.

3.1.2   Energy

Energy is needed for most of the essential chemical reactions which go on in a living cell; it is also needed e.g. for flagellar motility and for the uptake of various nutrients. All of this energy is obtained from sources in the environment. In phototrophic species, energy is derived mainly or solely from light, while chemotrophic species obtain energy by 'processing' chemicals taken from the environment. Some species can use both methods. Energy is discussed in Chapter 5.

3.1.3   Water

Some 80% or more of the mass of a bacterium is water, and, during growth, nutrients and waste products enter and leave the cell, respectively, *in solution*; hence, bacteria can grow only in or on materials which have adequate free (available) water. (Not all the water in a given material is necessarily available for bacterial growth; some, for example, may be 'tied up' by hydrophilic gels or by ions in solution.)

An extreme lack of water (desiccation) is tolerated to different degrees by different species of bacteria, though many species do not survive for long in the air-dried state. [Desiccation tolerance of prokaryotes: Potts (1994) MR *58* 755–805.]

3.1.4   Temperature

Generally, for a given type of bacterium, growth proceeds most rapidly at a particular temperature: the *optimum growth temperature*; the rate of growth tails off as temperatures increase or decrease from the optimum. For any given bacterium there are maximum and minimum temperatures beyond which growth will not occur.

Thermophilic bacteria are those whose optimum growth temperature is >45°C. These *thermophiles* occur e.g. in composts, hot springs and hydrothermal vents on the ocean floor; they include species of *Thermobacteroides* (opt. 55–70°C) and *Thermomicrobium* (opt. 70–75°C.) Among the Archaea, *Pyrodictium* has a remarkable optimum growth temperature of 105°C. Some thermophiles exhibit features associated with adaptation to growth at high temperatures; such adaptation can be seen in the composition of their cell components (e.g. cytoplasmic membrane) and, in some cases, even in their mode of energy metabolism (see section 5.3.6).

Thermoduric bacteria can survive – though not necessarily grow – at temperatures which would normally kill most other vegetative (i.e. growing) bacteria. In dairy bacteriology, 'thermoduric' bacteria are those which survive pasteurization (section 12.2.1.1).

Mesophilic bacteria grow optimally at temperatures between 15 and 45°C.

The *mesophiles* live in a wide range of habitats, and they include those bacteria which are pathogens in man and other animals.

Psychrophilic bacteria grow optimally at or below 15°C, do not grow above about 20°C, and have a lower limit for growth of 0°C or below. The *psychrophiles* occur e.g. in polar seas.

Psychrotrophic bacteria can grow at low temperatures (e.g. 0–5°C), but they grow optimally above 15°C, with an upper limit for growth >20°C.

3.1.5 pH

Most bacteria grow best at or near pH 7 (neutral), and the majority cannot grow under strongly acidic or strongly alkaline conditions. However, some (found e.g. in mine drainage and in certain hot springs) not only tolerate but actually 'prefer' acidic or highly acidic conditions; these *acidophiles* include *Thiobacillus thiooxidans*, a species whose optimal pH is 2–4. Among the Archaea, species of *Sulfolobus* grow at pH 1–5, while *Thermoplasma acidophilum* has an optimum pH of 0.8–3.

Alkalophiles grow optimally under alkaline conditions – typically above pH 8. *Thermomicrobium roseum* (opt. pH 8.2–8.5) occurs e.g. in hot springs, and *Exiguobacterium aurantiacum* (opt. pH 8.5 and 9.5) has been found in potato-processing effluents; other alkalophiles occur in natural alkaline lakes.

Acidophiles and alkalophiles may grow slowly – or not at all – at pH 7.

3.1.6 Oxygen

Some bacteria need oxygen for growth. Others need the *absence* of oxygen for growth. Yet others can grow regardless of the presence or absence of oxygen.

Bacteria which *must* have oxygen for growth are called 'strict' or 'obligate' *aerobes* to emphasize their absolute need for oxygen.

Strict or obligate *anaerobes* grow only when oxygen is absent; these organisms occur e.g. in river mud and in the rumen.

Bacteria which normally grow in the presence of oxygen but which can still grow under anaerobic conditions (i.e. in the absence of oxygen) are called *facultative anaerobes*. Similarly, those which normally grow anaerobically but which can grow in the presence of oxygen are called *facultative aerobes*.

Microaerophilic bacteria generally grow best when the concentration of oxygen is (usually much) lower than it is in air.

Some bacteria exhibit *aerotaxis* (section 2.2.15.2).

3.1.7 Inorganic ions

All bacteria need certain inorganic ions (e.g. those of chlorine and magnesium) in low concentrations, higher concentrations usually inhibiting growth. The ions have various functions – e.g. magnesium in the outer membrane (section

2.2.9.2), iron in cytochromes (section 5.1.1.2) and in a range of enzymes, and manganese and nickel in enzymes or enzyme systems.

Some bacteria (the *halophiles*) grow only in the presence of a high concentration of electrolyte (usually NaCl). Certain members of the Archaea (e.g. *Halobacterium salinarium*), which occur in salt lakes and salted fish etc., are examples of extreme halophiles: they need at least 1.5 M NaCl for growth, 3–4 M NaCl for good growth. The electrolyte serves to maintain the structure of e.g. ribosomes and the cell envelope; in dilute solutions the cells of some species break open due to weakening of the cell envelope.

Halotolerant bacteria are non-halophiles which can grow in electrolyte up to about 2.5 M; they include many strains of *Staphylococcus*.

Inorganic ions can also play a *regulatory* role in various aspects of bacterial physiology. For example, Ca^{2+} ions have known or suspected roles in e.g. transformational competence (section 8.4.1), chemotaxis (section 2.2.15.2) and sporulation in *B. subtilis* (section 7.8.6.1) [calcium signalling in bacteria: Norris *et al.* (1996) JB *178* 3677–3682].

3.1.8   Response to changed conditions: adaptation

Bacteria which live in highly specialized environments – e.g. at extreme temperatures, or within the cells of other organisms – usually cannot tolerate other types of environment. By contrast, many bacteria continue to grow (or, at least, survive) even when major changes occur in their environment; such continued growth (or survival) means that these organisms are either unaffected by the particular change(s) or are able to *adapt* (i.e. change themselves) so that they can then either tolerate or take advantage of the new conditions.

For many types of adaptation the cell needs to synthesize new types of protein (for specific functions) and to suppress the synthesis of other proteins. For example, *Salmonella typhimurium* appears to adapt to strongly acidic conditions in a two-stage process [Foster (1991) JB *173* 6896–6902]; thus, synthesis and suppression of particular proteins occur at a 'pre-shock' stage (at about pH 6) and again at a specific, lower pH as the pH continues to fall – the cell surviving in acidity as low as pH 3.3. (See also section 7.8.2.6.)

Increased osmolarity may tend to withdraw water from a cell (by osmosis), lowering its turgor pressure. In *Escherichia coli*, a significant up-shift in osmolarity affects many of the cell's *genes*, i.e. the cell alters its composition and behaviour in a number of ways. For example, it changes the relative proportions of certain pore-forming proteins in the outer membrane – thereby reducing the average pore size (section 7.8.6). It also rapidly takes up potassium ions (K^+) and synthesizes glutamate (as counter ion), thus increasing internal osmolarity. High osmolarity also promotes the transport (uptake) of the 'compatible solute' glycine betaine [Lucht & Bremer (1994) FEMSMR *14* 3–20]; the enhanced uptake of this small molecule (which is common in nature) increases internal osmolarity – so that glycine betaine can replace the accumulated potassium ions (which are physiologically less friendly). The halotolerant bacterium *Staphylococcus aureus* has two transport systems for glycine betaine [see e.g. Pourkomailian & Booth (1994) Microbiology *140* 3131–3138].

Many facultatively anaerobic bacteria (section 3.1.6) can adapt to anaerobiosis (or to a return to aerobiosis) by switching to an alternative form of energy metabolism (Chapter 5), this requiring e.g. the synthesis of new (protein) enzymes.

The synthesis and/or suppression of specific proteins for adaptation means that genes (Chapter 7) must be switched on or off – or their activity increased or decreased – according to the new conditions; some of the mechanisms involved are described in Chapter 7.

Genetic/regulatory aspects of the adaptations to heat shock, cold shock, starvation and acid shock are discussed in sections 7.8.2.3–7.8.2.6.

3.2   GROWTH IN A SINGLE CELL

In a growing cell there is a co-ordinated increase in the mass of the component parts. 'Co-ordinated' does not mean that all the parts are made simultaneously: some are synthesized more or less continually, but others are made in a definite sequence during certain fixed periods. The cycle of events in which a cell grows, and divides into two daughter cells, is called the *cell cycle*.

3.2.1   The cell cycle

Most studies on the bacterial cell cycle have been carried out on *Escherichia coli* (a Gram-negative bacillus), and the following summary is based mainly on such studies.

The cell's dimensions during growth. Rapidly growing cells are larger (in mass and volume) than slow-growing cells. When the doubling time (section 3.2.3) is less than 1 hour, the C and D periods of the cell cycle (Fig. 3.2, *lower part*) are practically constant, and, at the start of each C period, the cell's mass (per chromosome 'origin') is also practically constant; hence, under these conditions, the faster a cell grows (i.e. the shorter the I period) the more will its mass (and volume) increase during the constant $C + D$ interval before cell division.

When growing at a *constant* rate, the cell's increase in mass involves an increase in length only, the diameter remaining unchanged. If the growth rate increases the cells become larger: the cell's diameter increases, slowly, but its length increases more rapidly – and, in fact, initially 'overshoots' (i.e. increases more than is appropriate for the new rate of growth); subsequently, the diameter increases to its new value, and the length decreases to its final

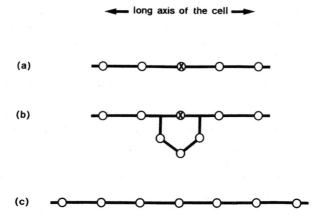

Fig. 3.1 Outline of the 3-for-1 model describing the mode of growth of the peptidoglycan sacculus in *E. coli* [Höltje (1996) Microbiology *142* 1911–1918]. In peptidoglycan, the glycan backbone chains (Fig. 2.7a) are *perpendicular* to the long axis of the cell; the diagram shows the *end-on* view of backbone chains (circles) joined by peptide bridges (straight lines).

(a) A peptidoglycan monolayer; the monolayer is a stress-bearing structure and is under considerable tension. The chain in the centre (⊗) – called the 'docking' chain – is to be replaced by three chains (hence 3-for-1).

(b) A triplet of three linked chains is located below (i.e. on the cytoplasmic side of) the peptidoglycan monolayer, with the flanking chains of the triplet covalently bound either side of the docking chain; for this to happen, a stem peptide (Fig. 2.7b) from each flanking chain has bound to the ε-amino group of a dimeric peptide bridge to form a *trimeric* bridge (Fig. 2.7b).

(c) The docking chain has been removed by enzymic action, and – because of the tension in the monolayer – the incoming triplet of chains has taken up its position in the sacculus; note that this contributes to cell *elongation*. The model assumes that the central chain in the triplet will, when incorporated in the sacculus, function as a new docking chain; for this reason, the dimeric bridge to each flanking chain is assumed to have a free ε-amino group located in the stem peptide of the flanking chain.

For cell length to *double* during the cell cycle, the model assumes that (i) docking and non-docking chains alternate in the sacculus, and (ii) only those docking chains which are present at the *start* of the new cell cycle will be replaced by an incoming triplet; that is, 'new' docking chains inserted *during* the current cell cycle will not be replaced until the next cell cycle. For this purpose, the growth mechanism must be able to distinguish 'old' from 'new' docking chains. This is possible because newly synthesized peptidoglycan is linked to the sacculus solely by tetra-tetra peptide bridges (Fig. 2.7b); it is therefore postulated that *new* chains (identified by their tetra-tetra links) are not recognized as docking chains, but that, at the start of each new cell cycle, all tetra-tetra links will have been changed enzymically (by an LD-carboxypeptidase) to tetra-tri links (Fig. 2.7b) and that this allows recognition of the docking chains.

The successive addition of triplets at the mid-point of the cell – perhaps in association with the FtsZ ring (section 3.2.1, septum formation) – might account for the ingrowth of peptidoglycan during the development of the septum.

value (which is still higher than the original value). A fall in growth rate causes an initial 'undershoot' in cell length, the cell's diameter, again, adjusting more slowly than its length to the new growth rate [Zaritsky et al. (1993) JGM 139 2711–2714].

Synthesis of the cell envelope. During continuous steady growth, synthesis of the cell envelope presumably continues throughout the cell cycle. According to one model [Cooper (1991) MR 55 649–674], peptidoglycan is incorporated diffusely in the lateral wall of the cell and preferentially at the poles; this seems to agree with a later study on peptidoglycan turnover in growing cells [Park (1993) JB 175 7–11] which suggests that the lateral wall contains a monolayer of peptidoglycan, with multilayered peptidoglycan probably occurring at the poles.

More recently, Höltje [Microbiology (1996) 142 1911–1918] has described a '3–for–1' model for the biosynthesis of the peptidoglycan sacculus; this model accounts for a number of the observed features of biosynthesis and for the different types of peptide bridge which are now known to occur in the sacculus (Fig. 2.7, caption). The model is outlined in Fig. 3.1. To co-ordinate the various synthetic and lytic processes involved, Höltje postulates the existence of two types of holoenzyme – one active at the cell's wall, the other active during septum formation; both types of holoenzyme would involve penicillin-binding proteins (PBPs – section 2.2.8) and 'lytic transglycosylases', i.e. enzymes which can degrade a glycan strand processively (that is, moving and cutting).

During growth, the turnover of peptidoglycan seems to involve the loss of disaccharide units; however, the peptides are re-cycled following uptake across the cytoplasmic membrane by a specialized transport system (section 5.4) called the oligopeptide permease (Opp) system.

It has been assumed that the cytoplasmic membrane develops passively, i.e. simply 'keeping in step' with the expanding peptidoglycan sacculus. However, recent work indicates that (conversely) peptidoglycan synthesis is dependent on the synthesis of membrane phospholipids – suggesting a novel form of regulation for the growth of the cell envelope [Rodionov & Ishiguro (1996) Microbiology 142 2871–2877].

Chromosome (DNA) replication. The necessary co-ordination between chromosome (DNA) replication (section 7.3) and cell division is described in section 3.2.1.1 and Fig. 3.2.

What controls the *initiation* of a round of replication in the cell cycle? The answer is not known, but initiation – though precisely timed – can occur at different precise times after the start of the cycle (depending on growth rate), and it occurs at all chromosomal origins in the cell at more or less the same time [Nordström & Austin (1993) MM 10 457–463]. The dependence of the timing of initiation on growth rate can be seen in Fig. 3.2: compare the top and lower sets of $I + C + D$ sequences, and note the timing of initiation (start of the

C period) at the different growth rates. (The reader may find it useful to draw a further set of $I + C + D$ sequences with I half the length of C, noting the occurrence of initiation.)

Certain factors seem to be important in the initiation of replication. One factor is the intracellular concentration of DnaA protein (or, perhaps, of an activated form of it); thus, before replication can begin, the strands of (double-stranded) DNA must separate at the chromosome's origin (section 7.3) – this requiring the prior binding of many molecules of DnaA in this part of the chromosome. Insufficient DnaA may therefore delay initiation.

Another factor is the degree of methylation (sections 7.4, 7.8.8) at the origin, *oriC*; the timing of initiation may involve delayed or gradual methylation at *oriC* – which could occur e.g. if the relevant sites were not freely accessible to the methylating enzymes.

Finally, initiation is linked to growth rate. One suggestion for a mechanism involves a certain small molecule, guanosine 5'-diphosphate 3'-diphosphate ('guanosine tetraphosphate'; ppGpp), whose intracellular concentration is inversely proportional to growth rate; at low rates of growth, the high level of ppGpp can e.g. inhibit synthesis of DnaA protein – which, in turn, may delay initiation (see above).

Chromosome decatenation and partition. During cell division, daughter cells each receive the same number of chromosomes. When synthesized, chromosomes are initially *catenated* (interlocked, like the links of a chain), and they have to be separated; separation involves certain enzymes (*topoisomerases*) which can break DNA strand(s), transiently, and pass DNA strand(s) through the gap. Once separated, each chromosome is compacted to form a nucleoid (section 2.2.1); this may involve e.g. the binding of many molecules of *HU protein*: a protein which binds preferentially to kinked/angular DNA [Pontiggia *et al.* (1993) MM *7* 343–350].

The mechanism by which nucleoids are partitioned to daughter cells is not known. It may involve e.g. (i) attachment of nucleoids to the cytoplasmic membrane (the nucleoids being drawn apart mechanically); (ii) nucleoids repelling each other; and/or (iii) a force-generating protein (product of the *mukB* gene) [chromosome partitioning in *Escherichia coli*: Løbner-Olesen & Kuempel (1992) JB *174* 7883–7889].

Septum formation. The growing cell eventually divides into two cells by the development of a cross-wall (*septum*). Septum formation begins with the development of a ring-shaped mass of FtsZ molecules (products of the *ftsZ* gene) circumferentially on the inner surface of the cytoplasmic membrane, mid-way along the cell; this seems to determine the plane of the forthcoming septum. Binding between FtsZ and a cytoplasmic membrane protein (ZipA) is apparently necessary for cell division [Hale & de Boer (1997) Cell *88* 175–185]. The FtsZ ring may assemble by an energy-dependent mechanism, but we do not know how assembly is initiated or understand the coupling between initiation and the cell cycle. Subsequently, at the site of the FtsZ ring, peptidoglycan grows inwards to form the septum; the EnvA protein then splits the septum into two layers – each layer forming the new pole (end) of one of the daughter cells. During septum formation, the outer membrane, too, grows inwards to complete the cell envelope of each daughter cell.

Following cell division in *E. coli*, daughter cells usually separate immediately. In other species, separation may not occur immediately – leading to one or other of the groupings mentioned in section 2.1.3.

3.2.1.1   *The Helmstetter–Cooper model*

During the cell cycle it's clearly essential that chromosome replication and cell division be properly co-ordinated. Chromosome replication within the cell cycle is described by the Helmstetter–Cooper model (Fig. 3.2). During slow growth, one complete duplication of the chromosome occurs for each division of the cell – so that each new daughter cell receives only one chromosome; during faster growth, a new round of chromosome replication starts before the previous round has been completed, so that each daughter cell receives more than one chromosome – about $1\frac{1}{2}$ chromosomes in Fig. 3.2. This helps to explain why it is that the rate of growth can affect the number of chromosomes in a cell (section 2.2.1).

3.2.1.2   *Genes involved in the cell cycle: the morphogenes*

Genetic control (Chapter 7) of the cell cycle involves appropriate functioning of the so-called *morphogenes* and their products. These genes have to ensure not only that appropriate products are formed at the right time but also that particular gene products are targeted to the correct positions within the cell. For example, during cell division, the septum must develop mid-way along the length of the cell – rather than at alternative *polar* sites in the cell. This means that the FtsZ protein (section 3.2.1) must somehow be 'guided' to the correct mid-cell division site. One model supposes that the products of genes *minC* and *minD* form a complex, MinCD, which binds at the alternative (polar) sites for the septum, thus excluding FtsZ and preventing the development of a polar septum. MinCD is prevented from binding at the mid-cell division site by the prior binding of the *minE* gene product (MinE) at a closely adjacent site [Huang, Coa & Lutkenhaus (1996) JB *178* 5080–5085] – allowing FtsZ to initiate the mid-cell septum.

The product of morphogene *envA* is involved in splitting the septum into the two new poles of the daughter cells.

Oher morphogenes include those encoding the penicillin-binding proteins.

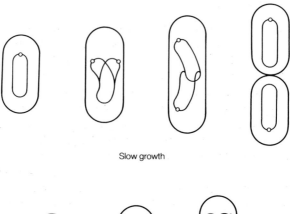

Slow growth

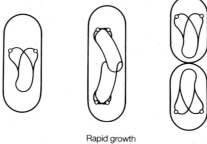

Rapid growth

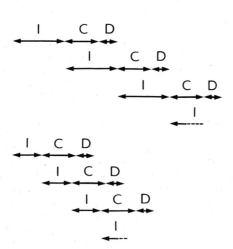

3.2.1.3   Control of the cell cycle

In the cell cycle, the occurrence of any given event could be directly dependent on the occurrence of the previous event. Alternatively, events could be controlled independently, and co-ordinated. Some [e.g. Nordström, Bernander & Dasgupta (1991) MM 5 769–774] argue for independent control; their reasons include the following. (i) Some cell-cycle events can be inhibited without inhibiting others. For example, cell division *can* occur without prior replication of the chromosome, or without proper partition of the nucleoids; in such cases there would be no initiating 'signal' from the (normally) preceding event. (ii) There is no evidence for *direct* coupling between chromosome replication and cell division, but indirect co-ordination clearly does occur. Thus, damage to DNA (which may inhibit chromosome replication) triggers the SOS system (section 7.8.2.2) – a *general* regulatory mechanism which, among other effects, inhibits cell division; such inhibition operates via the SOS gene *sulA* whose product inhibits FtsZ, a protein required for the initiation of septum formation (section 3.2.1).

Fig. 3.2 Chromosome replication during growth at different rates. Cells and chromosomes are represented diagrammatically. Replication begins at a specific location in the chromosome – the 'origin' (section 7.3) – which is shown as a small circle. During slow growth (*top*) each new daughter cell has exactly one chromosome because, following duplication of the chromosome, a new round of chromosomal replication does not begin until after completion of cell division. During faster growth (*below*) a new round of replication has begun before the previous round has been completed – i.e. well before cell division; consequently, each daughter cell in the diagram has about $1\frac{1}{2}$ chromosomes.

The cell division cycle may be represented as a linear sequence of three periods: *I*, *C* and *D*. *C* is the period during which chromosome replication occurs. *D* is the period in which the septum forms, cell division occurring at the end of the *D* period. *I* is the period between each successive *initiation* of chromosomal (i.e. DNA) replication; it is also equal to the doubling time (section 3.2.3). When one *I* period ends the next *I* period begins.

In each set of $I + C + D$ sequences in the diagram, the relationship between a given $I + C + D$ sequence and the one immediately above it is that of a daughter cell to its parent cell. The upper set of $I + C + D$ sequences shows successive cell cycles during slow steady growth; notice that when $I = C + D$, a new round of chromosome replication does not start until after the *D* period, i.e. until after completion of cell division. The lower set of $I + C + D$ shows successive cell cycles during faster steady growth; here, with *I* shorter than *C*, chromosome replication is initiated before the previous round has been completed. Hence, at the end of *D* the new round of replication is already well advanced. Thus, in faster-growing cells, the stage which chromosome replication has reached in a new daughter cell is determined by the timing of initiation – i.e. start of the *C* period – in the parent cell.

In rapidly growing cells of *Escherichia coli* the *C* and *D* periods are more or less constant at 42 minutes and 22–25 minutes, respectively.

3.2.2 Modes of cell division

The division of one cell into two (typically similar or identical) cells by the formation of a septum (as in *E. coli*) is called *binary fission*; it is the commonest form of cell division in bacteria. 'Asymmetrical' binary fission, in which daughter cells are not similar to the parent cell, occurs e.g. in *Caulobacter* (Chapter 4).

Multiple fission involves repeated binary fission of cells within a common bag-like structure; it occurs e.g. in certain cyanobacteria.

In *ternary fission*, three cells are formed from one; it occurs e.g. in *Pelodictyon* (which forms three-dimensional networks of cells).

Budding is a form of cell division in which a daughter cell develops from the mother (parent) cell as a localized outgrowth (*bud*); it occurs e.g. in species of *Blastobacter*, *Hyphomicrobium* and *Nitrobacter*.

3.2.3 Doubling time

The time taken for one complete cell cycle, the *doubling time*, varies with species and with growth conditions. For a minimum doubling time, optimum growth conditions are necessary. In *E. coli* the minimum doubling time is about 20 minutes, and in some species of *Mycobacterium* (for example) it is many hours; it has been estimated that *Mycobacterium leprae* (the causal agent of leprosy) has a doubling time of about 2 weeks in infected tissues.

3.3 GROWTH IN BACTERIAL POPULATIONS

Since, following cell division, each daughter cell can itself grow and divide, one cell can quickly give rise to a large population of cells if conditions are favourable. Given suitable conditions, such populations may develop either on solid surfaces or within the body of a liquid. In bacteriology, any solid or liquid specially prepared for bacterial growth is called a *medium* (section 14.2).

3.3.1 Growth on a solid medium

One common type of 'solid' medium, widely used in bacteriological laboratories, is a jelly-like substance (an *agar* gel) containing nutrients and other ingredients. Suppose that a single bacterial cell is placed on the surface of such a medium and given everything necessary for growth and division. The cell grows, divides into two cells, and each daughter cell does the same. If growth and division continue, the progeny of the original cell eventually reach such immense numbers that they form a compact heap of cells that is usually visible to the naked eye; this mass of cells is called a *colony*.

Typically, under given conditions, each species forms colonies of characteristic

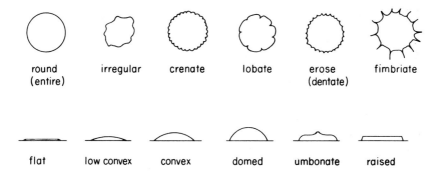

Fig. 3.3 Shapes of bacterial colonies. The upper row shows the outline or 'edge' of some colonies as seen from above. The lower row shows the 'elevation' of some colonies as seen from one side. A colony with, say, a round (entire) outline may have any one of the elevations, depending e.g. on the species of bacterium.

size, shape (Fig. 3.3), colour and consistency; different types of colony may be formed when growth occurs on different media or when other factors differ. The size of a colony may be limited e.g. by local exhaustion of nutrients (due to the colony's own growth); for this reason, crowded colonies are generally smaller than well-spaced ones. The rate at which a colony increases in size depends on temperature and other factors. Bacteria which produce pigments generally form brightly coloured colonies (e.g. red, yellow, violet) while colonies of non-pigmented bacteria usually look grey, whitish or cream-coloured. In consistency, a colony may be mucoid (viscous, mucus-like), butyrous (butter-like), friable (crumbly) etc. and its surface may be smooth or rough, glossy or dull etc.

Instead of starting with a single cell, suppose that a very large number of cells is spread over the surface of the medium. In this case there may not be enough space for individual colonies to develop; accordingly, the progeny of all these cells will form a continuous layer of bacteria which covers the entire surface of the medium. Such continuous growth is called *confluent growth*. Confluent growth may also result when one or a few cells of a *motile* bacterium are deposited on the medium; following growth, the numerous progeny of these cells may swim through the surface film of moisture and eventually cover the whole surface of the medium.

3.3.2 Growth in a liquid medium

Bacteria can move freely through a liquid medium either by diffusion or, in motile species, by active locomotion; thus, as cells grow and divide, the progeny are commonly dispersed throughout the medium. Usually, as the concentration of cells increases, the medium becomes increasingly turbid (cloudy). Certain bacteria are exceptional in that they tend to form a layer (a *pellicle*) at the surface of the medium; below the pellicle the medium may be almost free of cells. Some pellicles include bacterial products as well as the bacteria themselves; for example, a tough cellulose-containing pellicle is formed by strains of *Acetobacter xylinum*.

3.3.2.1 Batch culture

Suppose that a few bacterial cells are introduced into a suitable *liquid* medium which is then held at the optimum growth temperature for that species. At regular intervals a small volume of the medium can be withdrawn and a count made of the cells it contains (counting methods: section 14.8). In this way we can follow the development of a population, i.e. the increase in cell numbers with time. By plotting the number of cells against time we obtain a *growth curve* which, for a given species growing under given conditions, has a characteristic shape.

By growing bacteria in or on a medium we produce a *culture*; thus, a culture is a liquid or solid medium containing bacteria which have grown (or are still growing) in or on that medium. The process of maintaining a particular temperature (and/or other desired conditions) for bacterial growth is called *incubation*. The initial process of adding the cells to the medium is called *inoculation*.

When bacteria are introduced into a fresh liquid medium, cell division may not begin immediately: there may be an initial *lag phase* in which little or no division occurs. During lag phase the cells are adapting to their new environment – for example, by making enzymes to utilize the newly available nutrients. The length of the lag phase will depend largely on the conditions under which the cells existed *before* they were introduced into the medium. A long lag phase will often occur if the cells had previously existed under harsh conditions, or had been growing with different nutrients or at a different temperature; the lag phase will be short (or even absent) if the cells had been growing in a similar or identical medium at the same temperature.

During the (adaptive) lag phase, molecules are being synthesized, but the increase in total mass of the cell population is not matched by an increase in cell numbers; the cells are said to be undergoing *unbalanced growth*.

Once adapted to the new medium, the cells begin to grow and divide at a rate which is maximum for the species under the existing conditions; this is the *logarithmic* (= *log*) *phase* or *exponential phase* of growth. In this phase, cell numbers double at a constant rate (Table 3.1) – as does the mass of the population; this indicates *balanced growth*.

In the log phase of growth, a plot of cell numbers versus time gives a sharply rising curve on a simple arithmetical scale (Fig. 3.4a); clearly, such a scale would not be adequate for large numbers of cells. Is there a better way of plotting growth in the log phase? Table 3.1 (bottom row) shows that cell numbers can be expressed as powers of 2; for example, the 8 cells at 60

Table 3.1 Increase in cell numbers with time for *Escherichia coli* growing under optimal conditions in the logarithmic phase

Time (minutes)	0	20	40	60	80	100	200	300
Number of generations (i.e., rounds of cell division)	0	1	2	3	4	5	10	15
Number of cells	1	2	4	8	16	32	1024	32768
Number of cells as a power of 2	2^0	2^1	2^2	2^3	2^4	2^5....	2^{10}....	2^{15}

minutes can be written as 2^3 cells (in which 3 is the *index*). Each of the indices in Table 3.1 is, of course, the *logarithm* (to base 2, i.e. $\log_2$) of the corresponding number of cells. Now, if – instead of plotting cell numbers directly – we plot the $\log_2$ of each number, the result is a straight-line graph (Fig. 3.4b). In such a graph each unit on the $\log_2$ scale represents a doubling in cell numbers; the *doubling time* (here, the time, in minutes, needed for a doubling in cell numbers) can therefore be read off directly from the time-scale of the graph. The doubling time is also called the *generation time*.

Usually, it's more convenient to use $\log_{10}$ rather than $\log_2$ when constructing a growth curve. The $\log_{10}$ and $\log_2$ of any number can be interconverted by using the formula: $\log_{10}N = 0.301 \log_2 N$; the log phase of the graph will still be a straight line – only the *slope* of the graph will change.

As they grow, the cells use nutrients and they also produce waste products which accumulate in the medium. Eventually, therefore, growth slows down and stops due either to a lack of nutrients or to the accumulation of waste products (or both); the phase in which there is no overall increase in the number of living cells is called the *stationary phase*. (A minireview entitled 'Life after log' is worth reading [Siegele & Kolter (1992) JB *174* 345–348].) The stationary phase leads eventually to the *death phase* in which the number of living cells in the population progressively decreases.

Figure 3.5 shows the phases of growth during batch culture. 'Batch culture' is so-called because growth from lag phase to death phase occurs in the same batch of medium.

3.3.2.2   Cells in different phases of growth

Cells can undergo marked changes in their general biology and metabolism during the different phases of growth. For example, cells which – when growing – are normally killed by penicillin (section 15.4.1) are resistant to this antibiotic when growth ceases.

During periods of non-growth or growth-limitation, by-products of primary (growth-directed) metabolism may be used by the cell to synthesize so-called 'secondary metabolites' (*idiolites*) which are not used for growth. In some species, idiolites include important antibiotics or toxins. [Function of secondary metabolites: Vining (1990) ARM *44* 395–427.]

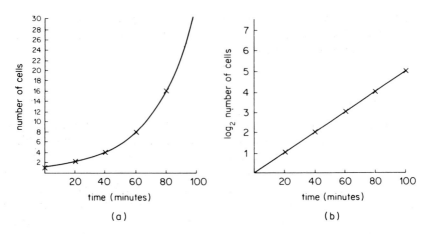

Fig. 3.4 The logarithmic (exponential) phase of growth in *Escherichia coli*. Cell numbers are plotted on (a) a simple arithmetical scale, and (b) a logarithmic ($\log_2$) scale. (Compare with Table 3.1.)

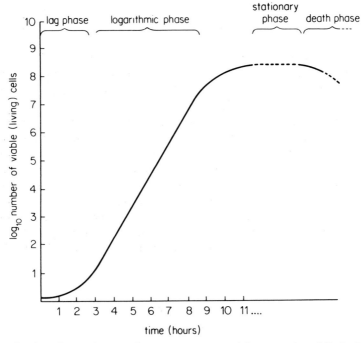

Fig. 3.5 Batch culture. A growth curve constructed for a strain of *Escherichia coli* growing in nutrient broth at 37°C.

The production of substances during a particular phase of growth can be optimized by adding nutrient(s) at appropriate times; such *fed batch culture* is used in some industrial processes.

3.3.2.3 Continuous culture

When bacteria are grown in a fixed volume of liquid medium (as in batch culture) the composition of the medium continually changes as nutrients are used up and waste products accumulate. Batch culture is suitable for many types of study, but sometimes it is preferable that cells be grown under constant, controlled, defined conditions. This is achieved by using *continuous culture (continuous-flow culture, open culture)*. In this process, bacteria are grown in a liquid medium within an apparatus called a *chemostat*; during growth, there is a continual inflow of fresh, sterile medium, and a simultaneous outflow – at the same rate – of culture (i.e. medium + cells). Constant and thorough agitation of the medium is necessary to ensure rapid mixing of the inflowing fresh medium with culture in the chemostat. Under these conditions, cells can exhibit continual logarithmic growth, i.e. balanced growth, for an extended period of time. Because growth is occurring under constant and defined conditions, this form of culture is useful e.g. for studies on bacterial metabolism.

In the chemostat, growth is normally kept at a *sub*maximal rate because instability tends to occur in the system at or near the maximum growth rate. The growth rate is controlled by controlling the concentration of an essential nutrient in the inflowing medium.

Under 'steady-state' conditions in a chemostat, the increase in cell numbers through growth is exactly balanced by the loss of cells from the chemostat; the mass of cells (the *biomass*) in the chemostat therefore remains constant. To achieve a steady state, the dilution rate is made equal, numerically, to the specific growth rate. The *dilution rate* (D) is given by F/V, where F is the rate at which the medium enters the chemostat (in litres per hour) and V is the volume of culture. The *specific growth rate* (μ) is the number of grams of biomass formed, per gram of biomass, per hour. The specific growth rate and the concentration of the growth-limiting nutrient often have the relationship predicted by the Monod equation:

$$\mu = \mu_{max}\frac{s}{k_s + s}$$

in which μ_{max} is the maximum growth rate, s is the concentration of the growth-limiting nutrient, and k_s is the concentration of the growth-limiting nutrient when $\mu = 0.5\ \mu_{max}$.

Clearly, D should not exceed the *critical dilution rate*, D_c, which corresponds to μ_{max}. If it does, the culture becomes progressively diluted to extinction and is said to have undergone 'wash-out'.

In practice, ideal (predictable) operation of a chemostat is not always achieved – due e.g. to less than perfect (instantaneous) mixing, or to factors such as 'wall growth' (organisms growing on the walls of the culture vessel and forming a separate, static (unmixed) population of cells). Another type of problem involves the emergence of *mutants* (Chapter 8) during the long periods of growth.

3.3.2.4 Synchronous growth

In a population of growing bacteria, all the cells do not divide at the same instant. However, in the laboratory, we can obtain a population in which all the cells divide at approximately the same time; in such *synchronous growth* the logarithmic portion of the growth curve (Fig. 3.5) appears as a series of steps, each step representing an abrupt doubling of cell numbers.

3.3.2.5 Growth rate versus temperature – Arrhenius plots and Ratkowsky plots

Changes in temperature markedly affect growth rate because they affect the cell's growth-directed chemical reactions (metabolism – Chapters 5 and 6).

Over a limited range of temperatures, the relationship between growth rate and temperature is similar to that between chemical reaction rates and temperature; this has been expressed in Arrhenius plots – in which the logarithm of growth rate is plotted against the reciprocal of absolute temperature (1/K). However, over an extended temperature range such plots are non-linear. Ratkowsky *et al.* [JB (1983) *154* 1222–1226] showed that, for various species of bacteria, a linear relationship could be demonstrated over the full biokinetic temperature range by plotting the square root of growth rate against absolute temperature.

3.4   DIAUXIC GROWTH

If a bacterium is given a mixture of two different nutrients, it may use one in preference to the other – utilization of the second beginning only after the first has been exhausted. For example, given a mixture of glucose and lactose, *Escherichia coli* will use the glucose first, starting on the lactose only when all the glucose has been used. During the transition from one nutrient to the other, growth may slow down or even stop; this pattern of growth is called *diauxie* (or diauxy). The mechanism of diauxie is discussed in section 7.8.2.1.

3.5   MEASURING GROWTH

Growth (change in cell numbers, or biomass, with time) can be measured e.g. by (i) counting cells (section 14.8); (ii) determining the increase in dry weight of biomass formed in a given time interval; (iii) monitoring the uptake (or release) of a particular substance; (iv) measuring the amount of a radioactive substance incorporated in biomass in a given time, and (v) *nephelometry*: measuring the increase in scattered light from a beam passing through a liquid culture (light scattering increases as cell numbers increase).

4 Differentiation

In most species of bacteria, major changes in form or function do not occur: progeny cells are more or less identical, both in appearance and behaviour, to their parental cells. In some bacteria, however, one type of cell can give rise to a markedly different type of cell; the timing of such *differentiation* is commonly related to conditions in the cell's environment. In the next few pages we look at some diverse examples of bacterial differentiation.

4.1 THE LIFE-CYCLE OF *Caulobacter*

Caulobacter is a Gram-negative, strictly aerobic bacterium found in soil and water. It forms two distinctly different types of cell, and the change from one type to the other is an essential part of the life-cycle (Fig. 4.1). Having a swarm cell in the life-cycle is advantageous since it allows the organism to spread to different locations. The motile daughter cell must lose its flagellum and develop a stalk (called a *prostheca*) before it can divide, and may thus be regarded as immature; the stalked (mature) mother cell can produce swarm cells but it cannot itself become a swarm cell.

Clearly, for *Caulobacter* to grow normally, the prostheca and flagellum must develop at the correct times in the cell cycle. How is this achieved? It appears that, at the appropriate times, signals from within the cell cause a temporary increase in the intracellular level of a particular protein [Brun & Shapiro (1992) GD 6 2395–2408]; this protein – a *sigma factor* (section 7.5) – promotes the activity of those genes required for the synthesis of the prostheca and flagellum.

A similar type of differentiation, from non-motile to motile cells, and vice versa, occurs in species of *Hyphomicrobium* and *Rhodomicrobium*.

4.2 SWARMING

Proteus is a Gram-negative bacillus found e.g. in the intestines of man and other animals. If cells of *P. mirabilis* (or of another species, *P. vulgaris*) are incubated on a suitable solid medium, the first progeny cells are short, sparsely flagellated bacilli about 2–4 μm in length; these cells form a colony in the usual way. However, after several hours of growth, some of the cells around the edge of the colony grow to lengths of 20–80 μm and develop

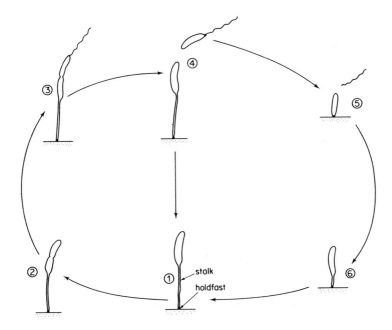

Fig. 4.1 The life-cycle of *Caulobacter*. **1.** A mature, stalked cell attached to a surface by an adhesive 'holdfast'. **2.** The stalked cell begins to divide. **3.** A flagellum develops at the free end of the daughter cell. **4.** Asymmetric binary fission is complete, and the motile daughter cell (*swarm cell*) swims away. The stalked cell can continue to grow and produce more swarm cells. **5.** The swarm cell loses its flagellum and becomes attached by its holdfast. **6.** A stalk develops, and the daughter cell matures into a new stalked mother cell.

numerous additional flagella; these cells are called *swarm cells*. The swarm cells swim out to positions a few millimetres from the colony's edge, and, there, each divides into several short bacilli – similar to those in the original colony; these cells grow and divide normally for a number of generations, forming a ring of heavy growth which surrounds (and is concentric with) the original colony. Later, another generation of swarm cells is formed at the outer edge of the ring and the cycle is repeated. In this way the entire surface of the medium becomes covered by concentric rings of growth. This phenomenon is called *swarming*.

Swarming is not essential for *Proteus*, and it doesn't happen on all types of medium. It occurs typically on rather moist surfaces; under such conditions, swarm cells help the organism to spread to new sources of nutrients.

Swarming also occurs in various other bacteria, both Gram-negative (including e.g. *Serratia marcescens* (Plate 4.1) and *E. coli*) and Gram-positive; it is often demonstrated by using an *agar* medium (section 14.2) containing a concentration of agar lower than that in normal growth media. Unlike *P. mirabilis*, which forms concentric rings of growth (due to discrete periods of swarming), some bacteria swarm continuously, forming a thin layer of growth, while in others swarming gives rise to individual microcolonies.

[Swarming (review): Harshey (1994) MM *13* 389–394.]

4.3   RESTING CELLS

In some bacteria differentiation can result in the formation of a resting cell – either a *spore* or a *cyst*. Resting cells may function as disseminative units (helping to spread the organism) and/or as dormant cells which are capable of surviving in a hostile environment. Under suitable conditions a spore or cyst *germinates* to form a new vegetative cell.

4.3.1   Endospores

Endospores have been studied more thoroughly than any other type of bacterial spore; they are formed by species of *Bacillus, Clostridium, Coxiella, Desulfotomaculum, Thermoactinomyces* and a few other genera. An endospore is formed *within* a cell as a response to starvation – specifically, a shortage of carbon, nitrogen and/or phosphorus. It exists in a state of dormancy: few, if any, of the chemical reactions in a vegetative (growing) cell take place in the mature endospore. Dormancy can persist for long periods, and the endospore is highly resistant to many hostile factors: extremes of temperature and pH, desiccation, radiation, various chemical agents and physical damage; in fact, irreversible inactivation of endospores can be guaranteed only by the harsh treatment of a *sterilization* process (section 15.1).

The formation of an endospore is shown diagrammatically in Fig. 4.2, and its structure is shown in Plate 4.1. The heat-resistance of the endospore is believed to be due to its low level of water; calcium dipicolinate (in the core) may act as a secondary stabilizer.

Under suitable conditions, an endospore *germinates*, i.e. it becomes metabolically active. The endospores of some species need to be 'activated' before they can germinate; activation may consist e.g. of sublethal heating or exposure to certain chemicals. Germination is promoted by chemicals called *germinants*; according to species, germinants may include e.g. L-alanine, some purine nucleosides, various ions or certain sugars. Germination may be initiated by the binding of a germinant to an inner membrane receptor. The transition from a germinated endospore to a vegetative cell is called *outgrowth*.

Genetic control of the initiation of endospore formation is described in section 7.8.6.1.

Note. 'Endospore' is often abbreviated to 'spore'. However, the endospore should not be confused with other types of bacterial spore (see section 4.3.2).

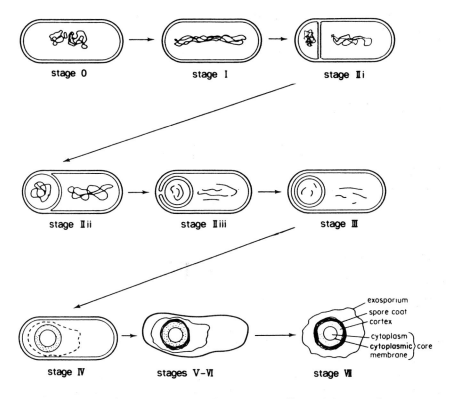

Fig. 4.2 Endospore formation. **Stage 0**. A vegetative cell containing two chromosomes is about to sporulate. **Stage I**. An *axial filament*, composed of the two chromosomes, is formed. **Stage II**. The development of an asymmetrical septum divides the protoplast into two unequal parts. The septum is considerably thinner than that formed during normal cell division because it contains much less peptidoglycan; subsequently the peptidoglycan is hydrolysed (removed). The smaller of the two protoplasts will become the endospore, and is called the *forespore* or *prespore*. The cytoplasmic membrane of the larger protoplast invaginates to engulf the forespore. Stage II is subdivided into parts i, ii and iii, as shown. When completely engulfed, the forespore is bounded by two membranes (**stage III**). **Stage IV**. Modified peptidoglycan is laid down between the two membranes of the forespore to form a rigid layer called the *cortex*. A loose protein envelope called the *exosporium* may begin to develop at about this time. **Stages V–VI**. A multilayered protein *spore coat* is deposited outside the outermost membrane (stage V), and the spore matures (stage VI) to develop its characteristic resistance to heat and its bright, refractile appearance under the light microscope; during this time, calcium dipicolinate accumulates in the 'core' – i.e. the spore's protoplast (bounded by the inner membrane). **Stage VII**. The completed spore is released by disintegration of the mother cell. (Note that an exosporium is not present on all endospores.)

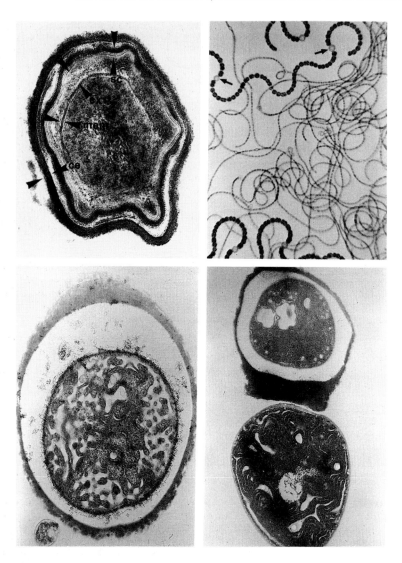

Plate 4.1 Differentiation.

Top left. Transmission electronmicrograph of an endospore of *Bacillus subtilis* within the mother cell (approx. 1 μm across). The 'core' (protoplast) of the endospore is bounded by the membrane (mem). Between the membrane and the multilayered spore coat (sc) is the cortex (cx). Surrounding the endospore is the cell envelope (ce) of the mother cell.

Top right. Light-micrograph of *Anabaena spiroides* (large filaments) and *Anabaena circinalis* (fine filaments) from Hebgen Lake, Montana, USA. Each arrow points to a heterocyst; these heterocysts have developed at intercalary positions among the vegetative cells of the filament.

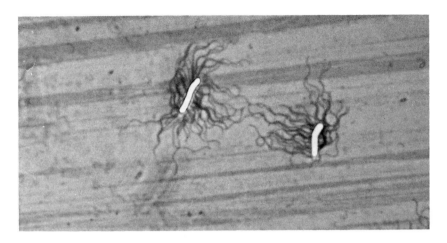

Bottom left. Transmission electronmicrograph of a heterocyst of *Anabaena oscillarioides*. Note the thick envelope. The very fine black dots (seen most clearly with a hand lens) indicate locations of the enzyme *nitrogenase*; each dot is a particle of colloidal gold (attached, as an electron-opaque label, to a molecule which binds specifically to nitrogenase).

Bottom right. Transmission electronmicrograph of an *Anabaena oscillarioides* heterocyst (upper) and vegetative cell (lower).

Top of this page. Swarm cells of *Serratia marcescens* (see section 4.2). Note the large number of flagella on each cell.

Endospore courtesy of Dr John Coote, University of Glasgow. Heterocysts courtesy of Associate Professor John C. Priscu, Montana State University, Bozeman, Montana, USA. Swarm cells courtesy of Dr Rasika M. Harshey, University of Texas at Austin, Texas, USA.

4.3.2 Other bacterial spores

In many of the hypha-forming actinomycetes, *exospores* are produced by septation and fragmentation of the hyphae (Fig. 4.3). These spores lack specialized structures (such as cortex and spore coat) but they do show some resistance e.g. to dry heat, desiccation and certain chemicals. The spores of *Streptomyces* are metabolically less active than the vegetative hyphae, though they are not completely dormant.

In the actinomycetes *Actinoplanes* and *Pilimelia*, motile (flagellated) *zoospores* are formed inside a closed sac called a *sporangium*; sporangia develop from the vegetative hyphae.

4.3.3 Bacterial cysts

Cysts are formed e.g. by the soil bacterium *Azotobacter vinelandii*. The desiccation-resistant cysts of this organism are dormant, and they can survive in

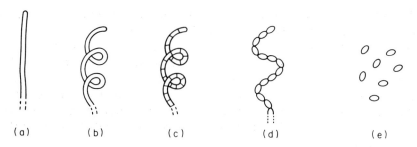

Fig. 4.3 Exospore formation in a species *Streptomyces* (an actinomycete). (a) The tip of a vegetative aerial hypha. (b) The tip of the hypha becomes coiled. (c) Septa develop along the length of the coiled hypha. (d) The walls of the developing spores thicken, and each spore becomes rounded. (e) The spores are released.

dry soil for years. Encystment may be triggered by changes in the levels of carbon and nitrogen in the environment; it involves loss of flagella and the development of a complex cyst wall which contains alginate, protein and lipid. Typically, PHB (section 2.2.4.1) accumulates in the cyst.

4.4   AKINETES, HETEROCYSTS AND HORMOGONIA

These structures are formed by various filamentous (trichome-forming) *cyanobacteria* – photosynthetic organisms which occur e.g. in soil, in natural waters and in symbiotic associations with certain eukaryotes.

4.4.1   Akinetes

Akinetes are differentiated cells produced by some species e.g. under starvation conditions; each has a thickened cell wall and a cytoplasm rich in storage compounds (such as glycogen). Akinetes are usually larger than vegetative cells, and they have a lowered rate of metabolism; they show some resistance to desiccation and cold, and may function as overwintering and/or disseminative units.

4.4.2   Heterocysts

Heterocysts are formed by some species when there is a shortage of usable nitrogen compounds. Under such conditions, some of the cells in a trichome undergo differentiation, each forming a *heterocyst*: a specialized compartment within which atmospheric (gaseous) nitrogen can be 'fixed' – i.e. converted to a usable nitrogen compound (section 10.3.2.1). The process of differentiation includes e.g. development of a thick envelope, re-arrangement of the thylakoids, cessation of (photosynthetic) oxygen evolution, and synthesis of *nitrogenase* (an enzyme used in nitrogen fixation). The thick envelope (see Plate 4.1) seems to protect (oxygen-sensitive) nitrogenase against atmospheric oxygen; *Anabaena flos-aquae* forms a thicker envelope when grown under higher partial pressures of oxygen [Kangatharalingam, Priscu & Paerl (1992) JGM *138* 2673–2678]. Communication between a heterocyst and an adjacent vegetative cell occurs via fine pores (*microplasmodesmata*) in their contiguous cytoplasmic membranes; during nitrogen fixation, fixed nitrogen is transferred to the vegetative cell which, in turn, transfers carbon and other materials to the heterocyst.

4.4.3 Hormogonia

A hormogonium is a short trichome, lacking both akinetes and heterocysts, formed e.g. from a vegetative trichome; the cells of a hormogonium may be smaller than those of the parent trichome. Typically, hormogonia exhibit gliding motility. In some species (e.g. *Nostoc muscorum*) only the hormogonia contain gas vascuoles – reinforcing the idea that these short trichomes have a primarily disseminative role.

5   Metabolism I: energy

'Metabolism' refers to the chemical reactions which occur in living cells: molecules are built up (*anabolism*), broken down (*catabolism*) or changed from one type to another, and various atoms are oxidized or reduced. Most of the reactions involve specific protein catalysts called *enzymes*.

A sequence of metabolic reactions, in which one substance is converted to another (or others), is called a *metabolic pathway*; in such a pathway the *substrate* (e.g. a nutrient) is converted, often via one or more *intermediates*, to so-called *end product(s)*. Many of the metabolic pathways in bacteria do not occur in eukaryotes.

Metabolic reactions are commonly *endergonic*, i.e. they require energy; energy is also needed for locomotion (in motile species) and for the uptake of various nutrients. Most bacteria obtain energy by 'processing' chemicals from the environment; such bacteria are called *chemotrophs*. Other bacteria, which use the energy in sunlight, are called *phototrophs*. However, neither chemicals from the environment nor sunlight can be used *directly* to fuel a cell's energy-requiring processes; consequently, the cell must have ways of converting these 'environmental' sources of energy into a usable form of energy. What *is* a 'usable form of energy'? Given certain chemicals, or sunlight, bacteria can make specific 'high-energy' compounds with which they can satisfy their energy demands; these compounds include adenosine 5'-triphosphate (ATP), phosphoenolpyruvate (PEP), acetyl phosphate and acetyl-CoA. Such compounds have been called 'energy currency molecules' because the cell can spend them (rather than 'environmental' energy) just as we spend coins and banknotes instead of gold bars. Some 'currency' molecules are shown in Fig. 5.1.

ATP (Fig. 5.1a) yields usable energy when its terminal phosphate bond is broken; accordingly, as molecules of ATP are used up (supplying the cell's energy needs) molecules of adenosine 5'-diphosphate (ADP) are formed. Hence, environmental energy must be harnessed in such a way that ATP can be re-synthesized by the phosphorylation of ADP.

Another type of energy currency molecule – nicotinamide adenine dinucleotide (NAD) – carries energy in the form of 'reducing power': it accepts and yields energy by being (respectively) reduced and oxidized (Fig. 5.1b). Other carriers of reducing power include NAD phosphate (NADP) and flavin adenine dinucleotide (FAD).

Environmental energy can also be converted to an electrochemical form of energy; this consists of a gradient of ions (usually protons: H^+) between the two surfaces of the cytoplasmic membrane. The energy in this ion gradient

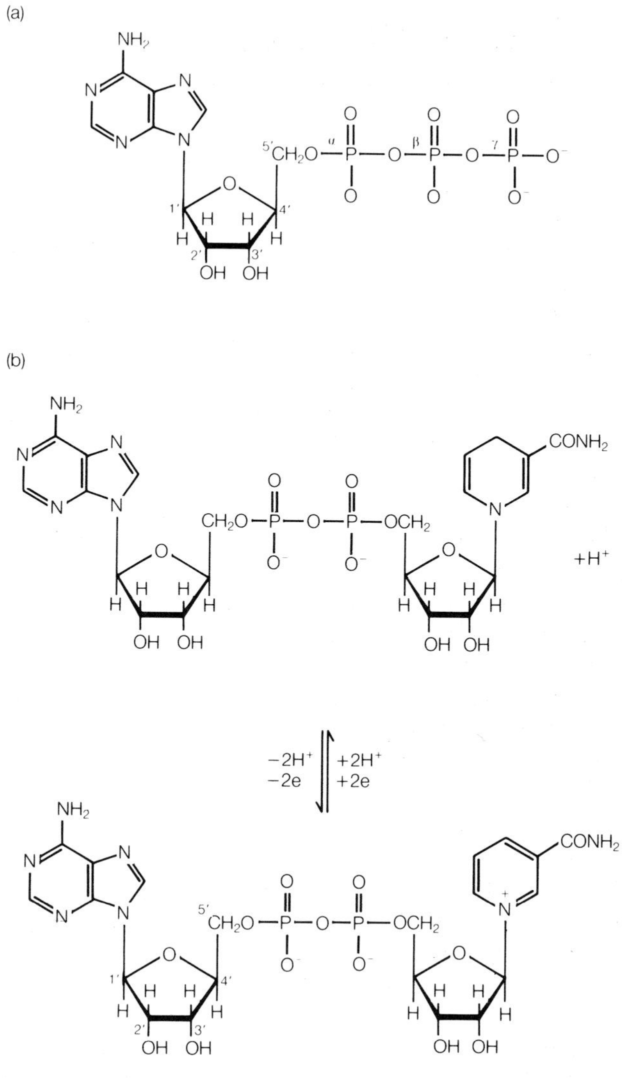

(a)

(b)

Fig. 5.1 Some energy currency molecules. (a) Adenosine 5'-triphosphate (ATP). When donating energy, the γ-bond is broken and the terminal phosphate group is lost; the resulting molecule, adenosine 5'-diphosphate (ADP), must be phosphorylated to regenerate ATP.

(b) Nicotinamide adenine dinucleotide (NAD) showing the reduced (upper) and oxidized (lower) forms; e = electron. Strictly, the oxidized form should be written NAD^+ but is often written NAD for convenience; similarly, the reduced form should be written $NADH + H^+$ but is often written NADH. NAD phosphate (NADP) is NAD 2'-phosphate – the phosphate group being at the 2'-position of the (left-hand) sugar (D-ribose) molecule.

can be used for transport (see later), for driving flagellar rotation (section 2.2.14.1) – and for the synthesis of high-energy compounds!

How *does* a bacterium form high-energy compounds, or an ion gradient, from 'environmental' energy? Different strategies are used by chemotrophs and phototrophs, as discussed below.

5.1   ENERGY METABOLISM IN CHEMOTROPHS

The chemotrophs 'process' chemicals for energy; those which use organic compounds are called *chemoorganotrophs*, and those which use inorganic compounds, or elements, are called *chemolithotrophs*. (Mechanisms used for the *uptake* of chemicals are discussed in section 5.4.)

5.1.1   Energy metabolism in chemoorganotrophs

Chemoorganotrophs use their organic substrates in one of two main types of energy-converting metabolism: *fermentation* and *respiration*. Some chemoorganotrophs can carry out only one of these processes; others can carry out either – depending on conditions.

5.1.1.1   Fermentation

Fermentation is a type of energy-converting metabolism in which the substrate is metabolized *without the involvement of an exogenous (i.e. external) oxidizing agent*. (*Note.* Fermentation typically – but not necessarily – occurs anaerobically, i.e. in the absence of oxygen, but this is *not* the distinguishing feature of fermentation: as we shall see later, respiration also can occur anaerobically.) Because no external oxidizing agent is used, the products of fermentation – collectively – are neither more nor less oxidized than the substrate; that is, the oxidation of any intermediate in a fermentation pathway is balanced by the reduction of other intermediate(s) in that pathway. This is illustrated diagrammatically in Fig. 5.2.

These ideas can be understood more easily by looking at some real metabolic pathways. In many bacteria the fermentation of glucose begins with a pathway known as *glycolysis* or as the *Embden–Meyerhof–Parnas pathway* (EMP pathway) (Fig. 5.3). In this pathway, 1 molecule of glucose yields (via a number of intermediates) 2 molecules of the end-product, pyruvic acid. At two places in this pathway (Fig. 5.3) energy from *exergonic* (i.e. energy-yielding) reactions is used to phosphorylate ADP – that is, to synthesize the 'energy currency molecule' ATP from ADP. When energy from a chemical reaction is used, directly, for the synthesis of ATP from ADP the process is known as *substrate-level phosphorylation*.

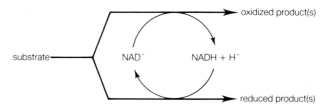

Fig. 5.2 A diagrammatic view of a fermentation pathway. Here, the (organic) substrate gives rise to intermediates which undergo *mutual* oxidation and reduction; taken together, the products have the same oxidation state as the original substrate. (Compare with Fig. 5.4.)

In terms of energy currency molecules, the EMP pathway can be summarized as follows:

$$2ATP + 4ADP + 2NAD \rightarrow 2ADP + 4ATP + 2NADH$$

Thus, each molecule of glucose metabolized gives 2NADH and a *net* yield of 2ATP.

For the metabolism of further molecules of glucose there is clearly a need for fresh supplies of both ADP and NAD. ADP is regenerated from ATP when the latter is used to supply energy. However, as no external oxidizing agent is used in fermentation, how is NAD regenerated from NADH? Earlier it was said that the fermentation of glucose may *begin* with the EMP pathway; in fact, the EMP pathway is only the 'front end' of a number of different pathways: what happens to the NADH (and the pyruvic acid) depends on subsequent reactions. In the simplest case, NADH donates its reducing power to the pyruvic acid, i.e. the oxidation of NADH to NAD is coupled with the reduction of pyruvic acid to lactic acid:

$$\text{pyruvic acid} + NADH \rightarrow \text{lactic acid} + NAD$$

This reaction completes one possible fermentation pathway: a so-called *lactic acid fermentation*; a pathway in which lactic acid in the only (or predominant) product is called a *homolactic fermentation*, while a *heterolactic fermentation* yields a mixture of lactic acid and other products. Lactic acid – a waste product – is released by the cells.

The so-called 'lactic acid bacteria' include e.g. species of *Lactobacillus* and *Leuconostoc* (see Appendix); they are widely used in the manufacture of fermented foods (e.g. some cheeses and other dairy products, salami, sauerkraut and sourdough bread). [Genetics, metabolism and applications of lactic acid bacteria (symposium): FEMSMR (1993) *12* 1–272.]

Notice that lactic acid ($C_3H_6O_3$) has the same oxidation state as glucose ($C_6H_{12}O_6$) – i.e. no net oxidation or reduction has occurred; although glyceraldehyde 3–phosphate undergoes oxidation, this is balanced by the reduction of pyruvic acid to lactic acid (Fig. 5.4).

Other fermentation pathways which have the EMP pathway as their 'front

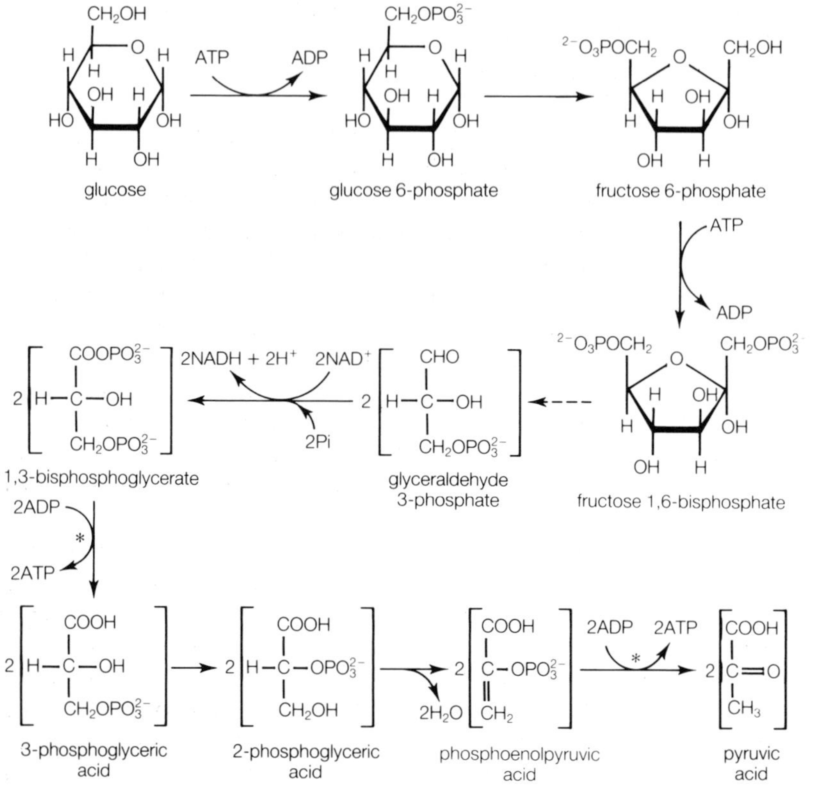

CH₂OH ... glucose → glucose 6-phosphate → fructose 6-phosphate

(The chemical structures of the Embden–Meyerhof–Parnas pathway are shown, including: glucose, glucose 6-phosphate, fructose 6-phosphate, fructose 1,6-bisphosphate, glyceraldehyde 3-phosphate, 1,3-bisphosphoglycerate, 3-phosphoglyceric acid, 2-phosphoglyceric acid, phosphoenolpyruvic acid, and pyruvic acid, with ATP/ADP and NADH/NAD⁺ conversions.)

Fig. 5.3 The Embden–Meyerhof–Parnas pathway. The broken arrow from fructose 1,6–bisphosphate indicates a simplified section of the pathway. Each of the two substrate-level phosphorylations is marked with an asterisk; in each case, phosphate is transferred to ADP from an energy-rich organic phosphate. Although two reactions in the pathway actually require ATP, there is nevertheless a *net* gain of 2ATP for each molecule of glucose metabolized. (*Note.* Pyruvic acid and phosphoenolpyruvic acid are often referred to as 'pyruvate' and 'phosphoenolpyruvate', respectively.) Most of the reactions in this pathway are reversible.

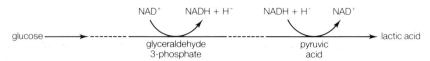

Fig. 5.4 Homolactic fermentation: a diagrammatic view of the fermentation of glucose to lactic acid via the Embden–Meyerhof–Parnas pathway. Note that the oxidation is balanced by an equivalent reduction. The homolactic fermentation is carried out by some of the so-called 'lactic acid bacteria' which are used in the food industry.

end' include the mixed acid fermentation (Fig. 5.5) and the butanediol fermentation (Fig. 5.6). In these pathways the pyruvic acid is metabolized to several end-products, the relative proportions of which can vary according to conditions of growth.

The *mixed acid fermentation* (Fig. 5.5) occurs e.g. in *Escherichia coli* and in species of *Proteus* and *Salmonella*. Under acidic conditions, *E. coli*, and other species which have the formate hydrogen lyase enzyme system, can split the formic acid into carbon dioxide and hydrogen; thus, these organisms carry out an *aerogenic* (gas-producing) fermentation. Organisms which lack the enzyme system carry out the fermentation *anaerogenically* (i.e. without forming gas); they include species of *Shigella*. (*Shigella* is a reminder that gas is not *necessarily* produced during fermentation.)

The *butanediol fermentation* (Fig. 5.6) occurs e.g. in species of *Enterobacter*, *Erwinia*, *Klebsiella* and *Serratia*. The amount of acid formed in this fermentation is generally much less than that formed in the mixed acid fermentation. Some strains form small amounts of diacetyl ($CH_3.CO.CO.CH_3$) from the acetolactic acid under certain conditions.

Although the mixed acid and butanediol fermentations are more complicated than the homolactic fermentation, they nevertheless conform to the same basic principles; while the relative proportions of end-products may vary, the formation of products more oxidized than glucose (e.g. formic acid) is always balanced by the formation of products more reduced than glucose (e.g. ethanol).

NADH and other compounds formed during fermentation can be used in various ways. Some of the NADH will be oxidized in biosynthetic reactions (rather than in the formation of waste products such as lactic acid); this also allows compounds such as pyruvate to be used as precursor molecules for biosynthesis (section 6.3.1 and Fig. 6.3). Both in respiration (section 5.1.1.2) and fermentation, there is (in chemoorganotrophs) a close link between energy metabolism and carbon metabolism (section 5.3.1).

Note. Before leaving the topic of fermentation it is worth mentioning that, in industry, the term 'fermentation' is commonly used for *any* chemical process mediated by microorganisms – even for those processes in which fermentation is not involved.

5.1.1.2 Respiration

Respiration is a type of energy-converting metabolism in which the substrate is metabolized *with the involvement of an exogenous* (i.e. *external) oxidizing agent* (compare fermentation: section 5.1.1.1). Respiration can occur in the presence of oxygen (oxygen itself serving as external oxidizing agent) – but it can also occur anaerobically (when other inorganic or organic oxidizing agents are used in place of oxygen). (As we are still talking about chemoorganotrophs,

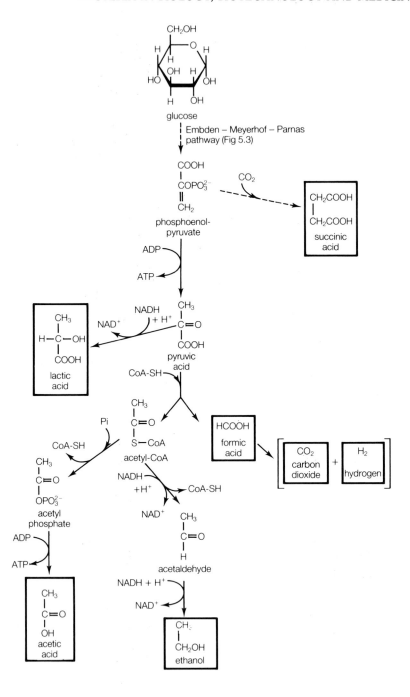

Fig. 5.5 The mixed acid fermentation. CoA = coenzyme A; Pi = inorganic phosphate.

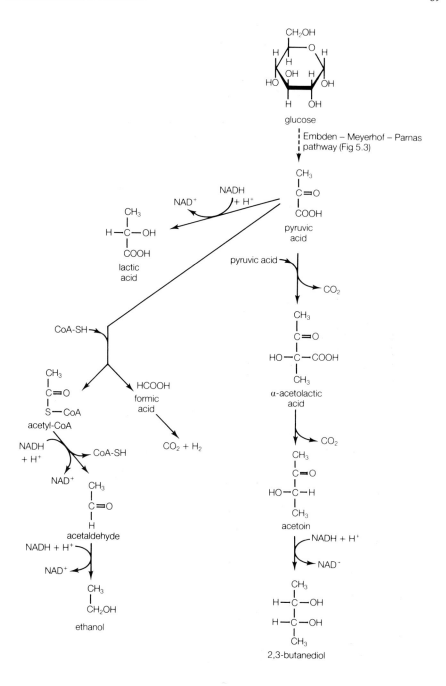

Fig. 5.6 The butanediol fermentation. (See also section 16.1.2.5 – the Voges–Proskauer test.)

the *substrate* is always an organic compound, even though the oxidizing agent may be inorganic or organic.)

Because an external oxidizing agent is used, the substrate undergoes a *net* oxidation (Fig. 5.7); glucose, for example, can be oxidized to carbon dioxide and water. The oxidation of a substrate provides more energy than that obtainable – from the same substrate – by fermentation.

How is the substrate oxidized, and how is usable energy obtained? The usual mode of oxidation of a typical *organic* substrate is shown in Fig. 5.7: oxidation is coupled with the reduction of NAD, the resulting NADH being oxidized by an external oxidizing agent. NADH and the external oxidizing agent usually interact indirectly via an *electron transport chain* (ETC) located in the cytoplasmic membrane. An ETC is a chain of specialized molecules (redox agents) which form a conducting path for electrons; the sequence of the (different) redox agents is such that electrons can flow down a redox gradient (towards the more positive end) in a series of oxidation/reduction reactions. Electrons from NADH flow down the gradient to an external oxidizing agent; when the latter is oxygen, the situation can be summarized as in Fig. 5.8. The final recipient of the electrons (in this case oxygen) is called the *terminal electron acceptor*.

Electron flow of this kind necessarily yields energy because the electrons are moving from high-energy to lower-energy locations. Typically, this liberated energy is used by the cell for pumping protons (hydrogen ions: H$^+$) across the cytoplasmic membrane – from the inner to the outer surface; this creates an imbalance of electrical charge (and pH) between the two surfaces of the membrane, and the tendency of protons to move back across the membrane (and thus abolish the imbalance) constitutes a form of energy known as *proton motive force*. Proton motive force (pmf) is one of the cell's most important and versatile forms of energy. It can be used – directly – to

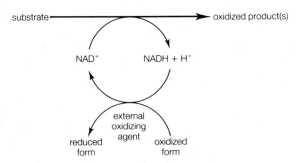

Fig. 5.7 Respiration (diagrammatic): the relationship between substrate, end-products and external oxidizing agent in the respiration of a typical *organic* substrate (compare Fig. 5.2). In this example, NADH – formed during metabolism of the substrate – is oxidized by the external oxidizing agent; NADH and the external oxidizing agent are generally coupled *indirectly* via an electron transport chain (see text).

satisfy several types of energy demand. Thus, it can drive flagellar rotation (section 2.2.14.1). It can provide energy for the transport (uptake) of various ions; ion uptake is an energy-requiring process because the cytoplasmic membrane is ordinarily impermeable to ions (section 2.2.8). It can provide energy for the transport of certain substrates across the cytoplasmic membrane – e.g. lactose uptake in *E. coli*. Pmf can also provide energy for the phosphorylation of ADP to ATP at enzyme complexes (*ATPases*) located in the cytoplasmic membrane. In respiration, when pmf is used as a source of energy for the synthesis of ATP from ADP the process is called *oxidative phosphorylation* (compare 'substrate-level phosphorylation': section 5.1.1.1). Interestingly, membrane ATPases can also catalyse the hydrolysis of ATP to ADP, the liberated energy being used to pump protons across the membrane – i.e. to augment pmf; thus, the energy in ATP and pmf is interconvertible! These ideas are summarized in Fig. 5.9.

In a bacterium, the electron transport chain involved in respiration (the *respiratory chain*) occurs in the cytoplasmic membrane (and, in some cases, in the thylakoids); its components vary from species to species, and variations may occur even in a given species growing under different conditions. Components found in respiratory chains include: (i) *cytochromes*: iron-containing proteins which receive and transfer electrons by the alternate reduction and oxidation of the iron atom; (ii) *iron–sulphur proteins* such as the *ferredoxins*; and (iii) *quinones*: aromatic compounds which can undergo reversible reduction. In the respiration of a typical organic substrate (Fig. 5.8), the oxidation of NADH and the reduction of the terminal electron acceptor (oxygen in Fig. 5.8) both appear to occur at the *inner* (i.e. cytoplasmic) face of the cytoplasmic membrane.

In Fig. 5.8 the source of NADH is given simply as 'NADH-producing metabolism'. We can now look at some actual pathways used in bacterial respiration. In many bacteria, the respiratory metabolism of glucose starts with the Embden–Meyerhof–Parnas pathway (EMP pathway, Fig. 5.3). Beyond pyruvic acid, however, the pathways of fermentation and respiration are completely different. In respiration, pyruvic acid is often converted to acetyl-CoA and fed into a cyclical pathway known as the *tricarboxylic acid cycle* (TCA cycle, Fig. 5.10) – also known as the *Krebs cycle* or the *citric acid cycle*.

Figure 5.10 shows that, in the TCA cycle, acetyl-CoA and oxaloacetic acid (OAA) combine to form citric acid; in the subsequent reactions, the original molecule of pyruvic acid is, in effect, oxidized to carbon dioxide. For each molecule of pyruvic acid oxidized, 4 molecules of NAD(P) and 1 of FAD are reduced, 1 molecule of ATP is synthesized, and 1 molecule of OAA is regenerated. NADH and $FADH_2$ can be oxidized via a respiratory chain, the resulting pmf being used e.g. for the synthesis of ATP at a membrane ATPase. In terms of energy yield, it should now be clear that respiration is much more efficient than fermentation: in respiration, the oxidation of NADH can lead, via pmf, to the synthesis of ATP, whereas in fermentation (where there is no

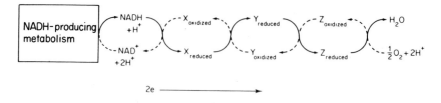

Fig. 5.8 Respiration (diagrammatic): an electron transport chain (ETC) consisting of three components (X, Y and Z), with oxygen as the terminal electron acceptor, in the respiration of a typical *organic* substrate. The cell's ongoing metabolism produces NADH, from which NAD must be regenerated by oxidation; some NADH is oxidized to NAD in biosynthetic reactions, and some is oxidized via the ETC. In the ETC, NADH is oxidized by transferring electrons to (and thus reducing) the oxidized form of X. The reduced form of X is oxidized by transferring electrons to the oxidized form of Y – and so on. The final step is the reduction of oxygen to water. The solid curved lines indicate the path of electron flow. Oxidation of NADH (enzyme: NADH dehydrogenase) and reduction of oxygen (at a cytochrome oxidase) both appear to occur at the *inner* (i.e. cytoplasmic) face of the cytoplasmic membrane.

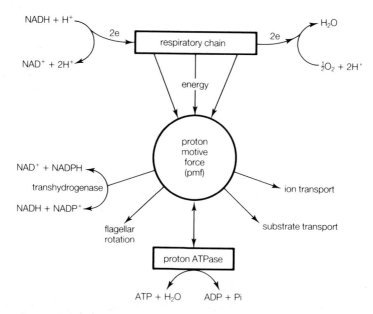

Fig. 5.9 Some of the roles of proton motive force. 'Respiratory chain' is the term used for an electron transport chain involved in respiratory metabolism (i.e. respiration). 'Proton ATPase' is an enzyme system (in the cytoplasmic membrane) which catalyses the pmf-dependent phosphorylation of ADP to ATP as well as the hydrolysis of ATP to ADP and inorganic phosphate (Pi); pmf is *used* for the synthesis of ATP but is *augmented* by the (energy-yielding) hydrolysis of ATP. Pmf also controls the (reversible) reduction of NADP by NADH; NADPH is used e.g. for some of the cell's biosynthetic reactions (e.g. ammonia assimilation – section 10.3.2).

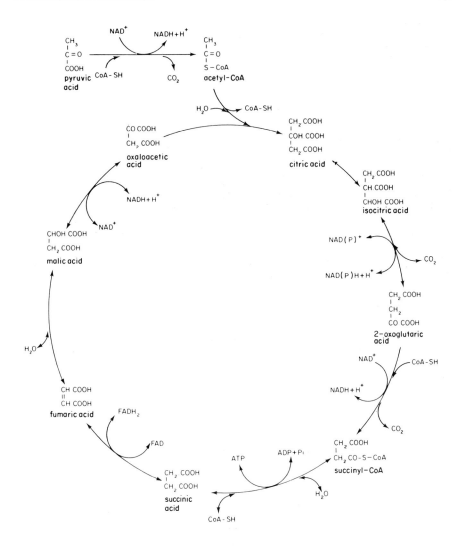

Fig. 5.10 The tricarboxylic acid cycle (TCA cycle): a common pathway in respiratory metabolism (section 5.1.1.2). The reaction isocitric acid→2–oxoglutaric acid is catalysed by the enzyme isocitrate dehydrogenase which, in most bacteria, is specific for NADP rather than NAD; NADPH is used e.g. in biosynthetic reactions. FAD (flavin adenine dinucleotide) – like NAD – carries energy in the form of 'reducing power'; $FADH_2$ can be oxidized via a respiratory chain, oxidation yielding pmf. In many bacteria the step from succinyl-CoA to succinic acid involves the synthesis of guanosine 5'-triphosphate (GTP) rather than ATP; GTP is used as an energy currency molecule e.g. in the binding of the aminoacyl-tRNA to the 'A' site in protein synthesis (Fig. 7.9).

external oxidizing agent) the cell has to get rid of NADH by synthesizing waste products such as lactic acid.

Anaerobic respiration. In principle, anaerobic respiration is the same as aerobic respiration: both use an external oxidizing agent. Under anaerobic conditions, however, oxidizing agents such as nitrate, sulphate and fumarate are used in place of oxygen; pmf can be generated by an anaerobic respiratory chain. The substrate (electron *donor*) used by chemoorganotrophs in anaerobic respiration may be any of various organic compounds, depending on species and conditions.

In *nitrate respiration* nitrate is used as the terminal electron acceptor, the nitrate being reduced to nitrite, nitrous oxide, nitrogen or ammonia – depending on species. When the nitrate is reduced mainly to nitrogen and/or nitrous oxide (i.e. gases) the process is called *denitrification* because it results in a loss of nitrogen to the atmosphere. This process can be a problem in agriculture because it can lower soil fertility (section 10.3.2.2). However, denitrification can be useful for eliminating the nitrogenous content of sewage in waste-water treatment plants (section 13.4).

Bacteria capable of denitrification include strains of *Alcaligenes faecalis*, *Bacillus licheniformis*, *Paracoccus denitrificans*, and *Pseudomonas stutzeri*. It was once thought that denitrification occurs only under anaerobic or microaerobic conditions. We now know that oxygen affects different denitrifying bacteria in different ways. For some of these bacteria anaerobiosis *is* necessary for denitrification. Others can denitrify in the presence of oxygen – in some cases consuming both oxygen and nitrate simultaneously – though the rate of denitrification typically decreases with increasing concentrations of oxygen.

Dissimilatory reduction of nitrate to ammonia (DRNA) is carried out by certain species under appropriate conditions (section 10.3.2).

Substrates used as electron *donors* by chemoorganotrophs in nitrate respiration include e.g. succinate (in *E. coli*).

The *sulphate-* and *sulphur-reducing bacteria* use sulphate and sulphur, respectively, as terminal electron acceptors; during anaerobic respiration, sulphate or sulphur is reduced to sulphide (Fig. 10.3). This type of anaerobic respiration is carried out e.g. by species of *Desulfovibrio* (sulphate reducers) and *Desulfuromonas* (sulphur reducer); these organisms typically occur in anaerobic mud and soil, and they are responsible for much of the hydrogen sulphide found in organically polluted waters.

In *fumarate respiration* the terminal electron acceptor is exogenous fumarate – which is reduced to succinate. Fumarate respiration is carried out by a range of bacteria, including *E. coli*, under appropriate conditions.

5.1.2 Energy metabolism in chemolithotrophs

Chemolithotrophs use *inorganic* substrates for energy metabolism – e.g.

sulphide, elemental sulphur, ammonia, hydrogen, ferrous ions etc. Metabolism usually involves aerobic or anaerobic respiration: electrons from the substrate are transferred to the external oxidizing agent, pmf being generated (see also section 5.3.4); NAD is not involved (compare respiration in chemoorgano-trophs). Obligate or facultative chemolithotrophs include (for example) the thiobacilli and the nitrifying bacteria.

Species of *Thiobacillus* occur e.g. in soil and marine muds. Typically, they oxidize substrates such as sulphide and sulphur aerobically, though *T. denitrificans* can metabolize these substrates anaerobically by nitrate respiration. A recent report indicates that *T. denitrificans* can use Fe (II) (ferrous ions) as electron donor in anaerobic nitrate-dependent growth (the nitrate being reduced primarily to dinitrogen) [Straub *et al.* (1996) AEM 62 1458–1460]. *T. ferrooxidans*, a strict aerobe, can oxidize ferrous ions as well as sulphur substrates. Thiobacilli are important in the sulphur cycle (section 10.3.3).

Nitrifying bacteria are obligately aerobic respiratory organisms which live in soil and aquatic environments. They obtain energy by *nitrification*: the oxidation of ammonia to nitrite (e.g. species of *Nitrosococcus* and *Nitrosomonas*) and nitrite to nitrate (e.g. species of *Nitrobacter* and *Nitrococcus*). These bacteria are important in the nitrogen cycle (section 10.3.2).

5.1.2.1 Inorganic fermentation

Until quite recently it was thought that no *inorganic* substrate could be fermented. Then, in 1987, Bak and Cypionka [Nature (1987) 326 891–892] described a type of energy metabolism in which the substrate (sulphite *or* thiosulphate) underwent 'disproportionation' to yield sulphide and sulphate. This 'chemolithotrophic fermentation' occurs e.g. in *Desulfovibrio sulfodismutans*.

5.1.2.2 Methane production

Methane (CH_4) is a by-product of an energy-yielding process in certain obligately anaerobic members of the Archaea – the *methanogens*; apparently, no member of the Bacteria produces methane. Methanogens live e.g. in river mud and in the rumen of cows and other ruminants (all environments characterized by an E_h below about -330 mV).

Some methanogens (e.g. *Methanobacterium*, *Methanobrevibacter*, *Methanococcus*, *Methanothermus*) produce methane by a complex series of reactions in which hydrogen is involved in the reduction of CO_2. Essentially, the carbon atom of CO_2 is bound, sequentially, to each of a series of C_1 carrier molecules and undergoes progressive reduction via $-CH_2-$ and CH_3 to methane (CH_4); the carrier molecules include methanofuran, tetrahydromethanopterin and several coenzymes. The final step of the reaction is thermodynamically favourable ($\Delta G^0 = -112.5$ kJ) and seems to be coupled to the generation of pmf.

Some methanogens (e.g. *Methanococcoides*, *Methanolobus*) can form methane from e.g. acetate or methanol. Growth on acetate is slower than that on carbon dioxide and hydrogen (the $\Delta G^{0'}$ of the reaction $CH_3COOH \rightarrow CH_4 + CO_2$ is about -36 kJ). Essentially, cleavage of acetate yields CO and CH_3, the latter being reduced to methane while CO is oxidized by water (enzyme: CO dehydrogenase) to CO_2 and 2H. In marshy soils, acetate may be the preferred substrate at lower temperatures [Wagner & Pfeiffer (1997) FEMSME 22 145–153].

5.2   ENERGY METABOLISM IN PHOTOTROPHS

Phototrophs obtain energy from sunlight – in most cases by *photosynthesis*.

5.2.1   Photosynthesis

In photosynthesis, the energy in light is absorbed by specialized pigments and is used to form energy currency molecules and/or pmf; in all cases, photosynthesis occurs in membranes containing *chlorophylls*, accessory pigments and electron transport chain(s). Chlorophylls are green, magnesium-containing pigments. Cyanobacteria contain chlorophyll *a* (which also occurs in algae and higher plants); the other photosynthetic bacteria contain one or more *bacteriochlorophylls* – pigments which are similar to the chlorophylls. The so-called *light reaction* of photosynthesis refers to all the (photochemical) events involved in the conversion of light energy to pmf or chemical energy; the *dark reaction* (= light-independent reaction) refers to the cell's use of its photosynthetically-derived energy for the synthesis of carbon compounds.

Photosynthetic bacteria can be divided into two main categories: (i) those which carry out photosynthesis aerobically (and which produce oxygen as a by-product), and (ii) those which carry out photosynthesis anaerobically (and which do not produce oxygen).

5.2.1.1   *Oxygenic (oxygen-producing) photosynthesis in bacteria*

This process – which closely resembles photosynthesis in green plants and algae – occurs in the cyanobacteria. Because the cyanobacteria carry out eukaryotic-type photosynthesis they were, for many years, regarded not as bacteria but as 'blue-green algae'; today, the bacterial nature of these organisms is not in doubt.

In almost all cyanobacteria photosynthesis occurs in the thylakoid membranes (section 2.2.7). (In strains of *Gloeobacter*, which lack thylakoids, the photosynthetic components appear to occur in the cytoplasmic membrane.)

Chlorophylls occur in so-called *reaction centres* to which light is channelled

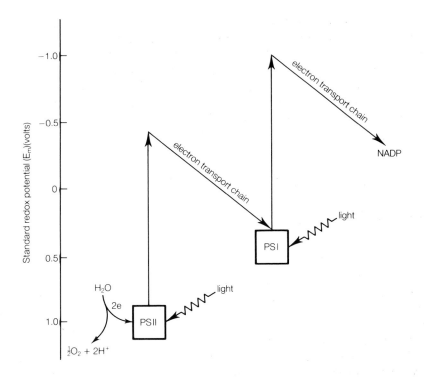

Fig. 5.11 Oxygenic photosynthesis: simplified Z scheme (scale approximate). Light energy causes the reaction centres (PSI, PSII) to eject energized electrons whose energy levels and destinations are shown by the arrows; electron flow from PSII to PSI generates pmf. Electrons ejected from PSII are replaced by the oxidation of water, oxygen being liberated. [How does PSII split water? (review): Rögner *et al.*(1996) TIBS 21 44–49.]

by specialized protein–pigment *light-harvesting complexes* (LHCs); the composition of the LHCs can be influenced by environmental factors such as light quality and the availability of nitrogen and sulphur [Grossman *et al.* (1993) JB *175* 575–582]. When the chlorophyll receives light energy it ejects highly energized electrons; these electrons can flow down an electron transport chain and provide energy for (i) pmf generation and/or (ii) the direct reduction of NADP. (The cell uses NADPH e.g. for various biosynthetic reactions.)

Oxygenic photosynthesis is generally represented by the Z *scheme* (Fig. 5.11). Energized electrons ejected from photosystem II (PSII) flow down an electron transport chain to photosystem I (PSI), and this electron flow generates pmf across the thylakoid membrane; when such photosynthetically-derived pmf is used for the synthesis of ATP (at a membrane-bound ATPase) the process is called *photophosphorylation*. Electrons ejected from PSI have sufficient energy to reduce NADP to NADPH. The flow of electrons (from left to right in Fig. 5.11) requires an input of electrons to PSII. This is achieved by the oxidation of water, oxygen being formed as a by-product. A recently revised eukaryotic Z scheme includes the reduction of NADP by PSII [see Prince (1996) TIBS *21* 121–122].

Some cyanobacteria (e.g. *Oscillatoria limnetica*) can – as an alternative – carry out photosynthesis anaerobically, using sulphide instead of water as an electron donor; the sulphide is oxidized to elemental sulphur. Some cyanobacteria can even grow as chemoorganotrophs.

A small group of prokaryotic organisms (Prochlorophyta; the prochlorophytes) – related to the cyanobacteria – contain both of the eukaryotic-type chlorophylls: *a* and *b*. Only three species are currently known: *Prochloron didemni*, *Prochlorothrix hollandica* and *Prochlorococcus marinus*.

5.2.1.2 Anoxygenic photosynthesis

Anoxygenic photosynthesis (in which oxygen is not produced) is carried out anaerobically by bacteria of the order Rhodospirillales. In the so-called 'purple' photosynthetic bacteria (suborder Rhodospirillineae), all the photosynthetic components occur in intracellular membranes which are continuous with the cytoplasmic membrane. In the 'green' photosynthetic bacteria (suborder Chlorobiineae), components of the light-harvesting complexes occur in *chlorosomes* (section 2.2.7) while the reaction centres occur in the cytoplasmic membrane.

In the 'purple' bacteria, electrons ejected from a reaction centre follow a *cyclic* path – via an electron transport chain – back to the reaction centre; the pmf which is generated can be used e.g. for the synthesis of ATP (i.e. photophosphorylation).

In 'green' bacteria the electron flow generates pmf (for e.g. ATP synthesis) and it can also be used for the direct reduction of NAD to NADH. Such *non-cyclic* electron flow needs a supply of electrons from an exogenous donor; electron donors used by the 'green' photosynthetic bacteria include e.g. sulphide and thiosulphate – but never water, because in these organisms photosynthesis is always anoxygenic.

Some bacteria of the Rhodospirillales (including many 'purple' bacteria) can live as chemoorganotrophs under aerobic or microaerobic conditions.

5.2.1.3 Electron donors in photosynthesis

Inorganic electron donors in photosynthesis include e.g. water (used only by cyanobacteria), sulphide, sulphur and hydrogen; organic donors include e.g. formic acid and methanol. Phototrophs which use inorganic electron donors are called *photolithotrophs*; those which use organic donors are called *photoorganotrophs*.

5.2.2 The purple membrane

In the Archaea, certain strains of *Halobacterium salinarium* (an extreme halophile – section 3.1.7) can use the energy in sunlight even though they cannot carry out photosynthesis. In these strains, a differentiated, purple-pigmented region of the cytoplasmic membrane – the *purple membrane* – develops under anaerobic or microaerobic conditions in the light; the main purple pigment is called *bacteriorhodopsin*. When a molecule of bacteriorhodopsin absorbs light energy it passes rapidly through a cycle of states (photointermediates), all within a few thousandths of a second; during this time protons are pumped outwards across the membrane, i.e. the 'photocycling' in bacteriorhodopsin generates pmf. Unlike photosynthesis, pmf generation in the purple membrane appears to involve the *direct* pumping of protons across the membrane, and chlorophyll is not used.

5.3 OTHER TOPICS IN ENERGY METABOLISM

5.3.1 The yield of ATP

In a living cell, no pathway or process occurs in isolation but rather forms part of a complex whole. Because of this, the maximum theoretical yield of ATP from the respiration of (say) one molecule of glucose may not be achieved owing to interactions between the respiratory pathway and other pathways in the cell. For example, not all the energy liberated (as pmf) by electron transport may be used for oxidative phosphorylation: some may be needed e.g. for ion transport or flagellar motility. Again, some of the NADH generated during respiration may be used for biosynthesis; this NADH is clearly not available for oxidation (via the respiratory chain) so that it cannot contribute to pmf and, hence, ATP synthesis. Finally, intermediates in energy metabolism may be drawn off to supply the cell with 'building blocks' for biosynthesis; thus, e.g. pyruvic acid is used in the synthesis of the amino acids alanine, valine and leucine. The withdrawal of intermediates necessarily sacrifices the energy which would otherwise have been obtained by their metabolism.

5.3.2 Reverse electron transport

Bacteria need reducing power – e.g. NADH and/or NADPH – for biosynthesis. For fermentative bacteria this is never a problem, while the cyanobacteria and 'green' photosynthetic bacteria obtain these reduced molecules by direct reduction using non-cyclic electron flow. However, the 'purple' photosynthetic bacteria cannot do this: electrons ejected from their reaction centres do not have enough energy to reduce NAD. Instead, these organisms use *reverse electron transport*; in this process, pmf is used to drive electrons 'uphill' to a membrane-bound enzyme, NAD dehydrogenase, where NAD is reduced to NADH. This requires an input of electrons from an external electron donor; the 'purple' bacteria typically use organic electron donors, but some can use inorganic donors such as sulphide. Reverse electron transport is also used by the nitrifying bacteria and other chemolithotrophs; note that chemolithotrophs (section 5.1.2) do not form NADH during metabolism of the energy substrate.

5.3.3   End-product efflux

Fermentative bacteria make a great deal of NADH; in fact, they have to get rid of some of it by synthesizing waste products such as lactic acid (section 5.1.1.1). Making a virtue of necessity, some bacteria (e.g. *Lactococcus (Streptococcus) cremoris*) gain energy by linking protons to their waste lactate; when lactic acid passes outwards across the cytoplasmic membrane its 'passenger' protons automatically augment pmf!

5.3.4   Extracytoplasmic oxidation

Complex energy substrates are generally transported across the cytoplasmic membrane before being metabolized. However, some simple energy substrates (such as H_2 and Fe^{2+}) appear to be oxidized at the *outer* face of the cytoplasmic membrane or in the periplasmic region; such *extracytoplasmic oxidation* of a substrate is characterized by: (i) release of protons from the substrate and / or water extracytoplasmically, (ii) a transmembrane flow of electrons (from the substrate) to the cytoplasmic (inner) face of the membrane, and (iii) interaction of these electrons with protons and a terminal electron acceptor (e.g. oxygen). The net result is pmf generation. For example, the strict aerobe *Thiobacillus ferrooxidans* gains energy from the oxidation of Fe^{2+} (to Fe^{3+}) at low pH; according to one scheme, Fe^{3+} (produced extracytoplasmically) reacts with water to yield ferric hydroxide and protons, while electrons (from Fe^{2+}) reduce a terminal electron acceptor (oxygen?) at the cytoplasmic face of the membrane.

Some bacteria (e.g. *Pseudomonas aeruginosa*) can generate pmf by the extracytoplasmic oxidation of glucose to gluconate; in these bacteria the enzyme glucose dehydrogenase – with its cofactor *pyrroloquinoline quinone* (PQQ) – occurs bound to the outer surface of the cytoplasmic membrane. The gluconate can be transported across the membrane and phosphorylated to 6-phosphogluconate, which can enter the Entner–Doudoroff pathway (Fig. 6.2).

E. coli (and many related bacteria) normally form a membrane-bound glucose dehydrogenase which lacks the necessary PQQ; such organisms can carry out extracytoplasmic oxidation only if they are provided with PQQ [Bouvet, Lenormand & Grimont (1989) IJSB 39 61–67]. However, a mutant strain of *E. coli* with a non-functional PTS system (section 5.4.2) can synthesize

PQQ and carry out extracytoplasmic oxidation – suggesting that PQQ-encoding genes may be present but not expressed under normal conditions [Biville, Turlin & Gasser (1991) JGM *137* 1775–1782].

5.3.5 Pmf and aerotaxis

Surprisingly, the respiratory-type organism *Azospirillum brasilense* – which uses oxygen as terminal electron acceptor – is microaerophilic (section 3.1.6); to reach optimal conditions, the organism exhibits aerotaxis (section 2.2.15.2), moving towards regions where the oxygen concentration is only 3–5 μM. Recent work on *A. brasilense* has shown that pmf increases when the cell swims towards the preferred concentration of oxygen, and decreases when the cell swims away from it; it has been suggested that this change in the level of pmf acts as the signal which regulates aerotactic movements [Zhulin *et al.* (1996) JB *178* 5199–5204].

The ability of *A. brasilense* to carry out aerobic respiratory metabolism under such low levels of oxygen is believed to be due to a highly efficient *oxidase* (oxygen-reducing enzyme). Interestingly, for *A. brasilense* (a nitrogen-fixing organism), these oxygen-poor conditions are useful for carrying out nitrogen fixation (section 10.3.2.1).

5.3.6 Sodium motive force (smf)

Smf is the energy associated with an electrochemical gradient of sodium ions between the inner and outer surfaces of the cytoplasmic membrane; it is analogous to pmf.

In some cases a cell can generate smf by using pmf: the entry of protons into the cytoplasm is linked to the exit of sodium ions across the cytoplasmic membrane (= Na^+/H^+ *antiport*). In *E. coli*, such antiport is used to provide a source of energy (smf) suitable e.g. for the uptake (= *transport*, section 5.4) of the sugar melibiose; thus, melibiose and sodium ions are jointly transported into the cytoplasm (= Na^+/melibiose *symport*), i.e. melibiose uptake consumes smf.

In some marine bacteria (e.g. *Vibrio alginolyticus*) energy from respiration can be used directly for the outward pumping of Na^+ via a *sodium pump*; this organism can use smf e.g. for the uptake of various amino acids.

At the higher range of growth temperatures, the inherent permeability of the cytoplasmic membrane for both protons and sodium ions tends to increase; however, the increase in permeability is greater for protons than it is for sodium ions. A pmf-based organism will need to use extra energy for generating a given level of pmf simply in order to compensate for the higher inward diffusion of protons at these higher temperatures. In some organisms this problem is minimized because the (special) composition of their cytoplasmic membrane is such that raised temperatures have less effect on proton permeability. However, some thermophiles (section 3.1.4) have abandoned the use of protons in their energy metabolism; for example, *Clostridium fervidus* uses Na$^+$ as the sole energy-coupling ion (i.e. pmf is not involved in this species). [Ion permeability at high temperatures: Driessen, van de Vossenberg & Konings (1996) FEMSMR *18* 139–148.]

5.4   TRANSPORT SYSTEMS

The cytoplasmic membrane (section 2.2.8) is an efficient barrier which stops most molecules (and all ions) from passing freely into and out of the cytoplasm. (This, of course, is necessary to allow the cell to control its own internal environment.) In Gram-negative bacteria there is also the extra barrier of the outer membrane (section 2.2.9.2). However, during ongoing metabolism the cell must be able to take up various substrates (for energy and other purposes) and get rid of waste products; this is the job of the various transport systems. Each transport system is typically specific for one or for a small number of substrates, though a given substrate may have more than one type of transport system, even in the same cell. Transport often requires energy – which is typically supplied by pmf or by a 'high-energy phosphate'.

Some transport systems are complex and are not fully understood; a few are outlined below.

Pmf can be used, directly, as energy for the transport of certain charged and uncharged solutes. In *E. coli*, for example, 'proton–lactose symport' allows the uptake of lactose at the expense of pmf, lactose and protons being *jointly* transported into the cytoplasm.

Energy for the transport of ions is sometimes provided by the hydrolysis of ATP at a membrane-bound ATPase. For example, in *E. coli*, K$^+$ can be transported by the so-called Kdp system which involves a specialized K$^+$-ATPase (*potassium pump*); hydrolysis of ATP at the pump allows uptake of K$^+$ against a concentration gradient.

Transport systems are also involved in the recycling of cell components. During ongoing growth in *E. coli*, for example, peptidoglycan (Fig. 2.7) is continually being degraded and is partly recycled – oligopeptides being transported from the periplasm back to the cytoplasm (for re-use) by the *oligopeptide permease* (Opp) system [Goodell & Higgins (1987) JB *169* 3861–3865].

5.4.1   ABC transporters

An ABC transporter is a transport system consisting of a multi-protein complex in the cell envelope, a given transporter being involved in the import or export/secretion of particular ions or molecules; at least one of the transporter's proteins has a binding site for ATP – the *ATP-binding cassette*

(ABC) – and it appears that hydrolysis of ATP at this site generally provides the energy for transport. Interestingly, *as well as* transporting specific solute(s), at least some ABC transporters seem to regulate channels through which *other* solutes are translocated; for example, in *Salmonella typhimurium*, the transporter encoded by genes *sapABCDF* seems to import certain toxic peptides (for detoxification) but *also* helps to regulate a K^+ ion channel which is necessary for bacterial resistance to these peptides (see e.g. Higgins [(1995) Cell *82* 693–606]).

5.4.1.1   ABC importers (binding-protein-dependent transporters; periplasmic permeases)

These systems mediate the uptake of e.g. certain amino acids, sugars and ions. Each system includes a periplasmic substrate-binding protein and a protein complex which forms part of the cytoplasmic membrane. One example is a histidine uptake system in *Salmonella typhimurium*; in this system, the periplasmic protein binds histidine and passes it to the protein complex which, in turn, transfers it to the cytoplasm.

5.4.1.2   ABC exporters

Collectively, these exporters mediate the export/secretion of a range of proteins (e.g. haemolysins (section 16.1.4.1) and enzymes), peptide antibiotics and (in some species) polysaccharide components of a capsule; however, a given exporter will transport only one type of molecule, or closely related molecules. One example is the α-haemolysin exporter of *E. coli* which mediates one-step transport, from cytoplasm to the cell's exterior (i.e. *secretion*), via an assembly of exporter proteins which spans the cell envelope. [ABC transporters (bacterial exporters): Fath & Kolter (1993) MR *57* 995–1017.]

   Proteins secreted by pathogens include some important *virulence factors* (section 11.5) – e.g. the cyclolysin of *Bordetella pertussis* (causal agent of whooping cough), the alkaline protease of *Pseudomonas aeruginosa* and the *E. coli* α-haemolysin. In Gram-negative bacteria, protein secretion via an ABC exporter involves the ABC protein (located in the cytoplasmic membrane), a so-called *membrane fusion protein* (MFP) (in the cytoplasmic membrane, but partly in the periplasm) and a component in the outer membrane. The MFP apparently links the cytoplasmic and outer membranes, facilitating transport. Typically, proteins transported by these exporters lack an N-terminal signal sequence (section 7.6) but have a C-terminal *secretion sequence* that may undergo direct interaction with the ABC protein. Exporters which transport molecules into the periplasm, or outer membrane – as the *final* target site – may have fewer protein components than those which *secrete* proteins.

   Recent *in vitro* work on the secretion of proteins by Gram-negative bacteria has suggested that the three components of an ABC exporter – the ABC protein, the MFP and the outer membrane component – assemble to form the complete transport system in a definite order, and that assembly is promoted by the binding of the substrate (i.e. the protein to be secreted) to the ABC protein. Thus, on substrate–ABC binding, the ABC protein interacts with MFP – the latter then binding to the outer membrane component [Létoffé, Delepelaire & Wandersman (1996) EMBO Journal 15 5804–5811].

5.4.2 PTS: the phosphoenolpyruvate-dependent phosphotransferase system

This type of transport system is used by various Gram-negative and Gram-positive bacteria for the uptake of certain sugars (e.g. glucose, fructose, lactose), sugar alcohols (e.g. mannitol) and substituted sugars (e.g N-acetylglucosamine), the source of energy being phosphoenolpyruvate (PEP, formula: Fig. 5.3). In the simplest case, PEP supplies phosphate and energy for the sequential phosphorylation of two energy-coupling proteins (designated I and H) in the cytoplasm; the phosphate and energy are transferred from H to a cytoplasmic-membrane-associated, sugar-specific permease (the so-called II complex) which binds the substrate and then phosphorylates and *concomitantly* transports it into the cytoplasm (Fig. 5.12).

Since sugars are often metabolized via their phosphate derivatives (e.g. glucose 6-phosphate in Figs 5.3 and 6.2), phosphorylation during uptake is a positive aspect of PTS transport.

PTS is involved not only in transport: the cell can also respond *chemotactically* to those sugars which are taken up by the PTS system. In one possible mechanism, phosphorylated PTS components (Fig. 5.12) interact – via unknown intermediate(s) – with the CheY protein of the MCP system (Fig. 7.13), thus influencing flagellar rotation. PTS-mediated chemotaxis differs from the MCP system in several ways [Titgemeyer (1993) JCB 51 1–6]; thus, (i) the substrate must be *transported* (rather than simply bound); (ii) adaptation does not involve methylation; and (iii) the PTS does not respond to repellents.

[PTS (review): Saier & Reizer (1994) MM 13 755–764.]

5.4.3 The general secretory pathway (*sec*-dependent pathway) in Gram-negative bacteria

Many proteins which are secreted (i.e. released to the cell's exterior) – e.g. toxins, certain enzymes – are transported from the cytoplasm to the exterior via the two-stage *general secretory pathway* (GSP). Such transport requires energy: pmf as well as ATP.

A secretory protein (i.e. one to be secreted) is synthesized with a special N-terminal *signal sequence* (section 7.6) that facilitates passage through the cytoplasmic membrane; this sequence is cleaved (by an enzyme known as a

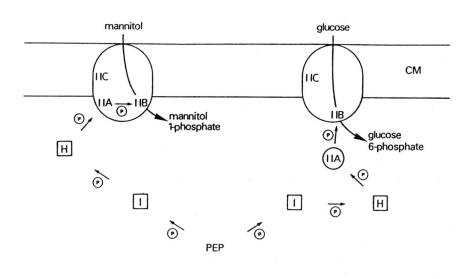

Fig. 5.12 Transport by the phosphoenolpyruvate-dependent phosphotransferase system (PTS) (simplified, diagrammatic). PTS (section 5.4.2) is shown here transporting mannitol and glucose across the cytoplasmic membrane (CM) into the cytoplasm in *Escherichia coli*.

The source of energy for transport is phosphoenolpyruvate (PEP). Energy is initially fed into the system by the sequential transfer of a phosphate group ℗ from the 'high-energy phosphate' PEP to enzyme I (designated I) and then to the Hpr protein (designated H). Phosphate is then transferred to a *permease* (= enzyme II complex, designated II). The permease for mannitol (left) is entirely membrane bound, and consists of three domains: IIA, IIB and IIC. IIC is a hydrophobic, trans-membrane domain which binds the substrate and appears to contain, or to contribute to, a trans-membrane channel; the IIA and IIB domains each contain a phosphorylation site, but phosphorylation of the substrate seems directly to involve only the IIB domain.

In the glucose permease the IIA domain is a separate protein, but the phosphorylation sequence is from IIA to IIB, and phosphorylation of the substrate involves IIB.

In all cases, phosphorylation of the substrate molecule is an integral part of its transfer into the cytoplasm.

In some PTS permeases the components are grouped in other ways. For example, the fructose permease of enteric bacteria consists of a membrane-bound protein, containing IIC and IIB, and a separate protein which includes both IIA and H functions. In the fructose permease of *Rhodobacter capsulatus*, a single, separate protein includes the functions of IIA, H and I.

According to a proposed nomenclature for PTS permeases [Saier & Reizer (1992) JB *174* 1433–1438], the mannitol permease of *E. coli* is designated IICBA^{Mtl,Eco}, and the glucose permease of *E. coli* is designated IICB^{Glc,Eco}+IIA^{Glc,Eco}.

signal peptidase) during passage of the protein from the membrane to the periplasm. This first stage of the GSP requires the products of the *sec* (secretion) genes.

In the second stage, the secretory protein is transported from the periplasm, through the outer membrane, to the cell's exterior. This stage, which requires the products of many genes – and which seems not to occur in *E. coli* – is not well understood.

[Reviews on protein secretion in Gram-negative bacteria: Salmond & Reeves (1993) TIBS *18* 7–12; Pugsley (1993) MR *57* 50–108.]

5.4.4   Transport through the peptidoglycan barrier

An intact cell envelope is needed to prevent osmotic lysis. However, the peptidoglycan layer cannot *completely* enclose the protoplast because various large molecules (e.g. the *E. coli* α-haemolysin – section 5.4.1.2) are transported through the envelope, while the assembly of fimbriae (and other appendages) requires the outward transport of protein components. In certain types of conjugation (section 8.4.2) a large DNA–protein complex is transported from donor to recipient cell via the envelopes of both cells. Moreover, the peptidoglycan layer must be breached during the assembly of the flagellar basal body, and adhesion sites (section 2.2.10) may involve the linkage of membranes at regions of discontinuity in the peptidoglycan.

The assembly of transport-specific or other structures which cross the cell envelope is likely to involve enzyme(s) which locally modify peptidoglycan while preserving structural integrity (cf. Fig. 3.1). The assembly of some of these structures has indeed been linked with peptidoglycan metabolism [review: Dijkstra & Keck (1996) JB *178* 5555–5562].

6 Metabolism II: carbon

Why carbon? Simply because virtually all the compounds which comprise – and which are formed by – living organisms are carbon compounds. In this chapter we consider (i) the carbon requirements of bacteria, (ii) the ways in which different carbon compounds are assimilated, and (iii) the synthesis, interconversion and polymerization of carbon compounds. The metabolism of nitrogen and sulphur is considered in Chapter 10.

What sort of carbon compounds do bacteria need for growth? Some bacteria are able to use carbon dioxide for most or all of their carbon requirements; such bacteria are called *autotrophs*, and the use of carbon dioxide as the sole (or main) source of carbon is called *autotrophy*. In some autotrophic bacteria autotrophy is optional; in others it is obligatory: *only* carbon dioxide can be used as a source of carbon – even when glucose or other substrates are freely available. The *obligate* autotrophs use simple energy substrates (Chapter 5) as well as a simple carbon source: they are either chemolithotrophs (i.e. *chemolithoautotrophs*) or photolithotrophs (i.e. *photolithoautotrophs*).

Chemolithoautotrophic metabolism occurs in relatively few bacteria (and it also occurs in some members of the Archaea); this type of metabolism is unique in the living world, being found only in these specialized prokaryotes and in no other type of organism. The chemolithoautotrophs have important roles e.g. in the nitrogen cycle (Chapter 10).

Photolithoautotrophic metabolism is found e.g. in the cyanobacteria – and, of course, in green plants and algae.

Most bacteria are *not* autotrophic: they cannot use carbon dioxide as a major source of carbon, and their growth depends on a supply of complex carbon compounds derived from other organisms; bacteria which need complex carbon compounds are called *heterotrophs*. Chemoorganotrophic heterotrophs are *chemoorganoheterotrophs*. Collectively, the heterotrophs can use a vast range of carbon sources – including sugars, fatty acids, alcohols and various other organic substances. Heterotrophic bacteria are widespread in nature, and they include (for example) all those species which cause disease in man, other animals and plants.

6.1 CARBON ASSIMILATION IN AUTOTROPHS

In autotrophs, carbon dioxide from the environment is used to form complex organic compounds; when carbon dioxide is incorporated into such compounds it is said to have been 'fixed'. Different autotrophs have different pathways for carbon dioxide fixation, but there are two very common pathways: the Calvin cycle and the reductive TCA cycle.

6.1.1   The Calvin cycle

This pathway (also called the *reductive pentose phosphate cycle*) is used by a wide range of autotrophs, including some anoxygenic photosynthetic bacteria and most or all cyanobacteria. Part of the Calvin cycle is shown in Fig. 6.1. Each turn of the Calvin cycle requires a considerable input of both ATP and reducing power, i.e. carbon dioxide fixation needs a lot of energy. The key enzymes in this pathway are ribulose 1,5–bisphosphate carboxylase-oxygenase (RuBisCO – see also section 2.2.6) and phosphoribulokinase; these enzymes are found only in cells which fix carbon dioxide via the Calvin cycle.

6.1.2   The reductive TCA cycle

This pathway is used for carbon dioxide fixation e.g. by phototrophic bacteria of the family Chlorobiaceae, and by the chemotrophic archaen *Sulfolobus*. Essentially, the pathway resembles the TCA cycle (Fig. 5.10) operating in reverse, with its one-way reactions (such as oxaloacetic acid→citric acid) being modified by different enzymes/reaction sequences. As in the Calvin cycle, carbon dioxide fixation requires a great deal of energy.

6.1.3   Carboxydobacteria

Carboxydobacteria can use carbon *monoxide* as the sole source of carbon and energy, i.e. they do not conform to the strict definition of 'autotroph'. However, they oxidize carbon monoxide to carbon dioxide, aerobically, and they assimilate carbon dioxide via the Calvin cycle. These organisms, which occur in soil, polluted waters and sewage, include e.g. *Bacillus schlegelii* and *Pseudomonas carboxydovorans*.

6.2   CARBON ASSIMILATION IN HETEROTROPHS

Collectively, the heterotrophs can assimilate a vast range of carbon sources; only a few examples are given below.

Many heterotrophs can use glucose, and its mode of assimilation will depend on the pathways and enzyme systems in any given organism. For example, many bacteria (including *E. coli*) can assimilate glucose via 'energy' pathways such as the EMP pathway (Fig. 5.3) and – subsequently – the TCA cycle (Fig. 5.10). In *Pseudomonas aeruginosa* glucose can be assimilated via

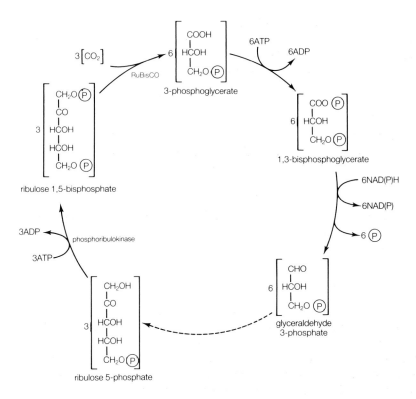

Fig. 6.1 Part of the Calvin cycle, showing the reaction in which carbon dioxide is 'fixed'. The purpose of the pathway is to enable carbon dioxide to be used for the synthesis of the complex carbon compounds which make up the cell itself. Essentially, carbon enters the cycle by the fixation of carbon dioxide, and carbon leaves the cycle when intermediates are withdrawn for use in biosynthesis. For example, 3-phosphoglycerate is used for the synthesis of certain amino acids (such as glycine and serine) and also for the formation of pyruvate – itself a precursor of alanine, leucine and other amino acids; note that obligate autotrophs must synthesize all the amino acids necessary for protein synthesis. Continued operation of the Calvin cycle demands a continual supply of ribulose 1,5-bisphosphate; hence, the amount of carbon removed from the cycle must not exceed that put into the cycle. (The encircled 'P' represents phosphate.) RuBisCO is the enzyme ribulose 1,5-bisphosphate carboxylase–oxygenase. The many interconnecting pathways in the Calvin cycle have been omitted for clarity.

extracytoplasmic oxidation (section 5.3.4). These pathways provide both energy and the compounds used as starting points in biosynthesis.

Some bacteria (e.g. species of *Acetobacter* and *Pseudomonas* can assimilate glucose via the *Entner–Doudoroff pathway* (Fig. 6.2); this yields NAD(P)H for biosynthesis as well as some useful precursor molecules. [Entner–Doudoroff pathway: history, physiology and molecular biology: Conway (1992) FEMSMR *103* 1–28.] The *hexose monophosphate pathway* (HMP pathway) provides a

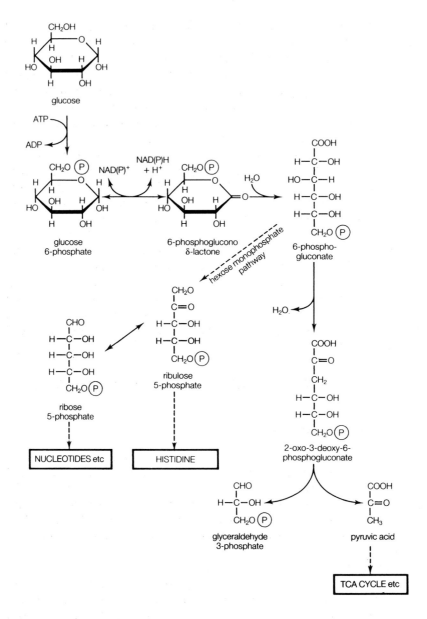

Fig. 6.2   The Entner–Doudoroff pathway: a pathway which can be used for the assimilation of carbon in some heterotrophic bacteria. The hexose monophosphate pathway is similar as far as 6-phosphogluconate; it continues (as shown) via ribulose 5-phosphate. Glyceraldehyde 3-phosphate may be converted to pyruvic acid (as in the EMP pathway: Fig. 5.3) or (e.g. in pseudomonads) re-cycled (via gluconeogenesis – section 6.3.2), re-entering the pathway as 6-phosphogluconate.

source of ribulose 5-phosphate (Fig. 6.2) – an important precursor of nucleotides (Chapter 7) and of the amino acid histidine. It also provides the NADPH needed as a source of reducing power for various biosynthetic reactions – such as the synthesis of fatty acids. Moreover, the HMP pathway is a source of erythrose 4-phosphate, a precursor molecule required in the synthesis of aromatic amino acids such as L-tryptophan and L-phenylalanine. The HMP pathway is widely distributed in nature, occurring also in eukaryotic microorganisms and in animals and plants.

In *E. coli*, the sugar lactose is split into glucose and galactose. The glucose is metabolized by the EMP pathway (Fig. 5.3), and the galactose is metabolized by the Leloir pathway. In the *Leloir pathway*, galactose is 1-phosphorylated (by ATP), and the galactose 1-phosphate is converted, enzymatically, via glucose 1-phosphate to glucose 6-phosphate – which enters the EMP pathway.

Cellulose, a polymer of glucose, is used as a carbon source by a number of actinomycetes (bacteria found in soil and compost heaps etc.) and e.g. by certain bacteria in the *rumen* (the grass-digesting part of the alimentary canal in cows and other ruminants). These bacteria produce enzymes which can degrade certain types of cellulose – outside the cell – into products which include e.g. the disaccharide *cellobiose*. Cellulolytic (i.e. cellulose-degrading) bacteria include species of *Cellulomonas* and e.g. *Clostridium thermocellum* and some strains of *Pseudomonas* and *Ruminococcus*.

6.3   SYNTHESIS, INTERCONVERSION AND POLYMERIZATION OF CARBON COMPOUNDS

6.3.1   Synthesis of carbon compounds

Synthesis of the vast range of molecules which form the structure of a living cell clearly demands an enormously complex (and rigorously controlled) network of chemical reactions. Moreover, as well as 'structural' molecules the cell needs energy with which to carry out the various anabolic reactions; carbon and energy metabolism are usually closely inter-linked (section 5.3.1). Here, we have space to look only briefly at some of the generalities of biosynthesis.

Many compounds in the initial assimilative pathways can be used more or less immediately as starting points for biosynthesis; from the cell's point of view this makes good sense: if lengthy metabolism were needed to make the carbon available, any early breakdown in the carbon pathway could be a serious problem for biosynthesis. Thus, for example, in the Calvin cycle (Fig. 6.1) the very first product of carbon dioxide fixation, 3-phosphoglycerate, can be used to synthesize the amino acids cysteine, glycine and serine (themselves components of proteins – section 7.6) as well as other compounds. Pyruvic acid (from the EMP and Entner–Doudoroff pathways) is a precursor of e.g. various amino acids, and ribose 5-phosphate (from the HMP pathway) is used e.g. to synthesize nucleotides (components of DNA and RNA – Chapter 7).

Intermediates in the TCA cycle (Fig. 5.10) can also be used for biosynthesis. For example, oxaloacetic and 2-oxoglutaric acids are precursors of a range of amino acids; acetyl-CoA can be used e.g. for fatty acid synthesis; and succinyl-CoA is used for the synthesis of porphyrins – components of chlorophylls and cytochromes. Since intermediates withdrawn from the TCA cycle cannot be used to regenerate oxaloacetic acid (OAA), this compound must be generated in some other way if the cycle is to continue. There are various reactions for achieving this – for example, under certain conditions, OAA may be formed directly by the carboxylation of pyruvic acid or of phosphoenolpyruvate; such 'replenishing' reactions are called *anaplerotic sequences.*

In general, the assimilative pathways in both autotrophs and heterotrophs yield a range of 3-, 4- and 5-carbon molecules which are useful precursors in biosynthesis.

Each step in a biosynthetic reaction is normally enzyme-mediated, very few reactions occurring spontaneously. As well as enzymes, there are commonly other requirements. Reductive steps typically involve NAD(P)H or FADH$_2$, oxidative steps NAD or FAD, and phosphorylation usually involves ATP, GTP or phosphoenolpyruvate. Coenzymes have important roles in bacterial metabolism; they include e.g. coenzyme A (a carrier of acetyl and other acyl groups) and thiamine pyrophosphate (TPP) (involved e.g. in various decarboxylation reactions). The way in which a diverse range of metabolic activities can depend on a given coenzyme is well illustrated by the roles of tetrahydrofolate (THF), a complex compound derived from a pteridine derivative, *p*-aminobenzoic acid and glutamic acid. Some of the roles of THF are shown in Table 6.1; notice that these THF-dependent functions involve essential aspects of cell growth: the synthesis of dTMP and purines (and, hence, the synthesis of DNA nd RNA – Chapter 7) and

Table 6.1 The coenzyme tetrahydrofolate (THF): examples of roles in bacterial metabolism

THF derivative	Function
N^{10}-formyl-THF	Initiation of protein synthesis (formation of N-formylmethionine – section 7.6)
N^5-methyl-THF	Biosynthesis of L-methionine
N^5, N^{10}-methylene-THF	Glycine $\leftrightarrow$ serine interconversion
N^5, N^{10}-methylene-THF	Biosynthesis of deoxythymidine monophosphate (dTMP) (Fig. 6.3)
N^5, N^{10}-methenyl-THF	Biosynthesis of purine bases (found in DNA and RNA)

the synthesis of proteins. THF itself is synthesized by the reduction of its precursor, dihydrofolic acid (DHF); in some strains of bacteria the synthesis of DHF (and hence THF) can be inhibited by sulphonamide antibiotics (section 15.4.9) – which thus inhibit the THF-dependent functions.

Some examples of simple biosynthetic pathways are shown in Fig. 6.3.

6.3.2 Interconversion of carbon compounds

In many pathways, individual reactions – or even sequences of reactions – can go in either direction, depending on conditions. Moreover, intermediates (as well as end-products) can pass from one pathway to another, so that the flow of carbon into various products can be regulated according to the cell's requirements. This versatility is indicated by the fact that, in many cases, the great diversity of structural molecules can be synthesized from a single carbon source. Only a few brief examples of interconvertibility can be mentioned here.

The interconvertibility of carbon compounds is well illustrated by *gluconeogenesis*: the synthesis of glucose 6-phosphate from non-carbohydrate substrates such as acetate, glycerol or pyruvate. Essentially, this is achieved by conversion of the substrate (where necessary) to an intermediate in the EMP pathway (Fig. 5.3) – followed by reversal of that pathway; non-reversible reactions in the EMP pathway are by-passed by other enzymes. Bacteria capable of gluconeogenesis (e.g *E. coli*) can, if necessary, use intermediates (as well as the end-product) in the reversed EMP pathway.

Where the hexose monophosphate and Entner–Doudoroff pathways occur in the same cell, the common intermediate 6-phosphogluconate (Fig. 6.2) may be metabolized by either pathway. Increased metabolism via the HMP pathway may be needed e.g. for chromosome replication – DNA synthesis requiring increased production of ribose 5-phosphate, a component of both purine and pyrimidine nucleotides (section 7.2).

6.3.3 Polymerization of carbon compounds

Polymerization involves the chemical linkage of small molecules to form a large, often chain-like molecule called a *polymer*; a *homopolymer* is formed when all the small molecules are similar, and a *heteropolymer* results when the small molecules are not all alike. In bacteria, polymerization is involved in the synthesis of certain storage compounds and in the formation of various cell wall and capsular structures. Two (hetero)polymers of major importance – proteins and nucleic acids – are considered in the next chapter.

6.3.3.1 Peptidoglycan synthesis

In *E. coli*, N-acetylglucosamine and N-acetylmuramic acid (Fig. 2.7) are synthesized separately, in the cytoplasm, as their UDP (uridine 5′-diphosphate)

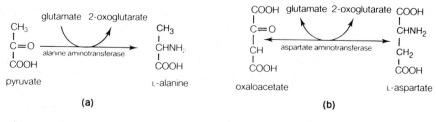

CH$_3$ — C=O — COOH
pyruvate

glutamate *2-oxoglutarate*

alanine aminotransferase

CH$_3$ — CHNH$_2$ — COOH
L-alanine

(a)

COOH — C=O — CH — COOH
oxaloacetate

glutamate *2-oxoglutarate*

aspartate aminotransferase

COOH — CHNH$_2$ — CH$_2$ — COOH
L-aspartate

(b)

CH$_2$O (P) — CHOH — COOH
3-phosphoglycerate

NAD$^+$ → NADH + H$^+$

phosphoglycerate dehydrogenase

CH$_2$O (P) — C=O — COOH
3-phosphohydroxy-pyruvate

glutamate → 2-oxoglutarate

phosphoserine aminotransferase

CH$_2$O (P) — CHNH$_2$ — COOH
3-phosphoserine

H$_2$O → Pi

phosphoserine phosphatase

CH$_2$OH — CHNH$_2$ — COOH
L-serine

(c)

uridine 5'-monophosphate
(UMP)

deoxyuridine 5'-monophosphate
(dUMP)

N^5,N^{10} – methylene-tetrahydrofolate → dihydrofolate

thymidylate synthase

deoxythymidine 5'-monophosphate
(dTMP)

(d)

derivatives – here abbreviated to UDP-GlcNAc and UDP-MurNAc, respectively. A chain of five amino acids is then added to UDP-MurNAc to form the so-called 'Park nucleotide'. The Park nucleotide is transferred (with release of UDP) to a long-chain lipophilic molecule (a *bactoprenol*) in the cytoplasmic membrane; subsequently, UDP-GlcNAc (with release of UDP) is added to form a bactoprenol–disaccharide–pentapeptide subunit. Subunits are then transferred (with release of bactoprenol) from the membrane to the periplasm; here, they are presumably joined together by *transglycosylation* reactions to form glycan chains (nascent strands of peptidoglycan). Incorporation of this new peptidoglycan into the existing sacculus may proceed according to the 3-for-1 mechanism (section 3.2.1, Fig. 3.1), peptide cross-linking involving *transpeptidation* reactions. Transglycosylation and transpeptidation are mediated by the enzymic action of penicillin-binding proteins (section 2.2.8).

Certain antibiotics inhibit peptidoglycan synthesis. β-Lactams (see section 15.4.1) inhibit the final stage. *Vancomycin* blocks the transfer of the subunit from the membrane to the periplasm by binding to the pentapeptide part of the bactoprenol–disaccharide–pentapeptide. Vancomycin, a complex glycopeptide, is bactericidal for a range of Gram-positive bacteria but is excluded from most Gram-negative cells; it is clinical useful e.g. against MRSA (section 15.4.11) and is also used for the treatment of pseudomembranous colitis associated with *Clostridium difficile*. *Teicoplanin* is structurally and functionally similar to vancomycin but is more active against enterococci.

6.3.3.2   Synthesis of poly-β-hydroxybutyrate

When non-carbon nutrients become scarce, many bacteria form intracellular granules of poly-β-hydroxybutyrate (PHB) – degrading these granules when conditions return to normal; PHB thus acts as a reserve of carbon and/or energy. Typically, PHB (Fig. 2.3) is synthesized from acetyl-CoA via acetoacetyl-CoA and β-hydroxybutyryl-CoA; the polymerizing enzyme, PHB synthetase, appears to occur at the surface of the granules.

Fig. 6.3   Some simple biosynthetic pathways: synthesis of the amino acids L-alanine(a), L-aspartate (b) and L-serine (c). (*Note*. Biochemists often give the *formula* of an organic acid in the un-ionized form but *name* the compound as though it were the salt – e.g. $CH_3.CO.COOH$ = pyruvate.) In these particular examples, the reaction with glutamate introduces nitrogen into the (non-nitrogenous) precursor molecule; glutamine carries out this function in the synthesis of L-tryptophan. Note that each reaction is catalysed by a specific enzyme. An encircled 'P' represents a phosphate group; 'Pi' is inorganic phosphate. (d) Uridine 5'-monophosphate (UMP) (top) has been synthesized via a series of reactions (not shown) from the simple initial precursor molecules aspartate and carbamoyl phosphate. Here, UMP first undergoes a reaction at the 2' position of the ribose (see Fig. 7.2), forming 2'-deoxyribose. In the next reaction, a coenzyme, N^5, N^{10}-methylene-THF (see section 6.3.1 and Table 6.1), donates a methyl group (-CH_3) to the 5-position of uracil (Fig. 7.3) – thus converting uracil to thymine.

The synthesis of PHB is described in a recent review on the production of polyhydroxyalkanoates [Poirier, Nawrath & Somerville (1995) Biotechnology 13 142–150].

6.4   METHYLOTROPHY IN BACTERIA

Methylotrophy is a type of aerobic metabolism characterized by (i) the obligate or facultative use of certain 'C$_1$ compounds' for carbon and energy, and (ii) the assimilation of formaldehyde – a product of methylotrophic metabolism – as a major source of carbon. The C$_1$ compounds used as substrates in methylotrophy include methanol, methylamine and methane; some methylotrophic bacteria can use methyl groups cleaved from certain larger molecules. The *methylotrophs* (organisms which are capable of methylotrophy) include species of *Hyphomicrobium*, *Methylococcus*, *Methylomonas* and *Methylophilus*. Those bacteria which use carbon monoxide as the sole source of carbon and energy (the carboxydobacteria – section 6.1.3) were once classified as methylotrophs, but are now excluded because they do not produce/assimilate formaldehyde.

The obligate methylotroph *Methylophilus methylotrophus* was once grown commercially (on methanol) for use as an animal feed; thousands of tons of the product ('Pruteen') were made each year – until the price of protein fell and the process was discontinued.

In methylotrophs, the formaldehyde produced metabolically is assimilated by the ribulose monophosphate pathway (RuMP pathway) or the serine pathway. Both are cyclic pathways. In the RuMP pathway, formaldehyde condenses with ribulose monophosphate to form hexulose-6-phosphate; subsequent reactions regenerate the acceptor molecule (RuMP) and allow pyruvate (for example) to be withdrawn for biosynthesis. In the serine pathway, formaldehyde reacts with glycine, forming serine; subsequent reactions regenerate glycine, 2-phosphoglycerate being withdrawn for biosynthesis.

Methanotrophic bacteria [review: Hanson & Hanson (1996) MR 60 439–471] are those methylotrophs which are able to use methane as sole source of carbon and energy; most are *obligately* methanotrophic. The methanotrophic bacteria, which include species of *Methylococcus* and *Methylomonas*, occur e.g. in soil, water and sediments; they use oxygen (O$_2$) and the enzyme *methane monooxygenase* (MMO) to oxidize methane to methanol and water. Further oxidation leads to the production of formaldehyde; some formaldehyde is assimilated by the RuMP or serine pathway, while some is oxidized to carbon dioxide to yield energy – common electron acceptors being NAD and NADP.

The oxidation of methane by methanotrophs significantly affects the global methane budget in that much less methane reaches the atmosphere; hence, methanotrophs help to reduce the contribution of methane to global warming.

7 Molecular biology I: genes and gene expression

7.1 CHROMOSOMES AND PLASMIDS

The chromosome (section 2.2.1) consists mainly of a polymer called *deoxyribonucleic acid* (DNA). DNA and a related polymer, ribonucleic acid (RNA), belong to a category of molecules (*nucleic acids*) which can carry information; in these molecules information is carried in the *sequence* with which subunits of the polymer are fitted together (see later). Chromosomal DNA carries all the information needed to specify both the structure and behaviour of a bacterium.

DNA dictates the life of a cell e.g. by encoding all the enzymes (thus controlling structure and metabolism) and by encoding various RNA molecules (involved in protein synthesis and certain control functions). Information carried by DNA regulates and co-ordinates growth and differentiation. Moreover, DNA controls its own replication and is also a self-monitoring system: there are various mechanisms for detecting and repairing damaged or altered DNA. All of this information is passed to daughter cells when the chromosome replicates and the parent cell divides (section 3.2.1). During replication the DNA is normally copied very accurately so that the characteristics of the species remain stable from one generation to the next; the tendency of daughter cells to inherit parental characteristics – heredity – is the main focus of *genetics*.

Clearly, nucleic acids determine both the 'routine' life of a cell and the process of heredity; both of these roles are studied in *molecular biology*.

Many bacteria contain one or more *plasmids*; a plasmid is an 'extra' piece of DNA which is usually much smaller than the chromosome and which can replicate independently. Some textbooks, and even research papers, 'define' a plasmid as '... a small circle of DNA ...'. Although many or most plasmids *are* circular (covalently closed, circular DNA, cccDNA), such a 'definition' is not correct because it ignores the existence of linear plasmids [see e.g. Hinnebusch & Tilly (1993) MM *10* 917–922]; linear plasmids were first reported over ten years ago.

Plasmids can encode various functions. Some encode enzymes which inactive particular antibiotics; such *R plasmids* (resistance plasmids) usually make the host cell resistant to the relevant antibiotics. Some plasmids encode structural elements – e.g. gas vacuoles (section 2.2.5) in certain strains of

Halobacterium. The 'Cit' plasmid encodes a transport system (section 5.4) for the uptake of citrate; strains of *E. coli* which contain this plasmid can use citrate as the sole source of carbon and energy – something which common or 'wild-type' strains of *E. coli* cannot do. (Other plasmid-mediated functions are mentioned elsewhere in the book.) However, although plasmids often encode useful functions they are typically not indispensable to their host cells.

Some (not all) plasmids encode the means to transfer themselves from one cell to another by the process of *conjugation* (section 8.4.2).

Plasmids are widely used in recombinant DNA technology (section 8.5). A recombinant (genetically engineered) plasmid is often indicated by the prefix 'p' (e.g. pBR322).

Prophages (section 9.2) which do not integrate with the host's chromosome, and which can replicate, are also called 'plasmids'.

7.2   NUCLEIC ACIDS: STRUCTURE

A nucleic acid is a polymer made of subunits called *nucleotides*; each nucleotide has three parts: (i) a sugar molecule, (ii) a nitrogen-containing base, and (iii) a phosphate group (Fig. 7.1). Nucleotides containing the sugar D-ribose are *ribo*nucleotides – which form *ribo*nucleic acid (RNA); those containing 2'-deoxy-D-ribose are *deoxyribo*nucleotides – which form *deoxyribo*nucleic acid (DNA) (Fig. 7.2). The nitrogen base in any given nucleotide is either a substituted *purine* or a substituted *pyrimidine* (Fig. 7.3); a ribonucleotide may contain adenine, guanine, cytosine or uracil, while a deoxyribonucleotide may contain adenine, guanine, cytosine or thymine. The names of the various nucleotides, and their corresponding nucleosides, are given in Table 7.1.

7.2.1   Deoxyribonucleic acid (DNA)

In DNA the nucleotides form an unbranched chain (a *strand*) in which sugar–base and phosphate residues are alternate links (Fig. 7.4). Note that a strand has *polarity*: there is a 5'-end and a 3'-end. DNA commonly consists of two strands which are held together by hydrogen bonding between their nitrogen bases; this double-stranded structure is called a DNA *duplex* (Fig. 7.5). Note that the two strands in a duplex are *antiparallel*, i.e. when read left to right (Fig. 7.5), one strand is 5'-to-3' while the other is 3'-to-5'.

Hydrogen bonding between the nitrogen bases is quite specific: adenine pairs with thymine, and guanine pairs with cytosine (Fig. 7.6); this specificity in *base-pairing* is referred to by saying that each of the two bases in a *base-pair* (bp) is *complementary* to its partner, and that (therefore) each strand in a DNA duplex is complementary to the other strand.

The ladder-like DNA duplex forms a *helix* (Fig. 7.7) – each turn of which

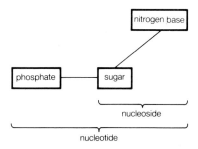

Fig. 7.1 A generalized nucleotide. Note the difference between a nucleotide and a nucleoside.

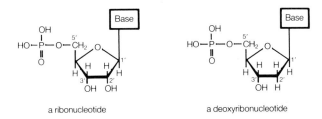

a ribonucleotide   a deoxyribonucleotide

Fig. 7.2 Nucleotides: a ribonucleoside monophosphate and a *deoxy*ribonucleoside monophosphate; these generalized molecules differ only in their sugar (ribose) residues: the deoxyribonucleotide lacks oxygen at the 2' position. (The numbers 1', 2' etc. refer to the positions of the carbon atoms in the ribose molecule.) The 'base' can be adenine, guanine or cytosine in either molecule, but uracil occurs only in ribonucleotides, while thymine is found only in deoxyribonucleotides. (In a *di*deoxyribonucleotide (not shown here) oxygen is missing at both the 2' *and* 3' positions; dideoxyribonucleotides are used e.g. in DNA sequencing – section 8.5.6.)

occupies about 10 base-pairs (10 bp) of distance along the duplex; this is the *double helix* worked out by Watson and Crick in the 1950s.

In *most* bacterial chromosomes the DNA duplex forms a closed *loop* (i.e. there are no free 5' and 3' ends): covalently closed circular DNA (cccDNA) [cf. *linear* chromosomes: Chen (1996) TIG *12* 192–196]. Circular DNA can exist in various states. A *relaxed* loop of DNA can theoretically lie in a plane, like a rubber band on a table, but suppose that the loop is cut, and that one end is twisted (the other held still) before the ends are rejoined; this will either increase or decrease the 'pitch' of the helix (forming an 'overwound' or 'underwound' helix, respectively) depending on the direction of twisting. Such a molecule is under strain: there is a tendency to restore the pitch to that of a relaxed molecule. To relieve the strain, the molecule contorts: the axis of the helix becomes helical, i.e. the helix itself coils up to form a helix! A molecule in this state is said to be *supercoiled*.

Naturally occurring supercoiled DNA is generally 'underwound' (i.e.

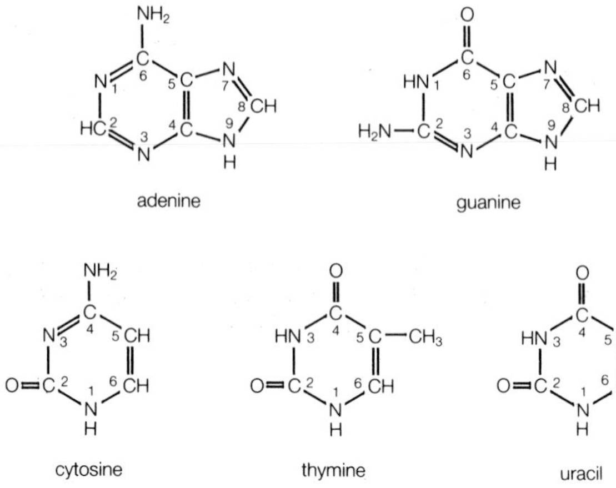

Fig. 7.3 Nitrogen bases found in DNA and RNA. Adenine and guanine are substituted *purines*; the others are substituted *pyrimidines*. Thymine occurs only in DNA, uracil only in RNA; adenine, guanine and cytosine occur in both DNA and RNA. Within a nucleotide (Fig. 7.2) a purine is linked via its 9-position to the sugar molecule, while a pyrimidine is linked via its 1-position.

Table 7.1 Nucleosides and nucleotides found in RNA and DNA

Sugar	Nitrogen base	Nucleoside (base + sugar)	Nucleotide[1] (base + sugar + phosphate)
Ribose	Adenine	Adenosine	Adenosine 5'-monophosphate (AMP) (or adenylic acid)
Ribose	Guanine	Guanosine	Guanosine 5'-monophosphate (GMP) (or guanylic acid)
Ribose	Cytosine	Cytidine	Cytidine 5'-monophosphate (CMP) (or cytidylic acid)
Ribose	Uracil	Uridine	Uridine 5'-monophosphate (UMP) (or uridylic acid)
Deoxyribose	Adenine	Deoxyadenosine	Deoxyadenosine 5'-monophosphate (dAMP) (or deoxyadenylic acid)
Deoxyribose	Guanine	Deoxyguanosine	Deoxyguanosine 5'-monophosphate (dGMP) (or deoxyguanylic acid)
Deoxyribose	Cytosine	Deoxycytidine	Deoxycytidine 5'-monophosphate (dCMP) (or deoxycytidylic acid)
Deoxyribose	Thymine	Deoxythymidine	Deoxythymidine 5'-monophosphate (dTMP) (or deoxythymidylic acid)

[1] Only the monophosphate is shown in each case. Diphosphates and triphosphates are named in a similar fashion, e.g. adenosine 5'-diphosphate (ADP) and adenosine 5'-triphosphate (ATP).

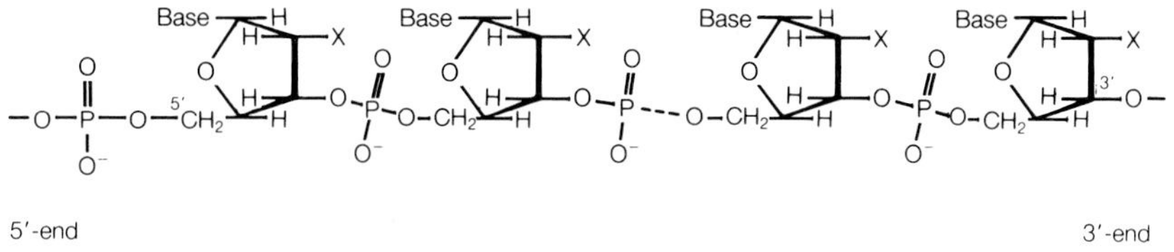

Fig. 7.4 The structure of a single strand of nucleic acid; note the *polarity* of the molecule. 'X' is a hydrogen atom (H) in DNA but a hydroxyl group (OH) in RNA. The sugar residues are linked together by *phosphodiester* bonds.

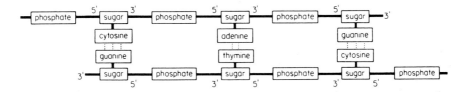

Fig. 7.5 A DNA duplex (diagrammatic). The dotted lines represent hydrogen bonds. Note that the strands are *antiparallel* (section 7.2.1).

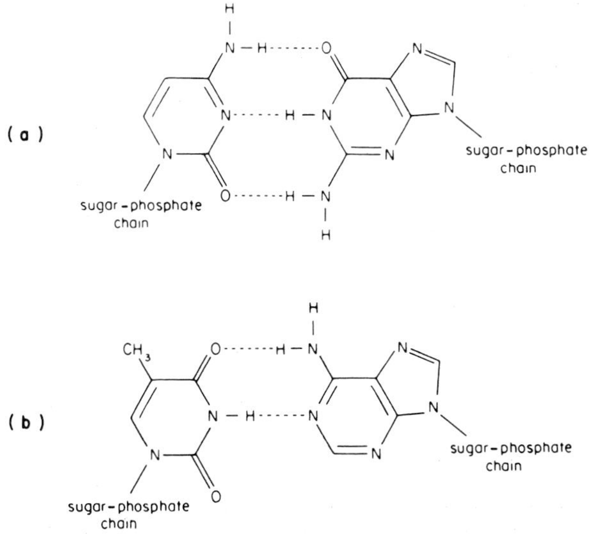

Fig. 7.6 Base-pairing between nucleotides. (a) Cytosine (left) pairing with guanine (right). (b) Thymine (left) pairing with adenine (right). (*Note.* In RNA synthesis (section 7.5) adenine pairs with uracil.) The dotted lines represent hydrogen bonds.

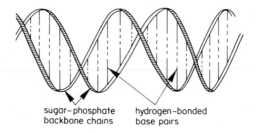

sugar–phosphate   hydrogen–bonded
backbone chains   base pairs

Fig. 7.7 The DNA 'double helix' (diagrammatic). Sugar residues are linked by phosphodiester bonds (Fig. 7.4). The (planar) nitrogen bases are stacked roughly perpendicular to the axis of the helix; chemical groups on these bases protrude into the grooves of the helical molecule – thus allowing e.g. recognition by enzymes.

'negatively supercoiled'). The degree of supercoiling is controlled by enzymes called *topoisomerases*. In *Escherichia coli* the general level of negative supercoiling seems to depend on the opposing activities of two such enzymes: topoisomerase II (= *gyrase*), which promotes negative supercoiling, and topoisomerase I, which prevents excessive supercoiling. However, supercoiling can also be affected by those environmental factors which affect the cell's energy status; the mechanism of this is not known, but it may involve environmental influence on the intracellular level of ATP because the activity of gyrase is known to be sensitive to the ratio ATP:ADP. For growth to occur normally the level of supercoiling must be kept within certain limits.

Topoisomerases III and IV may, together with gyrase, have a role in the decatenation of chromosomes (section 3.2.1).

[Control of bacterial DNA supercoiling: Drlica (1992) MM *6* 425–433.]

7.2.1.1   PNA (peptide nucleic acid)

PNA is an analogue of DNA in which the backbone chains are modified polypeptides (rather than chains of sugar–phosphate units); two strands of PNA can hybridize to form a DNA-like helical duplex [Wittung *et al.* (1994) Nature *368* 561–563]. This molecule may have evolutionary significance; it has been suggested that pre-biotic nucleic acids may not necessarily have had a sugar–phosphate backbone.

7.2.2   Ribonucleic acid (RNA)

RNA is a linear polymer of ribonucleotides (Fig. 7.4); although the ribonucleotides generally contain adenine, guanine, cytosine or uracil, modified bases occur in some RNA molecules. Bacterial RNA (unlike DNA) is typically single-stranded; however, double-stranded regions are formed in some cases by base-pairing between complementary sections within the same molecule.

7.3 DNA REPLICATION

As mentioned earlier, the *sequence* of bases in chromosomal DNA is a coded message specifying the cell's structure and behaviour. Before cell division occurs, the DNA must be duplicated precisely so that each daughter cell will receive an exact copy (replica) of the molecule. DNA synthesis (*replication*) is a complex process; the following is an outline, only, of the 'routine' replication of an *E. coli* chromosome during the cell cycle.

What starts (*initiates*) a round of DNA replication? We don't know (see section 3.2.1), but it begins in a specific region in the double-stranded DNA molecule, the *origin* (*oriC*), and requires a number of different proteins; additionally, the DNA must be supercoiled.

A number of molecules of DnaA protein (probably as DnaA·ATP) bind at specific sites (DnaA boxes) within the *oriC* sequence, inducing localized 'melting' (i.e. strand separation in DNA) in an *adjacent* part of *oriC*; in the melted region, single-stranded DNA may be stabilized by the binding of so-called single-strand binding proteins (SSB proteins). However, although strand separation is the essential first step in replication, replication apparently does not start within *this* melted region: it seems to begin at a site in that part of *oriC* containing the DnaA boxes. How strand separation occurs at the actual site where replication is to begin is not known; it may be that DnaA binding is involved in melting at this site as well as at the adjacent site. Another theory is that transcription (section 7.5) in the region of *oriC* causes a wave of transient melting to pass along the DNA duplex to the appropriate site; certainly, RNA polymerase (section 7.5) is needed for the initiation of DNA replication.

One factor involved in the initiation of DNA replication is the Fis protein (*fis* gene product). Fis has several binding sites in *oriC* near the DnaA boxes, and it also seems to regulate a certain cluster of genes (*dnaA–dnaN–recF*) [Froelich, Phuong & Zyskind (1996) JB *178* 6006–6012] – whose products include DnaA protein and the β subunit of DNA polymerase III (Fig. 7.8(b)). Fis binds at *oriC* in a cell-cycle-specific way [Cassler, Grimwade & Leonard (1995) EMBO Journal *14* 5833–5841] where it appears to form an 'initiation-preventive' complex [Wold, Crooke & Skarstad (1996) NAR *24* 3527–3532], and it also has a weak repressor effect on the *dnaAp2* promoter. The mechanism by which Fis influences the timing of initiation of DNA replication is not known.

Where replication begins, strand separation produces two *replication forks* (Fig. 7.8). Associated with each fork is a *primosome*: a complex of proteins which include DnaB and DnaG.

In *one* of the DNA strands, bases newly exposed by strand separation pair with the complementary bases of individual molecules of *ribo*nucleoside triphosphate, and the ribonucleotides are enzymically polymerized (covalently joined together) in the 5'-to-3' direction to form a short strand of RNA: a *primer* (Fig. 7.8a, *top*); in this process, the two terminal phosphate groups of

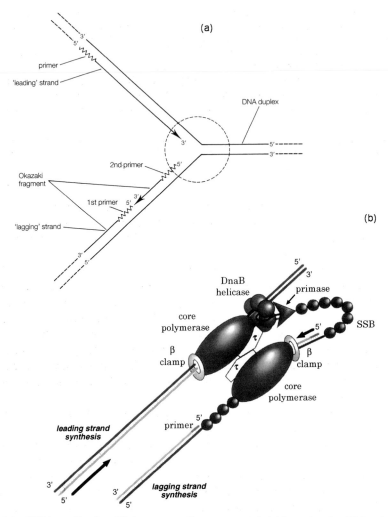

Fig. 7.8 DNA replication (Cairns-type, diagrammatic). (a) In a circular DNA duplex the molecule has split, locally, into its two component strands. The diagram shows only one half of the split region; in this half, the strands continue to separate in a left → right direction. On one exposed DNA strand (top) an RNA primer is synthesized (see text); the subsequent addition of deoxyribonucleotides forms the 'leading' strand of DNA – which is synthesized continuously in the 5'-to-3' direction. On the other DNA strand (bottom) the first primer is extended by synthesis of a DNA fragment (an *Okazaki fragment*) – again in the 5'-to-3' direction (because synthesis cannot occur in the 3'-to-5' direction); only when the duplex has opened out further is the second primer formed and the next Okazaki fragment synthesized. (All primers are subsequently replaced with DNA.) The encircled region (*replication fork*) moves to the right as replication continues, while the other replication fork (not shown) moves to the left.

(b) A diagrammatic view of the replication fork in *E. coli* (not drawn to scale). The two polymerases (one forming the leading strand, the other the lagging strand) work in opposite directions, and are shown physically linked via the τ subunit of each polymerase; a polymerase–helicase link is also shown (the helicase being a hexamer of DnaB proteins). The single-stranded DNA is protected by so-called *single-strand binding proteins* (SSBs) prior to serving as template for the lagging strand; the primase synthesizes primers on this template. The β-clamp (= β subunit of DNA polymerase III, *dnaN* gene product) may affect *processivity*, i.e. the coupled movement and function of the polymerase.

Drawing (b) reproduced from Cell (1996) *86* 177–180 by courtesy of the author, Dr Stephen C West, Imperial Cancer Research Fund, UK, and with permission from Cell Press, Cambridge MA 02138, USA.

---

each ribonucleoside triphosphate are cleaved in a reaction which provides the energy for polymerization. It is not known whether this polymerization is mediated by the DnaG protein (a *primase*) or by RNA polymerase; both enzymes can synthesize RNA.

The primer forms one strand of a short RNA/DNA hybrid duplex (Fig. 7.8a, *top*). On further strand separation, newly exposed bases pair with molecules of *deoxy*ribonucleoside triphosphate so that the primer is extended in the 5'-to-3' direction by a new strand of DNA; this polymerization, which is mediated by the enzyme *DNA polymerase III*, forms the start of the so-called *leading* strand of DNA (Fig. 7.8a, *top*). Note that the leading strand, together with one strand of the original duplex, is the beginning of a new duplex; the original strand determines the base sequence in the new strand and is therefore called a *template* strand.

An RNA primer is needed to start the process because a DNA polymerase cannot initiate a strand – it can only extend an *existing* strand.

The DnaB protein, present in each primosome, is a *helicase*: an enzyme which, in the presence of gyrase (section 7.2.1) and ATP, *unwinds* DNA – i.e. separates the strands of the DNA duplex; thus, DnaB allows DNA synthesis to continue by opening up the duplex (like a zip) so that the replication forks move in opposite directions, away from *oriC*. At 37°C, a replication fork moves at about 1000 nucleotides per second.

So far, we have considered only that strand of the DNA duplex on which the leading strand is synthesized. What of the other strand? That, too, acts as a template. However, because the strands of a duplex are antiparallel (section 7.2.1), and because DNA can be synthesized only in the 5'-to-3' direction, the direction of DNA synthesis on this strand is opposite to that of synthesis on the complementary strand (Fig. 7.8a, *bottom*). In fact, this new strand of DNA (the 'lagging' strand) is synthesized as a series of short fragments (*Okazaki fragments*), as follows. When strand separation in the DNA duplex has reached a certain extent, the DnaG protein (primase) synthesizes the first RNA primer near the replication fork; this primer is extended by the addition and polymerization of deoxyribonucleotides, in the 5'-to-3' direction, to form the first fragment of the new lagging strand (Fig. 7.8a, *bottom*). On further strand separation, the second primer is formed and the second fragment synthesized – and so on; each of these Okazaki fragments is about 2000 bases long (= 2 kilobases, 2 kb).

Co-ordination of leading/lagging strand synthesis at the replication fork, and the rapid rate of DNA synthesis, appear to involve (i) a physical link between the two DNA polymerases, and (ii) a coupling between polymerases and the helicase – loss of polymerase–helicase interaction resulting in a drastic reduction in the efficiency of the helicase [Kim *et al.* (1996) Cell *84* 643–650]. One view of the replication fork is shown in Fig. 7.8(b). Note that the helicase (which unwinds the duplex) is a hexamer (a ring-shaped structure comprising six DnaB proteins) [West (1996) Cell *86* 177–180]; in the figure, both strands of duplex DNA are shown passing through the ring, but whether both (or only one) pass through is currently unknown.

The two replication forks, moving in opposite directions, meet at the *ter* site on the chromosome, 180° from *oriC*. Before completion of replication all primers are replaced with DNA. Replication yields two supercoiled DNA duplexes which are initially catenated (section 3.2.1).

Since each duplex consists of one strand of the original (parent) duplex and one new (daughter) strand, replication is said to be *semi-conservative*.

Other modes of DNA replication. The replication described above is 'routine' replication in an *E. coli* chromosome, and it should be stressed that replication can occur in other ways. Even in *E. coli*, chromosomal replication can be mechanistically different under certain conditions; for example, in cells induced for the SOS system (section 7.8.2.2), DNA replication – inducible stable DNA replication (iSDR) – begins at a different origin (*oriMs*) and is independent of DnaA protein [Asai & Kogoma (1994) JB *176* 1807–1812].

In certain *linear* genomes (e.g. that of phage ϕ29 – Table 9.1), replication begins at one end of the duplex.

Some bacteria have linear chromosomes (and plasmids) in which DNA replication starts at an internal site; in *Streptomyces coelicolor*, for example, replication runs in both directions from an internal origin. One unexplained aspect of DNA replication in a *linear* duplex follows from the 5' → 3' direction of synthesis of a new strand, and the likelihood of leaving the 3' end of the template strand (in each daughter duplex) single stranded. Several models for completing synthesis on the template strand have been suggested [see Chen (1996) TIG *12* 192–196]. Two models envisage the 3' end of the template strand folding over and base-pairing with itself (to provide a double-stranded starting point for replication); in one of these models, synthesis of a complementary strand is then started from the 3' end of the template strand. A third model is based on the finding that all linear chromosomes and plasmids of *Streptomyces* so far studied have *terminal inverted repeats* (see section 8.3). DNA synthesized *on the new strand* at the *fully duplexed end* of a daughter duplex can displace the (5') terminal nucleotide sequence of the template strand and, because of the terminal inverted repeats, this displaced nucleotide sequence will be complementary to the single-stranded gap at the other end of the chromosome; the displaced nucleotide sequence can then be paired with the single-stranded region by homologous recombination.

The rolling circle mechanism (Fig. 8.5) is used for replicating the DNA of certain phages – e.g. M13 (section 9.2.2) and ϕX174.

7.3.1 Replication of plasmid DNA

A plasmid controls its own replication; however, the cell's biosynthetic machinery is needed to synthesize any plasmid-encoded protein etc. involved in replication.

The frequency of replication depends on the rate at which replication is *initiated*, and this is controlled in different ways by different plasmids. In plasmid ColE1, for example, control involves the synthesis of two, free (non-duplexed) strands of plasmid-encoded RNA, one of which (RNA II) can bind at the plasmid's origin and act as a primer, thus permitting replication; the other strand (RNA I) can interfere by base-pairing with RNA II, and the outcome of RNA I–RNA II interaction determines the occurrence of replication. Thus, in ColE1, initiation is controlled at the primer level (see also section 7.5.2).

Replication of plasmid R1162 depends on the binding of a plasmid-encoded initiator, the RepIB protein, to certain 'direct repeat' nucleotide sequences (*iterons*) in the plasmid's origin of replication; it has been shown that RepIB–iteron interaction brings about localized 'melting' (i.e. strand separation) in the DNA duplex [Kim & Meyer (1991) JB *173* 5539–5545]. Here, initiation seems to be controlled at the level of strand separation at the replicative origin.

In the F plasmid, control of replication involves the plasmid-encoded E protein.

To understand replication control in a given plasmid it is necessary to know whether control molecules (such as the RepIB and E proteins) are synthesized continually or periodically, how their synthesis/activity is regulated, and whether other factors/molecules are involved. In general, the replication of plasmids (and their partitioning to daughter cells at cell division) seems likely to depend on some kind of association between the plasmid and the cytoplasmic membrane [Firshein & Kim (1997) MM *23* 1–10].

Once initiated, replication may proceed as in the *E. coli* chromosome (Cairns-type DNA replication – see above). This occurs e.g. in ColE1, although in this plasmid replication proceeds *uni*directionally from the origin. Replication is bidirectional in the F plasmid.

In a bacterium, the number of copies of a given plasmid, per chromosome, is called the *copy number*. Copy number depends on the plasmid's replication control system, on the bacterial strain and on growth conditions. Some plasmids (e.g. the F and R6 plasmids) have a copy number of 1–2; in *multicopy plasmids* (e.g. ColE1, R6K) it may be e.g. 10–30.

7.3.1.1 *Stringent and relaxed control of plasmid replication*

Some plasmids (e.g. the F plasmid) fail to replicate if their host cells are treated with an antibiotic (e.g. chloramphenicol) which inhibits protein synthesis; the replication of such plasmids is said to be under *stringent control*. Other plasmids (e.g. ColE1) continue to replicate under these conditions – and can achieve higher-than-normal copy numbers; the replication of such plasmids is said to be under *relaxed control*.

7.4 DNA MODIFICATION AND RESTRICTION

In many (perhaps all) bacteria, DNA replication is followed by *DNA modification*: methylation of certain bases in *specific sequences of nucleotides* in each of the newly synthesized strands; in these sequences, the methyl group (CH_3-) is added at the N-6 position of adenine and/or the C-5 position of cytosine (see Fig. 7.3 for positions). Modification involves specific enzymes, methyltransferases (= 'methylases'), which use e.g. S-adenosylmethionine as a source of methyl groups.

Why is DNA methylated? Methylation protects the DNA from the cell's own *restriction endonucleases* (REs): enzymes which 'restrict' (i.e. cut) any DNA which lacks methylation in the appropriate sequence of nucleotides. A DNA duplex is not cleaved if at least one of the strands is methylated; immediately after replication, the newly synthesized (*un*methylated) daughter strand is normally protected from restriction by the (methylated) template strand until methylation occurs.

Different sequences of nucleotides are methylated in different strains of bacteria. Thus, each strain has its own strain-specific methylases and REs, both of which recognize a given sequence of nucleotides; for example, in *E. coli* the *Eco*RI methylase methylates a specific adenine residue:

$$CH_3$$
$$|$$
5'-GAATTC-3'

while the RE *Eco*RI cuts an (unmethylated) strand at:

5'-G|AATTC-3'

where '|' indicates the cut. Note that, in this example, the *complementary* strand is cut as follows: 3'-CTTAA|G-5'. This results in a *staggered* cut, i.e. each of the cut ends of the duplex has a single-stranded region (in this case 5'-AATT) (see Fig. 8.11).

Many strains of bacteria contain more than one type of methylase and RE.

The sequence recognized by a given RE is called its *recognition sequence*. Most REs can be classified as type I, II or III according e.g. to the relationship between their recognition sequence and their cutting site; for example, type I REs cut at random sites distant from their recognition sequences, and type II REs cut between a specific pair of bases within the recognition sequence.

Quite recently a new type of RE has been discovered; these enzymes make a staggered cut, at a specific site, on *both* sides of the recognition sequence – thus cutting out a small sequence of DNA which includes the recognition sequence. Cleavage by these REs typically requires Mg^{2+} and *S*-adenosyl-methionine. A new member of this family of REs has been reported by Sears *et al.* [(1996) NAR 24 3590–3592]: *Bae*I (from *Bacillus sphaericus*); cleavage by this enzyme is shown in Fig. 8.11.

REs are usually named as in the following example. *Eco*RI: RE from *Escherichia coli* strain R, 'I' indicating a particular RE from strain R. Other examples of REs are given in Table 8.1.

What is the purpose of restriction? Its main role seems to be to protect the cell from 'foreign' DNA – particularly phage DNA – which has entered the cell [restriction (review): Bickle & Krüger (1993) MR 57 434–450]. However, this protective mechanism is not always successful against phages: see section 9.6. Moreover, antirestriction proteins (offering protection against restriction) are known to be encoded by some plasmids [Belogurov, Delver & Rodzevich (1992) JB 174 5079–5085]; such proteins may be needed to overcome the restriction of the recipient cell during conjugation (section 8.4.2). Antirestriction also operates in conjugative transposition (section 8.4.2.3).

'Non-classical' forms of restriction. Some REs (e.g. *Eco*RII) are active only if they bind simultaneously to *two* recognition sequences – which need not necessarily be on the same DNA molecule. (The need for two sites resembles the requirement of the enzymes involved in site-specific recombination – section 8.2.2.)

The RE *Dpn*I from *Streptococcus* (incorrect name: '*Diplococcus*') *pneumoniae* is one example of an enzyme which cleaves only at a *methylated* site, i.e. it is a methyl-dependent RE (in contrast to the REs described above); *Dpn*I cleaves 5'-GATC-3' between A and T only if the adenine is methylated.

Restriction and modification in recombinant DNA technology. Type II REs (which need Mg^{2+} but not ATP) are widely used for cutting DNA at specific sites (section 8.5.1.3).

7.5 RNA SYNTHESIS: TRANSCRIPTION

In a cell, functions encoded by DNA are carried out by various types of RNA molecule – which are copied (transcribed) from the DNA.

Ribosomal RNA (rRNA) is a major component of ribosomes (section 2.2.3) – the minute bodies on which proteins are synthesized. Protein synthesis also involves other types of RNA molecule. Thus, synthesis of a given protein requires that the corresponding *gene* – i.e. the DNA sequence encoding that protein – be first copied in the form of a single strand of RNA; this strand is called *messenger RNA* (mRNA) because it carries the 'message' of the gene. During protein synthesis, molecules of *transfer RNA* (tRNA) carry amino acids and present them for polymerization in the order specified by the mRNA (section 7.6).

Other types of RNA molecule have specific control functions; an *antisense* RNA molecule can e.g. inhibit the activity of another nucleic acid molecule by binding to a complementary sequence in it (see e.g. RNA I in section 7.3.1).

Transcription of a gene. RNA is copied from DNA by base-pairing between individual *ribo*nucleotides and the exposed bases on a single-stranded DNA template, the ribonucleotides being polymerized in the 5'-to-3' direction. As in DNA synthesis, nucleoside *tri*phosphates base-pair with the template strand, the two terminal phosphate groups being cleaved to provide energy for polymerization. Note that, in RNA synthesis, adenine pairs with uracil, not with thymine. The synthesis of RNA on a DNA template is called *transcription*, and the single-stranded RNA product is called a *transcript*.

Polymerization of ribonucleotides is carried out by the enzyme *RNA polymerase*. In order to transcribe a gene, the polymerase first binds at a specific sequence of nucleotides – a *promoter* – in (double-stranded) DNA at the beginning of the gene. Usually within the promoter sequence is the so-called *start site* where transcription begins.

Before binding to a promoter, the RNA polymerase must first bind to another protein called a *sigma factor*. The cell encodes more than one type of sigma factor, and a given type confers on the polymerase the ability to bind to promoters of a particular class. In *E. coli*, an RNA polymerase complexed with sigma factor σ^{70} can initiate transcription from most promoters in the cell. Some of the cell's sigma factors are synthesized only when it is necessary to transcribe particular genes (see section 7.8.9).

Following binding of the RNA polymerase *holoenzyme* (= RNA polymerase + sigma factor, often symbolized $E\sigma^{70}$, $E\sigma^{32}$ etc.) to the promoter, the strands of DNA separate in the region of the start site, allowing transcription to begin. The first ribonucleotide base-pairs with a complementary deoxyribonucleotide on one of the DNA strands at the start site; this deoxyribonucleotide is designated +1, and subsequent deoxyribonucleotides – in the direction of synthesis ('downstream') – are designated +2, +3 ... etc. (Nucleotides in the upstream direction are designated –1, –2 . . . etc.) The RNA strand grows as more ribonucleotides are polymerized; shortly, the sigma factor is released and the RNA polymerase continues on its own to synthesize the remainder of the strand.

Elongation involves progressive unwinding of the DNA duplex and the sequential addition of ribonucleotides to the growing strand of RNA. As the process continues, the RNA strand peels away from the template strand and the DNA duplex re-forms.

Transcription stops at a specific sequence of nucleotides (a termination signal, or *terminator*) in the template strand. In so-called *rho-independent* termination, parts of the transcript of the terminator region base-pair with one another, and the resulting structure may cause the release of the transcript and/or polymerase. In *rho-dependent* termination, the *rho factor* (in *E. coli*: a helicase comprising six 46-kDa subunits) may stop transcription after recognizing a particular site on the transcript.

7.5.1 Regulation of transciption: other factors

Section 7.5 is a simplified account of transcription. Normally, the transcription of a gene, or operon (section 7.8.1), involves the activity of proteins called *transcription factor(s)* which bind to specific sequence(s) in the DNA, either close to or distant from the promoter, and which exert either a positive or negative regulatory influence on transcription; an example of these *DNA-binding proteins* is the repressor (regulator) protein of the *lac* operon (Fig. 7.11).

To study these (and other) protein–DNA interactions, use is made of methods such as Southwestern blotting and footprinting (section 8.5.12).

7.5.2 Polyadenylation of bacterial mRNA

Although it's well known that most *eukaryotic* mRNAs have a 3' polyadenylate 'tail' (i.e. -A-A-A-A-3'), the existence of polyadenylated *bacterial* mRNAs is less widely acknowledged; however, such mRNAs do occur, and in *E. coli* there are at least two enzymes (poly(A)polymerases, PAPs) which carry out polyadenylation.

The function of polyadenylation is not known, but it may help e.g. in regulating the half-life (stability) of those RNA molecules whose 3' terminus is a stem–loop structure encoded by a rho-independent transcription terminator (section 7.5); a stem–loop tends to resist the action of certain RNA-degrading enzymes, and one suggestion is that polyadenylation at the 3' end of the stem–loop may facilitate degradation of these molecules by providing a suitable binding site for degradative enzymes such as the 3'-exonuclease polynucleotide phosphorylase. (See also section 7.6.1.)

Interestingly, mutations in the *E. coli* gene *pcnB* (which encodes the poly(A)polymerase PAP I) reduce the *copy number* of e.g. plasmid ColE1. It appears that the antisense molecule RNA I (see section 7.3.1) is normally polyadenylated, and that this facilitates its degradation; in *pcnB* mutants the absence of polyadenylation tends to stabilize RNA I and thus to enhance its

Table 7.2 The 'universal' genetic code: amino acids and 'stop' signals encoded by particular codons[1-5]

First base (5' end)	Second base				Third base (3' end)
	U	C	A	G	
U	Phe	Ser	Tyr	Cys	U
	Phe	Ser	Tyr	Cys	C
	Leu	Ser	*ochre*	*opal*	A
	Leu	Ser	*amber*	Trp	G
C	Leu	Pro	His	Arg	U
	Leu	Pro	His	Arg	C
	Leu	Pro	Gln	Arg	A
	Leu	Pro	Gln	Arg	G
A	Ile	Thr	Asn	Ser	U
	Ile	Thr	Asn	Ser	C
	Ile	Thr	Lys	Arg	A
	Met	Thr	Lys	Arg	G
G	Val	Ala	Asp	Gly	U
	Val	Ala	Asp	Gly	C
	Val	Ala	Glu	Gly	A
	Val	Ala	Glu	Gly	G

[1] A = adenine; C = cytosine; G = guanine; U = uracil.
[2] Standard abbreviations are used for the amino acids: Gln = glutamine, Ile = isoleucine etc.
[3] As with other nucleotide sequences, codons are conventionally written in the 5'-to-3' direction; as anticodons also are written in this way, the *first* base in a codon pairs with the *third* base in its anticodon.
[4] The codon UAA (*ochre*), UGA (*opal*) and UAG (*amber*) are normally 'stop' signals (see text).
[5] Although called 'universal', a number of exceptions to the code have been reported. In *Escherichia coli*, for example, one codon base-pairs with a tRNA carrying serine, but the serine is converted to selenocysteine before incorporation in the polypeptide (*co-translation*) [Leinfelder *et al.* (1988) Nature 331 723–725].

inhibitory effect on RNA II – the effect being to inhibit replication of ColE1 (i.e. to reduce its copy number).

[Polyadenylation of bacterial mRNA (review): Sarkar (1996) Microbiology *142* 3125–3133.]

7.6 PROTEIN SYNTHESIS

All the cell's proteins are encoded by DNA; by studying protein synthesis we can see *how* DNA carries this information – and how the genetic code works.

A protein consists of one or more *polypeptides*, a polypeptide being a chain of amino acids covalently linked by peptide bonds (–CO.NH–). Each polypeptide is folded into a three-dimensional structure; this structure is stabilized mainly by hydrogen bonds or disulphide bonds formed between amino acids in different parts of the chain. The specific three-dimensional structure of a given polypeptide – essential for biological activity – is determined by the nature, number and sequence of its amino acids.

For any given polypeptide, the nature, number and sequence of amino acids are dictated by a particular sequence of bases in a DNA strand; this sequence of bases conforms to one definition of a *gene*.

How does a gene bring about the synthesis of a polypeptide? That is, how is the gene *expressed*? Unlike nucleotides, amino acids cannot simply 'line up' (undergo polymerization) on a DNA template strand. In fact, protein synthesis involves several stages. First, the gene is *transcribed* (section 7.5); the (RNA) transcript of the gene, which carries the 'message' from DNA, is called *messenger RNA* (mRNA). Now, along the length of the mRNA molecule, groups of three consecutive bases each encode a particular amino acid; each of these three-base groups is called a *codon*. Thus, e.g. the codon UCA (uracil–cytosine–adenine) encodes the amino acid serine (Table 7.2). Hence, the sequence of codons in mRNA encodes the sequence of amino acids in a polypeptide. For the synthesis of a polypeptide, each amino acid must first bind to an 'adaptor' molecule which is specific for that particular amino acid and for its codon; these 'adaptor' molecules are small molecules of RNA – *transfer RNA* (tRNA) – and each binds to its specific amino acid to form an *aminoacyl-tRNA*. (Such binding requires ATP.) As well as a binding site for the amino acid, a given tRNA molecule also contains a sequence of three bases (an *anticodon*) which is complementary to the codon of its particular amino acid; hence, a given tRNA molecule can bind its particular amino acid and can link it to the specific codon in mRNA through codon–anticodon base-pairing.

The synthesis of a polypeptide on mRNA (the *translation* process) takes place on a ribosome (section 2.2.3); a simplified version of this process is shown diagrammatically, and explained, in Fig. 7.9.

As shown in Fig. 7.9, the *initiator codon* (i.e. the first codon to be translated) is commonly AUG. The alignment of AUG with the ribosomal P site is generally promoted by a particular sequence of nucleotides (the *Shine–Dalgarno sequence*) located 'upstream' of AUG on the mRNA (i.e. to the left of AUG in Fig. 7.9); this sequence base-pairs with part of a 16S rRNA molecule in the ribosome.

When acting as an initiator codon, AUG encodes the modified amino acid N-formylmethionine in most or all bacteria (including *E. coli*). The initiator codon base-pairs with a specialized *initiator tRNA* which carries N-formyl-methionine; formylation of the methionine occurs in a reaction requiring N^{10}-formyltetrahydrofolate (Table 6.1). Note that, after completion of translation, either the formyl group or formylmethionine itself is enzymically removed from the free polypeptide. In *E. coli*, formylmethionine is cleaved from about 50% of proteins; whether or not cleavage occurs in a given protein depends apparently on the identity of the *penultimate* N-terminal amino acid – the longer the side-chain on the penultimate amino acid the smaller the probability of cleavage of formylmethionine. Cleavage of formylmethionine is effected by *methionine aminopeptidase* (MAP), one of a family of physiologically important enzymes [properties and functions of aminopeptidases: Gonzales

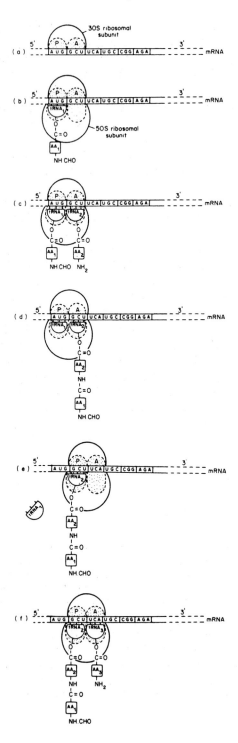

& Robert-Baudouy (1996) FEMSMR *18* 319–344]. Cleavage of the formyl group (only) is effected by the enzyme *methionine deformylase*.

The initiation of protein synthesis, outlined above, requires the participation of three protein *initiation factors* – IF-1, IF-2 and IF-3 – but the timing and mode of their involvement are not yet fully understood, and the current literature contains several different models for this phase of protein synthesis. It is generally agreed, however, that IF-2 and GTP are needed for the binding of tRNA–N-formylmethionine to AUG at the P site; that this tRNA binds to the 30S subunit *before* addition of the 50S subunit (i.e. between steps (a) and (b) in Fig. 7.9); and that all three initiation factors have been released by completion of step (b) in Fig. 7.9.

Elongation of the polypeptide chain proceeds by sequential addition of amino acids according to the codons which follow AUG. The first step in elongation is the binding of an aminoacyl-tRNA to the vacant A site (Fig. 7.9, step c); in *E. coli* this involves GTP and the *elongation factor* EF-T (which comprises two protein subunits: EF-Ts and EF-Tu). The binding of the aminoacyl-tRNA is followed by *transpeptidation* and *translocation* (Fig. 7.9, d–e); steps c–e are repeated throughout translation, codons generally being translated according to the *genetic code* (Table 7.2).

Base-pairing between mRNA codons and tRNA anticodons is actually not stable enough, in itself, to guarantee the degree of accuracy and efficiency needed in translation. The ribosome must therefore provide help, and this appears to be given by the rRNA [see e.g. Schroeder (1994) Nature *370* 597–598].

Termination of translation (the end of polypeptide synthesis) is signalled by

---

Fig. 7.9 Protein synthesis in bacteria (simplified, diagrammatic). (a) The 30S ribosomal subunit binds to mRNA; the *decoding region* of 16S rRNA (in the subunit) coincides with the 'P' and 'A' sites. The *initiator codon* (AUG) of the mRNA occupies the P site.

(b) The first amino acid (AA_1), on its tRNA ($tRNA_1$), occupies the P site, i.e. the anticodon of $tRNA_1$ base-pairs with codon AUG. The 50S ribosomal subunit binds, completing the ribosome.

(c) The second aminoacyl-tRNA occupies the A site, i.e. the anticodon of $tRNA_2$ base-pairs with codon GCU.

(d) A peptide bond is formed between the carboxyl group of AA_1 and the α-amino group of AA_2; this is called *transpeptidation*. (Note that AA_1 will form the 'amino end' or 'N-terminal' of the polypeptide chain.) The enzyme which catalyses transpeptidation is *peptidyltransferase*; this enzyme is actually part of the 50S ribosomal subunit (and, incidentally, is the site of action of certain antibiotics – see section 15.4.4).

(e) The ribosome moves along the mRNA, by one codon, in the 5'-to-3' direction (*translocation*). $tRNA_1$ is released. The dipeptidyl-tRNA now occupies the P site, and there is a vacant A ('acceptor') site opposite the third codon.

(f) A third aminoacyl-tRNA binds at the A site. Steps d–f are repeated for each codon in turn. When a stop codon (e.g. UAA) is reached, a protein *release factor* hydrolyses the ester bond between the last tRNA and the polypeptide chain – thus releasing the completed chain.

the presence, at the A site, of one of the *stop codons* (= *nonsense codons*): UAA, UAG and UGA (*ochre, amber* and *opal* codons, respectively). Termination involves the intervention of a protein *release factor* which must recognise the stop signal and interact directly with the mRNA; this causes hydrolysis of the polypeptide–tRNA bond at the P site, thus releasing the polypeptide. Although the 'stop' signal is traditionally regarded as a *triplet* of bases (as above), there is evidence that the base *following* a stop codon influences the efficiency of termination, and it may be that the actual stop signal is a 4-base (or even longer) sequence [Tate & Mannering (1996) MM *21* 213–219].

As one ribosome translocates along the mRNA another can fill the vacated initiation site and start translation of another molecule of the polypeptide; thus, a given mRNA molecule may carry a number of ribosomes along its length – forming a *polyribosome* (*polysome*) (Fig. 7.10).

Signal sequences. In many cases, a protein destined to form part of a membrane, or to pass through a membrane, is synthesized with a special N-terminal sequence of amino acids (a *signal sequence*); this helps passage into, or through, the hydrophobic region of the membrane, and is enzymically cleaved as the protein passes through the membrane. [Signal sequences (review): Izard & Kendall (1994) MM *13* 765–773.]

Protein folding. As mentioned earlier, polypeptide chains must be folded correctly to form functional proteins, and this often involves disulphide bond formation between amino acids in the chain; disulphide bonds are particularly common e.g. in periplasmic proteins (though rare in cytoplasmic proteins).

In *E. coli*, disulphide bonds in periplasmic, membrane and secreted proteins appear to be catalysed in the periplasmic region; this is inferred from the location of the (periplasmic) Dsb proteins (products of genes *dsbA, dsbB*) which are involved in the catalysis of such bonds. Mutations in the *dsb* genes are *pleiotropic*, i.e. they have multiple effects; this is because the proteins which are stabilized by disulphide bonds have various roles – e.g. flagellar proteins, membrane proteins and various secreted proteins. In pathogenic bacteria, mutations in *dsb* genes may affect secreted protein virulence factors, and this may reduce virulence.

[Disulphide bond formation (review): Bardwell (1994) MM *14* 199–205.]

The correct and efficient folding of newly synthesized proteins, and their translation across membranes (in a partly folded or unfolded state), requires the presence of specialized proteins called *molecular chaperones* (or simply 'chaperones'); chaperones appear to stabilize nascent proteins and promote correct folding by inhibiting incorrect modes of folding.

Chaperones in *E. coli* include the products of genes *groES, groEL* and *dnaK*. A given folding or translocation event may involve the joint action of two or more types of chaperone.

Chaperones are produced constitutively (i.e. under normal conditions), but

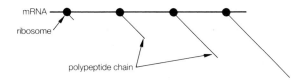

Fig. 7.10 A polysome (diagrammatic). As a given ribosome travels further along the mRNA molecule, its associated polypeptide chain increases in length.

under certain stress conditions (e.g. heat shock: section 7.8.2.3) they are produced in greater quantities; under these circumstances they are likely to be needed for tasks such as the refolding of denatured proteins.

Peptide synthesis without ribosomes. A number of very short polypeptides are synthesized by a multi-enzyme complex instead of by ribosomes and RNA molecules. This process is used e.g. for the synthesis of certain antibiotics such as *tyrocidins* (cyclic peptides containing 10 amino acid residues) and *gramicidins* (linear peptides); these antibiotics (produced by *Bacillus brevis*) act against certain Gram-positive bacteria, apparently by altering the permeability of the cytoplasmic membrane. [Non-ribosomal biosynthesis of peptide antibiotics (review): Kleinkauf & von Döhren (1990) EJB *192* 1–15.]

7.6.1 The fate of mRNA

After use, many *bacterial* mRNAs are short-lived, being rapidly degraded to re-cyclable components; this is essential: the cytoplasm would otherwise quickly fill up with 'used' mRNA molecules. (In some cases, however, degradation of mRNA is delayed – see section 7.8.4.) Degradation involves the action of *nucleases* (enzymes which cleave nucleic acids): *endonuclease(s)* (which cleave phosphodiester bonds in non-terminal regions), and *exonuclease(s)* (which sequentially cleave nucleotides from the 3' end). In *E. coli*, the exonucleases RNase II and polynucleotide phosphorylase jointly degrade mRNA to pieces of about 10 nucleotides in length – but no smaller; how, then, is mRNA re-cycled? Apparently, a previously unknown enzyme, RNase*, degrades these pieces to single nucleotides, thus completing the scheme for mRNA degradation [Cannistraro & Kennell (1991) JB *173* 4653–4659].

Degradation of mRNA can be affected by growth conditions. In *E. coli*, for example, the half-life of at least some mRNAs is greatly increased during slow anaerobic growth (doubling time: 700 minutes); under these conditions, the rate of *synthesis* of mRNA is much lower, so that the increased stability of mRNA may be a mechanism for maintaining gene expression during anaerobiosis [Georgellis *et al.* (1993) MM *9* 375–381].

More recent work indicates that polyadenylation may be an important factor in the stability of mRNAs (see section 7.5.2).

7.6.2　Proteins within proteins: inteins

In 1990 it was discovered that certain proteins contain an internal sequence of amino acids – an *intein* – which apparently catalyses a *self-splicing* reaction in the protein: the intein is excised (forming a separate protein) and the two terminal parts of the original polypeptide (the *exteins*) are joined to form a functional protein. Thus:

$$\text{extein–intein–extein} \rightarrow \text{extein–extein} + \text{intein}$$

This phenomenon was first detected in the *VMA1* ($=TFP1$) gene of the (eukaryotic) yeast *Saccharomyces cerevisiae*, but was later found in bacteria (e.g. the *recA* gene in *Mycobacterium tuberculosis*) and in members of the Archaea (e.g. the DNA polymerase gene in *Pyrococcus* sp and *Thermococcus litoralis*).

Inteins are typically about 350–550 amino acids long. Their essential role in autocatalytic 'self-splicing' is inferred from (i) loss of splicing following deletions in the intein-coding DNA, and (ii) 'heterologous expression' of an intein-coding gene, i.e. expression of such a gene in an organism which does not normally contain that gene; thus, expression of the *VMA1* gene in *E. coli* produced both mature protein and intein.

During self-splicing, an intermediate product is a *branched* polypeptide (with two N terminals and one C terminal); this is reminiscent of the branched mRNA formed, in eukaryotes, during the processing of pre-mRNA to mature mRNA (see Fig. 8.8), and there are indeed parallels between the roles of inteins and those of certain introns.

Some excised inteins have been shown to act as site-specific endonucleases (as have some translated introns). Moreover, the excised *VMA1* intein, of *S. cerevisiae* will specifically cleave a *VMA1* gene lacking an intein-coding sequence, cleavage occurring at the site normally occupied by intein DNA. In a strain of *S. cerevisiae* with one intein-coding *VMA1* gene and one 'intein-less' copy of the gene, cleavage of the latter gene was followed by insertion of intein DNA by a process (called *gene conversion*) in which intein DNA was copied from that in the intein-containing gene.

The ability of an intein to promote the insertion of its coding sequence into an 'intein-less' copy of the given gene (as in *S. cerevisiae*, above) is called *intein homing*. Like transposons (section 8.3), inteins are referred to as *mobile genetic elements*.

Inteins have been reviewed by Cooper & Stevens [TIBS (1995) *20* 351–356]. A new (cyanobacterial) intein has been reported by Pietrokovski [TIG (1996) *12* 287–288] who points out that we do not know whether inteins confer any advantage, but suggests that some may regulate the activity of their host proteins.

7.7 DNA MONITORING AND REPAIR

Abnormal DNA can result e.g. from the insertion of abnormal nucleotides during replication; it may be recognized and repaired immediately through *proof-reading*: DNA polymerase III can cleave a 'wrong' nucleotide from the 3' end of a growing strand, allowing replacement with a normal one.

DNA is also vulnerable to chemical change owing to the reactivity of its bases [Lindahl (1993) Nature *362* 709–715]. For example, aberrant, non-enzymatic, spontaneous methylation by *S*-adenosylmethionine (a normally legitimate methyl donor – see section 7.4) can produce 3–methyladenine and/or 7–methylguanine; each of these aberrantly methylated purine bases can interfere with DNA function. In *E. coli*, these aberrant bases are excised by *N*-glycosylases which cleave the sugar–base linkage. DNA glycosylase I (= Tag protein), which is synthesized constitutively, excises 3–methyladenine. DNA glycosylase II (= AlkA protein), which is inducible, excises not only both of these aberrantly methylated purines but also certain aberrantly methylated pyrimidines; work on the structure of AlkA has suggested that the broad specificity of this enzyme is associated with an electron-rich cleft, rich in aromatic side-chains, which may recognize each of the aberrantly methylated, electron-deficient bases [Labahn *et al.* (1996) Cell *86* 321–329].

After excision of aberrant bases, the site is repaired by an excision repair system.

7.7.1 Excision repair systems

7.7.1.1 *Mismatch repair*

Errors which escape proof-reading may be corrected shortly afterwards by the *mismatch repair system*. In *E. coli*, an enzyme (encoded by genes *mutH*, *mutL* and *mutS*) can recognize a single mismatched base-pair; the new (daughter) strand, identified by its transient under-methylation (section 7.4), is 'nicked' either side of the error, and an enzyme, apparently helicase II, removes the nicked section – which may be 1000 or more nucleotides in length. A DNA polymerase closes the gap by synthesizing on the parent (template) strand, and the junction is sealed by a ligase.

Mismatch repair can also correct mismatch in a heteroduplex (section 8.2.1); if the heteroduplex is fully methylated, then *either* strand can apparently be corrected (i.e. made complementary to the other strand).

7.7.1.2 *The UvrABC endonuclease*

In *E. coli*, an ATP-dependent enzyme, UvrABC endonuclease (= 'ABC excinuclease'), recognizes DNA which has been damaged/distorted by ultraviolet radiation and by other causes; the enzyme cleaves a phosphodiester bond (Fig. 7.4) on each side of the damaged section (the incision stage), and the damaged section is removed – apparently by helicase II. The gap is then closed and sealed as in mismatch repair, above. In *E. coli*, only about 10 nucleotides in and around the damaged site are involved – hence the name *short patch repair*. [Nucleotide excision repair in *E. coli*: van Houten (1990) MR *54* 18–51.]

7.7.1.3   *Other excision systems*

Some systems operate primarily to repair the damage caused by particular types of agent – e.g. oxidative damage [Demple & Harrison (1994) ARB *63* 915–948] or aberrant methylation (see above). Often the first stage of repair is the removal of a chemically aberrant base by a glycosylase; the result is an *apurinic* (purine-less) or *apyrimidinic* (pyrimidine-less) site termed an *AP site* (= abasic site). Next, the phosphodiester bond on one side of the AP site is cleaved by an 'AP endonuclease'; in *E. coli* the major AP endonuclease is exonuclease III, a multifunctional enzyme whose structure suggests that cleavage of phosphodiester bonds involves a nucleophilic attack on the P–3'O link [Mol *et al.* (1995) Nature *374* 381–386]. A DNA polymerase can then replace the damaged nucleotide, together with a few adjacent nucleotides, and a ligase completes the repair.

The spontaneous, low-rate deamination of cytosine (Fig. 7.3) to uracil is repaired by the initial removal of uracil by uracil-*N*-glycosylase (UNG) followed by replacement of the nucleotide. (UNG is useful in recombinant DNA technology – section 8.5.4.3.) However, in some cases a cytosine residue will be methylated at the 5-position owing to modification (section 7.4); deamination of 5-methylcytosine produces *thymine* (Fig. 7.3) – a normal base which is not removed by any enzyme. Hence, the thymine remains, and at the next round of DNA replication it pairs with adenine; thus, a *point mutation* (section 8.1.1; Fig. 8.1) is created at this site: CG in the original duplex is replaced by TA in one of the daughter duplexes. (The replacement of one pyrimidine with another, or one purine with another, is called a *transition mutation*.) Accordingly, 5-methylcytosines are 'hotspots' of spontaneous mutation.

7.7.2   **Photolyase**

One type of damage to DNA caused by ultraviolet radiation is the formation of *thymine dimers*; a thymine dimer is produced by chemical bonding between two adjacent thymine residues in a given strand. Some bacteria, including *E. coli*, encode an enzyme, photolyase, which can use the energy in (visible) light to repair the damage.

7.7.3 Priority repair of actively transcribed genes

Actively transcribed genes are repaired more rapidly than others. One model proposes that, during transcription, aberrant DNA blocks the movement of RNA polymerase and that a transcription-repair coupling factor (TRCF), encoded by gene *mfd* (mutation frequency decline), binds at the affected site, causing release of the polymerase and the unfinished mRNA; TRCF may then interact with the UvrABC system – which effects repair. [Minireview: Selby & Sancar (1993) JB *175* 7509–7514.]

7.8 REGULATION OF GENE EXPRESSION

A cell does not express all of its genes all of the time. For example, if a particular substrate were *not* available the cell would be wasting energy if it synthesized those proteins (e.g. enzymes) needed to metabolize that substrate. In fact, many genes can be 'switched on' (*induced*) or 'switched off' (*repressed*) – or their expression increased or decreased – according to conditions. Such regulation often involves a mechanism which promotes or inhibits synthesis of the gene's mRNA, that is, gene expression is often regulated 'at the level of transcription'; however, there are other modes of regulation, and some of these are described in the following pages.

Although gene expression commonly involves protein synthesis (section 7.6), it should be remembered that genes can also encode e.g. tRNAs and rRNAs.

Apart from the effects of external influences, a gene is inherently regulated by the *strength* of its promoter: strong promoters facilitate transcription, while weak ones make for low-level expression. However, the strength of a promoter is not necessarily fixed; in many cases the level of transcription from a given promoter can be altered by the binding of a specific regulator protein within or near the promoter.

Normal gene expression is also dependent on the correct degree of supercoiling of the DNA (section 7.2.1); for example, less-than-normal negative supercoiling tends to inhibit transcription from many promoters.

7.8.1 Operons

In many cases a *sequence* of genes is transcribed – as a single unit – from a single promoter; a sequence of genes which is subject to co-ordinated expression in this way is said to form an *operon*. Genes in a given operon often encode functionally related products – e.g. the enzymes for a particular metabolic pathway. Operons are controlled by various mechanisms – often at the level of transcription (sections 7.8.1.1 and 7.8.1.2), but sometimes at a translational level (section 7.8.1.3).

7.8.1.1 Operons under promoter control

In these operons, control involves a *regulator protein* which may be formed more or less continually. Operons which are expressed unless 'switched off' by the regulator protein are said to be under *negative control*; those which are not expressed unless 'switched on' by the regulator protein are under *positive control*.

The *lac* operon in *Escherichia coli* (Fig. 7.11) encodes proteins which promote the uptake and metabolism of β-galactosides such as lactose (a disaccharide composed of glucose and galactose residues); the presence of *allolactose* can 'induce' the *lac* operon (see Fig. 7.11). The *lacY* gene encodes β-galactoside permease, which promotes lactose uptake; *lacZ* encodes the enzyme β-galactosidase which can split lactose into glucose and galactose – but which also converts a small amount of lactose to allolactose. The *lacA* product (thiogalactoside transacetylase) appears not to be needed for lactose metabolism. The *lac* operon, which is under negative control, is explained in Fig. 7.11.

The term '*ara* operon' generally refers to the *araBAD* operon – genes *araB*, *araA* and *araD* which encode enzymes concerned with the metabolism of L-arabinose to D-xylulose 5-phosphate. In *E. coli* there are – additionally – genes involved with the uptake (transport: section 5.4) of arabinose, and these genes (*araE*, *araFG*) occur at two further loci on the chromosome. Genes at all three loci are controlled by a common regulator protein: AraC (encoded by *araC*). As genes at different loci are controlled by a common regulator protein, the complete set of genes (uptake + metabolism) conforms to the definition of a *regulon* (section 7.8.2).

In the absence of arabinose, AraC acts as a repressor (i.e. *negative* control, as in the *lac* operon). When present, arabinose converts AraC to an *activator* which initiates transcription; the (modified) regulator protein thus 'switches on' the system – an example of *positive* promoter control.

7.8.1.2 Operons under attenuator control

These operons are concerned e.g. with the synthesis of amino acids. In attenuator control, the initial sequence of nucleotides in the mRNA transcript (the *leader sequence*) encodes a small peptide (*leader peptide*) which is rich in the particular amino acid whose synthesis is governed by that operon; the leader sequence also includes a rho-independent terminator (section 7.5) – the *attenuator* – located between the leader-peptide-encoding region and the first gene of the operon.

If a cell contains adequate levels of the given amino acid (i.e. no synthesis required) transcription is stopped by the attenuator. (The leader peptide is synthesized because a ribosome follows closely behind the RNA polymerase, translating the newly formed transcript.) With inadequate levels of the amino acid (synthesis required) the ribosome synthesizing the leader peptide 'stalls'

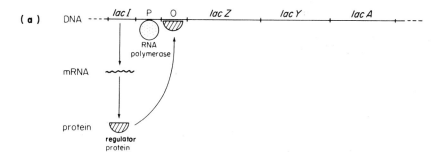

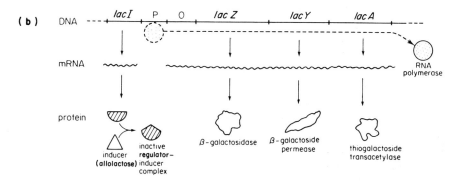

Fig. 7.11 The *lac* operon in *Escherichia coli*. (a) In the absence of an inducer, the regulator protein (product of the *lacI* gene) binds to the operator, O, and minimizes transcription of genes *lacZ*, *lacY* and *lacA* by the RNA polymerase (seen here at the promoter, P); there is some 'leakage': a few molecules of e.g. the *lacZ* product (β-galactosidase) are formed under these conditions. (b) In the presence of lactose (taken up e.g. by proton–lactose symport: section 5.4), the enzyme β-galactosidase converts some of the lactose to *allolactose* (section 7.8.1.1); allolactose acts as an inducer of the *lac* operon by binding to – and thereby inactivating – the regulator protein. Once the regulator protein has been inactivated, the three genes can be transcribed and translated. (For each of the three genes, the single mRNA transcript contains an initiator codon and a stop codon – see section 7.6.)

When the lactose has been used up, transcription of the *lac* operon is no longer needed; the inducer (allolactose) is no longer formed, and the (now active) regulator protein 'switches off' the operon. The mRNA is rapidly degraded (section 7.6.1).

The *lac* operon is subject to catabolite repression (section 7.8.2.1).

While the above account gives the essential operation of the *lac* operon, the actual process is rather more complex. For example, the regulator protein can bind at two other sites, albeit with lower affinity; these sites are called O-2 and O-3. During repression of the operon, it seems that a tetramer of regulator protein binds at the main operator (O) *and* at O-2 (or O-3) – thus forming a loop of DNA – and that this stabilizes regulator–operator binding [Reznikoff (1992) MM 6 2419–2422].

Isopropyl-β-D-thiogalactoside (IPTG) is a gratuitous inducer of the *lac* operon, i.e. it induces the operon but is not metabolized by the cell; IPTG is used in recombinant DNA technology (see Fig. 8.16).

when it reaches codon(s) specifying the given amino acid. If this happens, it allows 'downstream' parts of the transcript to base-pair with one another in such a way that the attenuator cannot form – so that the genes will be transcribed.

Attenuator control occurs e.g. in the *his* (histidine) operon in *E. coli*.

The *trp* (tryptophan) operon in *E. coli* is under both negative promoter control and attenuator control.

7.8.1.3   *Operon regulation by translational control*

An example of the translational control of genes in an operon is given by the IF3–L35–L20 operon (genes *infC-rpmI-rplT* respectively) in *E. coli*; this operon encodes two ribosomal proteins (L35, L20) and a protein factor involved in protein synthesis (IF3: translation initiation factor 3).

Translation of L35 and L20 is repressed by L20, i.e. the concentration of L20 in the cell regulates gene expression – high levels of L20 repressing translation. Repression by L20 appears to involve the binding of L20 to mRNA at a site upstream of *rpmI*. The mechanism of repression by L20 is unknown – L20 could simply block mRNA–ribosome binding, or it could bind the ribosome to form an inactive complex. However, an alternative explanation has been suggested by *in vitro* studies which indicate that the regulation of translation may involve an RNA 'pseudoknot', a structure formed by base-pairing between mRNA sites upstream of *rpmI*; it has been suggested that L20 may stabilize the pseudoknot – which includes the Shine–Dalgarno sequence (section 7.6) and initiator codon (section 7.6) of *rpmI* – so as to inhibit ribosome binding and, hence, prevent translation of both *rpmI* and *rplT* [Chiaruttini, Milet & Springer (1996) EMBO Journal 15 4402–4413].

7.8.2   Regulons and other 'global' and multi-gene control systems

A regulon is a system in which two or more typically non-contiguous genes and/or operons (each with its own promoter) are controlled by the same regulator molecule – all the genes/operons having similar regulatory sequences that are recognized by the regulator molecule.

7.8.2.1   *Catabolite repression*

Diauxic growth (section 3.4) shows that the presence of lactose does not *necessarily* induce the *lac* operon (section 7.8.1.1) – i.e. the effect of allolactose is overridden in the presence of glucose. Similarly, glucose will repress the *ara* operon (section 7.8.1.1) in the presence of arabinose. These are only two examples of *catabolite repression* (the 'glucose effect'): a common phenomenon

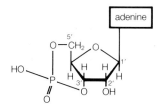

Fig. 7.12 Cyclic AMP (cAMP): a molecule involved e.g. in catabolite repression (section 7.8.2.1) – and, interestingly, in the pathogenesis of cholera (section 11.3.1.1). cAMP (adenosine 3',5'-cyclic monophosphate) is synthesized from ATP by the enzyme adenylate cyclase, and is degraded to AMP by cAMP phosphodiesterase. [Various roles of cyclic AMP in prokaryotes (review): Botsford & Harman (1992) MR 56 100–122.]

(in bacteria) in which a cell uses some substrates in preference to others. The mechanism is complex and not fully understood.

The overall regulatory molecule is the *catabolite activator protein* (CAP), also called CRP. To be functional, CAP must bind to (i.e. be activated by) cyclic AMP (cAMP) (Fig. 7.12) to form the cAMP–CAP complex.

cAMP–CAP allows the *lac, ara* and certain other operons to be expressed in the presence of their respective inducers. In the *lac* operon, it was once thought that cAMP–CAP might act by displacing RNA polymerase from a preferred, but weak, promoter site to the normal promoter site of the operon; this is now thought unlikely because e.g. inactivation of the alternative promoter site does not activate the normal promoter [Reznikoff (1992) MM 6 2419–2422]. Apparently, cAMP–CAP binds to DNA at the CAP (= CRP) site near the P1 promoter region – this facilitating the binding of RNA polymerase to the P1 promoter in preparation for transcription of the *lac* operon. (A study of the binding of CAP (= CRP) to the P1 promoter region of the *gal* operon is described in Plate 8.3.)

The presence/absence of cAMP–CAP in the cell depends e.g. on the level of cAMP. cAMP levels are low (cAMP–CAP absent) when glucose levels are high, and vice versa. The reason is not known. The level of cAMP seems to be determined mainly by the level/activity of its synthesizing and degradative enzymes, but we do not know how glucose affects these factors. Moreover, glucose-regulated CAP levels may also be important [Ishizuka *et al.* (1993) MM *10* 341–350]. For some sugars, catabolite repression is lower or absent in very low levels of glucose [Lendenmann & Egli (1995) Microbiology *141* 71–78].

7.8.2.2 The SOS system in Escherichia coli

This system of about 20 unlinked genes is expressed when the DNA is damaged and/or cannot replicate – due e.g. to the effects of ultraviolet radiation and/or certain chemicals; expression of the SOS system can (for example) stop cell division, increase DNA repair activity, affect energy metabolism and suppress restriction.

Control is exercised by the LexA protein (*lexA* gene product); under normal conditions (DNA not damaged), LexA (probably as a dimer) binds close to the promoters of the SOS genes and inhibits their transcription. For each SOS gene, the binding site of LexA (the 'SOS box') appears to be the consensus sequence 5'-CTGTN$_8$ACAG-3' (in which N$_8$ is an 8-nucleotide sequence). Damage to DNA activates the RecA protein (by an unknown mechanism). Activated RecA (designated RecA*) functions, non-enzymatically, as a *co-protease* which triggers the autocatalytic cleavage of LexA; this allows expression of the SOS genes.

The product of the SOS gene *sulA* represses septum formation (section 3.2.1) [SulA–FtsZ interaction: Huang, Cao & Lutkenhaus (1996) JB *178* 5080–5085], thus inhibiting cell division; cells may continue to grow as septum-less filaments. A physiological advantage of the inhibition of cell division may be that cells are given time to carry out repairs to their damaged DNA. DNA replication may continue (see end of section 7.3).

DNA repair, including that mediated by the UvrABC endonuclease (section 7.7.1.2), shows enhanced activity. A repair process involving genes *umuC* and *umuD* (so-called 'error-prone repair') operates only when the SOS system has been induced; as well as repairing DNA, error-prone repair results in an increased number of *mutations* (section 8.1) – this being referred to as *SOS mutagenesis*. In this process, RecA*, again acting as a co-protease, brings about the autocatalytic cleavage of UmuD to form the active fragment, UmuD', which is involved in SOS mutagenesis. The mechanism by which mutations are generated in this process is unknown, but it is generally believed to involve DNA (repair) synthesis on a template strand containing a 'lesion' – e.g. a pyrimidine dimer (two covalently linked pyrimidines); such synthesis (called *translesion* synthesis) would be mediated by a modified form of DNA polymerase (produced as a result of SOS induction) which, owing to the local lack of base-pairing specificity, is likely to introduce incorrect base(s). Subsequent excision repair of the lesion, and DNA replication, is therefore likely to give rise to a mutant genome. Translesion synthesis *in vitro* on a template strand containing an *abasic* site (section 7.7.1.3) has been found to require DNA polymerase III, RecA UmuD' and UmuC.

Once DNA has been repaired, RecA is inactivated and LexA again represses the SOS system.

Note that *lysogenic* bacteria may undergo lysis when the SOS system is expressed (see section 9.2.1).

7.8.2.3　The heat-shock response

Heat shock (a sudden rise in temperature) causes a characteristic adaptive response in organisms ranging from bacteria to plants and animals; the response includes increased synthesis of the so-called *heat-shock proteins* (HSPs).

In *E. coli*, the heat-shock response follows e.g. a $30° \rightarrow 42°C$ shift in temperature; it can also be triggered by stress factors such as ultraviolet radiation, ethanol, and the intracellular accumulation of abnormal/heterologous proteins (as in overproduction: section 8.5.11.4(d)).

Some of the 17 known HSPs in *E. coli* cope with stress-induced damage; thus e.g. the molecular *chaperones* GroES, GroEL and DnaK (section 7.6) appear to prevent mis-folding/aggregation of unfolded proteins, and the Lon protease (product of the *lon* gene) degrades damaged or abnormal proteins. Other HSPs include the products of genes *dnaJ*, *rpoD* and *lysU*.

Following heat shock, the synthesis of HSPs rises to a maximum within minutes and then decreases to a new steady state above that of the lower temperature.

In *E. coli* the heat-shock regulon is controlled by the product of gene *rpoH* (previously called *htpR*). RpoH is a sigma factor (section 7.5), σ^{32}, which is (transiently) synthesized in greater quantities following heat shock; compared with the 'routine' sigma factor (σ^{70}), σ^{32} permits more efficient transcription from the promoters of the HSP genes and so causes increased synthesis of the HSPs.

The increase in σ^{32} following heat shock is due mainly to its increased translation and stabilization (rather than to increased transcription of *rpoH*). Heat-induced translation from the *rpoH* mRNA appears to involve an effect on a 'secondary structure' in the mRNA formed by base-pairing between a region immediately downstream of the start codon (a 'downstream box' – see section 8.5.11.2) and another region in the coding sequence; it is believed that this secondary structure represses translation under normal conditions, and that repression is relieved during heat shock [Yuzawa *et al.* (1993) NAR *21* 5449–5455].

σ^{32} is synthesized under normal conditions but it has a short half-life (about 1 minute): it is complexed by DnaK, DnaJ and GrpE and subsequently degraded by the FtsH protease; during heat shock, DnaK binds preferentially to denatured proteins, and this sequestration of DnaK leaves σ^{32} free to function as a sigma factor – leading to increased synthesis of the HSPs [see e.g. Gamer *et al.* (1996) EMBO Journal *15* 607–617].

Heat shock in other bacteria. Different mechanisms for regulating the induction of the heat-shock response occur in some bacteria. For example, in some cases a regulatory sequence is located upstream of heat-shock genes (indicating control at the transcriptional level); this inverted repeat sequence, termed CIRCE (controlling inverted repeat of chaperone expression), has been found e.g. in strains of *Bacillus* and *Clostridium* and in some Gram-negative bacteria. Control in *Bradyrhizobium japonicum* consists of at least two distinct regulatory systems which involve CIRCE and a σ^{32}-like sigma factor [Narberhaus *et al.* (1996) JB *178* 5337–5346].

7.8.2.4   The cold-shock response

A sudden fall in temperature can bring about an altered pattern of gene expression in prokaryotes and eukaryotes. The following refers to cold shock in *E. coli*.

The cold-shock response is triggered e.g. by a 37° → 10°C down-shift (or, in general, a down-shift of at least 13°C). Growth stops, resuming (at a lower rate) after a lag of about 4 hours. The lag period is apparently due to a selective inhibition of translation; resumption of growth coincides with the resumption of protein synthesis.

Cold shock is characterized by e.g. (i) induction/increased synthesis of *cold-shock proteins*; (ii) ongoing synthesis of certain proteins involved in transcription and translation (despite the generalized inhibition of protein synthesis); (iii) repression of heat-shock proteins. This is believed to be an adaptive response.

The cold-shock proteins include CspA: a small (70 amino acid) protein, inducible (immediately on temperature down-shift) at the level of transcription, which appears to interact with nucleic acids; its suggested functions include (i) a low-temperature translational activator; (ii) a low-temperature 'RNA chaperone' which unfolds RNA molecules; and (iii) an agent which inhibits the translation of particular mRNAs.

Other cold-shock proteins include RecA; initiation factor 2 (IF-2), which mediates the binding of *N*-formylmethionine-charged tRNA to the 30S ribosomal subunit at the start of translation (section 7.6); NusA, involved in the termination stage of transcription; and the α-subunit of DNA gyrase (a topoisomerase – section 7.2.1). No cold-inducible sigma factors have so far been identified.

The mode of regulation of the cold-shock genes is not known. Suggestions include CspA as a transcriptional activator (CspA itself being regulated possibly by a repressor which loses activity at low temperatures). The small molecules ppGpp and pppGpp (in which G is guanosine, and p is phosphate) may have a regulatory role: following temperature down-shift, the concentration of these molecules decreases, and this fall in concentration has been associated with increased synthesis of some cold-shock proteins.

The cold-shock response, as described, can also be triggered by certain inhibitors of translation – e.g. the antibiotics chloramphenicol, erythromycin and tetracycline. That these agents (many of which also depress (p)ppGpp levels) have an effect similar to that of a temperature down-shift has suggested that the common factor – a decrease in translational capacity – may be important in the induction of the cold-shock response. Hence, according to one model, the temperature down-shift causes the cell's translational capacity to become too small in relation to the supply of charged tRNAs (i.e a state resembling a *nutritional* up-shift), and that this physiological state signals a fall in the concentration of (p)ppGpp with consequent induction of the cold-shock response [Jones & Inouye (1994) MM *11* 811–818].

7.8.2.5 The stringent response

In bacteria, starvation (e.g. lack of an essential amino acid) elicits the *stringent response* – which includes decreased synthesis of proteins, rRNA and various cell components, and inhibition of DNA synthesis; this conserves energy and material.

The response is triggered by an uncharged tRNA at the A site of a ribosome (see Fig. 7.9). This stimulates a ribosome-bound enzyme, pyrophosphotransferase (= RelA, the *relA* gene product; stringent factor), to synthesize ppGpp (in which G is guanosine, and p is phosphate) from GTP (or GDP) and ATP. ppGpp (guanosine 5'-diphosphate 3'-diphosphate) is an example of an *alarmone*: a small molecule which accumulates under certain stress conditions and which serves as a signal for re-directing the cell's metabolism.

ppGpp may inhibit the initiation of protein synthesis e.g. by interacting with a protein *initiation factor*, IF-2, and blocking the binding of the tRNA complex to the ribosome. ppGpp has been shown to inhibit both protein synthesis and rRNA synthesis [Svitil, Cashel & Zyskind (1993) JBC *268* 2307–2311].

ppGpp can apparently enhance transcription from some of the operons governing the biosynthesis of particular amino acids (e.g. histidine). ppGpp may thus help to maintain a correct balance of amino acids in the cell; for example, given a level of histidine low enough to trigger the stringent response, increased synthesis of ppGpp should help to restore the balance by (i) slowing protein synthesis, and (ii) stimulating the *his* operon.

Cells having a null mutation in *relA* do not exhibit the stringent response when starved of an essential amino acid and are said to be *relaxed*.

7.8.2.6 The acid tolerance response (ATR)

Acid shock elicits a response (ATR) involving the synthesis of so-called *acid-shock proteins* (ASPs) – which are believed to deal with acid-induced damage and/or promote survival at low pH. In *Salmonella typhimurium* the response appears to be a two-stage process (section 3.1.8); in this organism, adaptation to, or survival in, low pH is necessary for a pathogenic role because infection normally occurs via the (highly acidic) stomach.

In *S. typhimurium* at least 50 ASPs are synthesized in the ATR; expression of the ASP genes is controlled by at least two regulatory proteins: (i) the sigma factor σ^s (RpoS; *rpoS* gene product), and (ii) the Fur protein (*fur* gene product). RpoS, itself an acid-inducible ASP, regulates a subset of ASP genes; mutants lacking the σ^s function can still exhibit a transient ATR in which Fur regulates (σ^s-independent) ASP genes. Interestingly, Fur also regulates the iron-uptake

Table 7.3   Assembly of the flagellum of *E. coli/Salmonella typhimurium*: some of the genes and functions involved

Gene (product)	Function of gene product
Class I	
flhC, flhD (FlhC, FlhD)	Transcriptional activators of class II genes
Class II	
flgE (FlgE)	Hook
flgH (FlgH)	L ring
flgI (FlgI)	P ring
flgM (FlgM)	Anti-sigma factor; represses FliA until completion of hook
fliA (FliA)	Sigma factor regulating class III genes
fliF (FliF)	MS ring
fliG (FliG)	C ring
fliM (FliM)	C ring
fliN (FliN)	C ring
motA (MotA)	Torque generation (with MS/C rings)
motB (MotB)	Torque generation (with MS/C rings)
Class III	
fliC (FliC)	Filament subunit (the protein *flagellin*)

function in *S. typhimurium* (section 11.5.5) – but the roles of Fur in ATR and iron-uptake are physiologically and genetically distinct [Hall & Foster (1996) JB *178* 5683–5691].

7.8.2.7   Assembly of the bacterial flagellum

The structure and assembly of the flagellum (section 2.2.14.1, Fig. 2.8) involves about 50 different types of protein; assembly is highly organized – particular components being added in strict sequence – and the corresponding genes are expressed in a way which reflects this sequence. Some of the relevant genes are listed in Table 7.3.

In *E. coli*, the first genes to be expressed (class I genes) are those encoding certain transcriptional activators which seem to be necessary for expression of (class II) genes encoding components of the basal body and hook. Expression of class I and II genes is necessary for the expression of class III genes (whose products include the filament subunit protein *flagellin*).

Transcription of class III genes depends on a σ factor (section 7.5) encoded by a class II gene. This σ factor (protein FliA, *fliA* gene product) is temporarily inactivated by an anti-σ factor (FlgM, *flgM* gene product) until completion of the basal body and hook; completion of the hook acts as a signal which seems to allow release of FlgM (via the axial channel) – this allowing FliA to mediate transcription of the class III genes. Thus, genes encoding the final phase of construction are governed directly by the level of assembly of the partly completed flagellum.

Genetic control of flagellar assembly has been reviewed by Shapiro [(1995) Cell *80* 525–527].

7.8.3 Recombinational regulation of gene expression

See site-specific recombination (section 8.2.2).

7.8.4 Regulation of gene expression by the rate of decay of mRNA

In some cases, the expression of a bacterial gene is regulated by the rate of decay of its mRNA – 'decay' meaning enzymatic degradation (section 7.6.1). Such regulation has been studied e.g. in the *puf* operon of a photosynthetic bacterium, *Rhodobacter capsulatus*; the *puf* genes (encoding components of the photosynthetic apparatus) are transcribed together as a single mRNA transcript (i.e. *polycistronic* mRNA). (An example of a polycistronic mRNA has already been seen in Fig. 7.11.) Interestingly, different parts of *puf* polycistronic mRNA decay at different rates – i.e. some sequences survive for longer than others; in this way, the correct genes are expressed at appropriate times, so that photosynthesis and cell growth occur optimally. Special decay-promoting and decay-inhibiting regions in the *puf* polycistronic mRNA may be responsible for these differential rates of decay [Klug (1993) MM *9* 1–7].

7.8.5 Regulation of gene expression by translational attenuation

In some Gram-positive bacteria, the presence of chloramphenicol (section 15.4.4) induces the *cat* gene; *cat* encodes chloramphenicol acetyltransferase: an enzyme which inactivates chloramphenicol by acetylating it. To understand this inducible resistance to chloramphenicol we look at the regulatory process of *translational attenuation*.

The transcript of the *cat* gene consists of an initial short 'leader' sequence of nucleotides followed – in the same mRNA transcript – by the *cat* coding sequence. In the absence of chloramphenicol, the *cat* gene is transcribed but not translated because the ribosome-binding site is 'distorted' by local base-pairing between ribonucleotides. However, the leader sequence (which has its own ribosome binding site) seems to be translated continually. In the presence of chloramphenicol, the ribosome translating the leader sequence stalls; this results in a loss of the 'distortion' at the *cat* ribosome binding site – permitting translation of the *cat* coding sequence. Why is *cat* translation itself not prevented by chloramphenicol (which inhibits protein synthesis)? The *cat* induction mechanism is triggered by levels of chloramphenicol even lower than those which inhibit protein synthesis – possibly because the nascent leader peptide itself (as well as chloramphenicol) contributes to ribosome stalling [Gu, Rogers & Lovett (1993) JB *175* 5309–5313].

Another example of translation attentuation is the regulation of gene *pyrC* in *E. coli; pyrC* encodes the enzyme dihydroorotase – involved in the synthesis of pyrimidines and, hence, nucleic acids. Transcription of *pyrC* can begin at any of several (adjacent) nucleotides on the DNA template – transcription from a given nucleotide being influenced by the availability of pyrimidines in the cell (as 'sensed' by the CTP/GTP ratio). Transcripts of *pyrC* synthesized in the presence of adequate pyrimidines are mainly of a kind in which RNA–RNA base-pairing can occur in the region of the Shine–Dalgarno sequence; such base-pairing blocks ribosome binding and gene expression. With low levels of pyrimidines, most transcripts are synthesized from a slightly different start point, and, in these transcripts, base-pairing does not occur so that the enzyme is synthesized.

7.8.6   Regulation of gene expression via signal transduction pathways

Bacteria can detect various environmental signals (such as changes in osmolarity) by means of sensory systems located in the cell envelope; within the cell, these signals act via signal transduction pathways involving so-called *two-component regulatory systems*. The first component is a *histidine kinase*: an enzyme which can undergo ATP-dependent autophosphorylation (at an active site containing a histidine residue) and transfer the phosphate to another molecule. When appropriately influenced by environmental signals the histidine kinase transfers phosphate to a site containing an aspartate residue in the second component – a separate regulator protein (= response regulator); when phosphorylated, the regulator protein modifies the expression of certain genes, e.g. by controlling their transcription, thus eliciting a response to the environmental signal. [Histidine kinases and signal transduction (review): Alex & Simon (1994) TIG *10* 133–138.]

In some cases the environmental signal acts directly on the kinase, regulating its activity; that is, the kinase *is* the sensor. For example, in *E. coli*, increased osmolarity stimulates the kinase activity of a sensor protein, EnvZ, located in the cell envelope. EnvZ transfers phosphate to the regulator protein, OmpR, which then (i) enhances transcription of the gene encoding OmpC porin (section 2.2.9.2) and (ii) inhibits transcription of the gene encoding OmpF porin; thus, at high osmolarity the outer membrane contains more OmpC (which forms a slightly smaller pore) than OmpF. (Other responses to increased osmolarity are mentioned in section 3.1.8.)

In other cases the sensor (= 'receptor') and kinase are different molecules: environmental signals are detected by the sensor which regulates the activity of the kinase; this occurs e.g. in the chemotaxis pathway in *E. coli* (see Fig. 7.13) in which the kinase regulates *two* different regulator proteins.

In *Bacillus subtilis*, the initiation of sporulation involves a *phosphorelay* (see section 7.8.6.1 and Fig. 7.14).

In pathogens, the expression of certain virulence factors (e.g. the cholera toxin) is regulated by environmental conditions via signal transduction pathways.

7.8.6.1 *Initiation of endospore formation in* Bacillus subtilis

When growth becomes limited by a shortage of nutrients, *B. subtilis* re-organizes its metabolism: genes are induced and repressed to reflect the new conditions – but there is no abrupt change from active growth to active sporulation. Instead, exponential (log-phase) growth is followed by a *transition state* in which both growth-related and survival-related activities occur simultaneously [transition state in *B. subtilis* (review): Strauch (1993) PNARMB 46 121–153]; during this period the decision is made to either maintain the vegetative state, with low-level metabolism and no cell division, or to initiate sporulation.

Clearly, sporulation is a major step, and before a final commitment is made the cell must integrate and interpret signals from various sources, both environmental and intracellular. For example, one relevant signal is the level of calcium; thus, sporulation is *inhibited* if levels of calcium drop to 2 μM. (The value of this particular constraint can be seen by recalling that the core of an endospore contains an accumulation of calcium dipicolinate (section 4.3.1); were sporulation to begin with insufficient calcium it may abort at an intermediate stage.)

Signals which promote sporulation appear to be recognized by one or more sensor molecules; these sensors are *kinases*. (The nature of the signals which activate these kinases is unknown; for some years it has been thought that sporulation is triggered by a shortage of certain metabolite(s) – particularly guanine nucleotides – but experimental confirmation is lacking.)

Receipt of the appropriate environmental and/or internal signals apparently causes ATP-dependent autophosphorylation of the kinase(s) – which transfer the phosphate to the first component of a *phosphorelay* system (Fig. 7.14); this component, the protein Spo0F, thus acts as a main junction through which are channelled signals from various (external and internal) sources. The phosphorelay involves sequential phosphorylation of proteins – the final protein being Spo0A. The intracellular level of Spo0A~P (i.e. activated (phosphorylated) Spo0A) seems to be the key factor which determines the decision between sporulation and continued vegetative growth.

7.8.7 Regulation of gene expression by translational frame-shifting

In some cases, the synthesis (or composition) of a polypeptide is regulated during *translocation* (Fig. 7.9e): a specific sequence of nucleotides in the transcript causes the ribosome to 'slip' along the mRNA – commonly either a

a

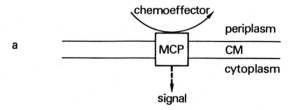

b

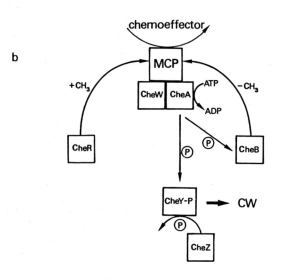

c

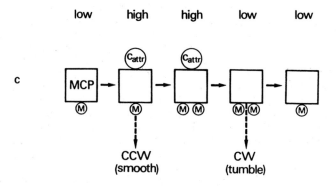

Fig. 7.13   Chemotaxis in *Escherichia coli*: a model of receptor-mediated chemotaxis (section 2.2.15.2) involving signal transduction (section 7.8.6).

(a) The receptors – called *methyl-accepting chemotaxis proteins* (MCPs) – occur in the cytoplasmic membrane (CM) and have functional regions in both the periplasm and cytoplasm. There are different types of MCP, each type recognizing its own range of chemoeffectors/environmental stimuli, and there are several hundred molecules of each type of MCP in a given cell. (The MCPs are encoded by genes *tap*, *tar*, *trg* and *tsr*.) Some chemoeffectors (e.g. aspartate, serine) bind directly to their MCPs, but ribose, and some other sugars, first complex with certain periplasmic proteins before binding. The binding/release of chemoeffectors by an MCP regulates an intracellular signal which controls the frequency of tumbling.

(b) The intracellular signal originates at the protein CheA (encoded by gene *cheA*). CheA is a *histidine kinase* (see section 7.8.6). In the diagram, CheA is bound to the MCP; CheW is a coupling protein. CheA can transfer phosphate from ATP to a regulator protein, CheY; the phosphorylated (activated) form of CheY (CheY~P) enhances CW flagellar rotation (and, hence, increases the frequency of tumbling) by interacting with the C ring of the flagellar motor (apparently with the FliM component – see Fig. 2.8). The basic signal thus consists of the transfer of phosphate from CheA to CheY; this signal is modulated (regulated) by (i) the binding/release of chemoeffector at the MCP, and (ii) the degree of methylation of the MCP. We look at each of these two factors in turn and then consider how, jointly, they regulate signal transduction and the frequency of tumbling.

The binding of a chemoattractant *inhibits* the kinase activity of CheA, thus inhibiting the transfer of phosphate to CheY; the release of a chemoattractant greatly *stimulates* CheA, promoting phosphate transfer to CheY and (hence) encouraging tumbling.

The cytoplasmic side of an MCP is subject to ongoing methylation by a methyltransferase, CheR, which uses S-adenosylmethionine as a methyl donor. Opposing this, a methylesterase, CheB, removes the methyl groups; the activity of CheB is enhanced when it is phosphorylated by CheA. The degree of methylation of an MCP thus depends on the activities of both CheR and CheB; increased methylation tends to *enhance* the activity of CheA.

(c) In a low, uniform concentration of chemoattractant (C_{attr}), tumbling occurs at a given rate. On entering the high concentration, C_{attr} binds to MCP – inhibiting CheA (inhibiting phosphorylation of CheY) and promoting CCW rotation (smooth swimming). Inhibition of CheA also inhibits CheB, allowing increased methylation of the MCP (because CheR methylates continually). Hence, the inhibition of CheA, due to C_{attr} binding, is subsequently offset by the stimulatory effect of increased methylation of the MCP – so that the frequency of tumbling returns to its original value; the cell has thus *adapted* to the new, higher uniform concentration of C_{attr} (MCP in the centre).

On re-entering the low concentration, release of C_{attr} stimulates CheA and promotes CW rotation (more frequent tumbling). CheB is also stimulated, demethylating the MCP until CheA activity has been reduced to its original level: adaptation to the new, low, uniform concentration (MCP at the right-hand side). Such adaptation requires efficient dephosphorylation of CheY and CheB appropriate to the adapted level of CheA activity; CheB undergoes autodephosphorylation, and the regulation of CheY~P levels involves the CheZ protein.

Note that the degree of methylation of an MCP in an *adapted* cell reflects the extracellular concentration of the chemoattractant.

Control of chemotaxis has been reviewed by Eisenbach [(1996) MM *20* 903–910].

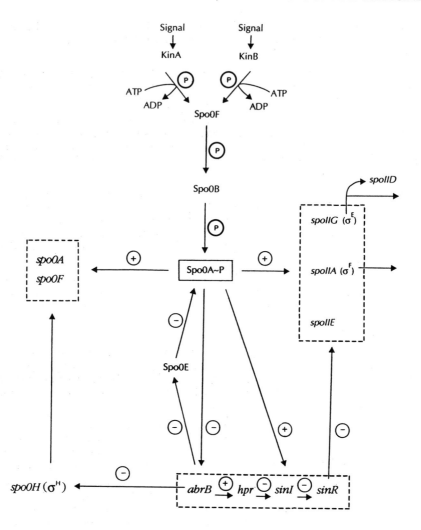

Fig. 7.14 Endospore formation in *Bacillus subtilis*: simplified scheme for the regulation of gene expression during the initiation of sporulation. The scheme is based on information from several sources.

Endospore formation (section 4.3.1) involves a range of proteins which are not synthesized in the vegetative (growing) cell, and this requires co-ordinated expression of various sporulation-specific genes. Not only must these genes be switched on, but other genes – which have been repressing sporulation genes during vegetative growth – must be switched off.

The mechanism for initiating sporulation appears to recognize both external (environmental) and internal (intracellular) signals. These signals are believed to cause autophosphorylation of certain kinases (section 7.8.6.1) – shown in the diagram as KinA and KinB. The kinases transfer phosphate (symbol ⓟ) to a sequence of proteins: Spo0F→Spo0B→Spo0A – the so-called *phosphorelay* [Strauch & Hoch (1993)

COGD 3 203–212; Hoch (1993) JCB 51 55–61]. The intracellular concentration of the phosphorylated form of Spo0A – Spo0A~P – appears to be the key factor in initiating sporulation.

In the diagram, inhibitory influences are indicated by the symbol ⊖, while ⊕ indicates that the expression of a given gene is promoted. In most cases, control is exercised at the level of transcription.

Spo0A~P promotes the expression of certain sporulation-specific genes (shown in the dashed box, centre right). SpoIIG (encoded by *spoIIG*) is a sigma factor (section 7.5), σ^E; this sigma factor is needed for the transcription of (i) the *spoIID* gene and (ii) certain genes in the mother cell. The sigma factor encoded by *spoIIA*, σ^F, is needed for the transcription of certain genes in the forespore.

During vegetative growth, SinR (encoded by *sinR* in the lower dashed box) represses expression of the *spoII* genes in the dashed box at centre right. In sporulation, Spo0A~P represses *abrB* (lower dashed box) – thus inhibiting expression of SinR and helping to promote the expression of the *spoII* genes; SinR is repressed at the protein–protein level, i.e., the protein SinI inhibits the protein SinR. Note that the transcription of *sinI* is also promoted by Spo0A~P.

Spo0A~P promotes the expression of genes *spo0F* and *spo0A* (dashed box, left) by a positive feedback loop [Strauch *et al.* (1993) MM 7 967–974]. Also, by repressing *abrB*, Spo0A~P promotes the expression of *spo0H* – whose product (Spo0H; = σ^H) is required e.g. for the transcription of *spo0F* and *spo0A* (σ^H being produced at low level during vegetative growth).

Production of the sigma factor σ^H is greatly enhanced following the initiation of sporulation. During sporulation, σ^H is needed for the transcription of gene *ftsZ* (see FtsZ, section 3.2.1) from a special promoter, p2, which is distinct from the *ftsZ* promoter used during exponential growth; FtsZ is required for the formation of the asymmetrical septum.

Spo0E is a negative regulator of the phosphorelay [Ohlsen *et al.* (1994) PNAS 91 1756–1760]; it is an enzyme which de-phosphorylates (and thus inactivates) Spo0A~P and which may therefore help to prevent sporulation until the cumulative effect of the various (extracellular and intracellular) signals dictates that sporulation is necessary.

In the study of gene control in sporulation, much information has been obtained from mutants which are blocked at certain stages (Fig. 4.2) of the process.

Sporulation can be prevented by mutations in the *spo0* genes. Thus, e.g. a null mutation (i.e. a mutation causing *total* loss of gene product/function) in *spo0B* will block the phosphorelay.

So far, no mutations have been found to affect stage I (Fig. 4.2); some workers do not regard stage I as a distinct stage.

Mutations in the *spoII* genes can prevent development beyond the formation of the asymmetrical septum. Recent work on *spoIIE* mutants has indicated that SpoIIE has two distinct roles: regulation of sigma factor σ^F, and involvement in the development of a normal, functional asymmetrical septum (SpoIIE may even be incorporated in the septum) [Barák & Youngman (1996) JB 178 4984–4989].

'+1 slip' (a 1-nucleotide shift downstream, i.e. in the direction of translation) or a '−1 slip' (a 1-nucleotide shift upstream, i.e. in the opposite direction). Such a shift, of course, affects all subsequent codons in the transcript (compare the frame-shift mutation, Fig. 8.1b). Frame-shifting may result e.g. in a new stop codon, thereby ending translation prematurely (and producing a shorter polypeptide). Frame-shifting can also nullify the effect of an existing stop codon within the transcript; in one such case, frame-shifting (leading to translation of the whole transcript) occurs when levels of the polypeptide are low – but is prevented (stop codon functional) by adequate levels of the polypeptide. [Translational frame-shifting (review): Engelberg-Kulka & Schoulaker-Schwarz (1994) MM *11* 3–8.]

7.8.8 DNA methylation as a control mechanism

After DNA replication, methylation (modification: section 7.4) protects the chromosome from the cell's own restriction enzymes. It has been suggested that, following replication, the chromosome's origin (*oriC*) may be left initially hemi-methylated (i.e. only the template strand methylated), and that the time required for further methylation (apparently needed for a functional *oriC*) may be a factor in regulating the initiation of chromosome replication in the cell cycle (section 3.2.1).

In *E. coli*, the Dam methylase (encoded by the *dam* gene) methylates the N-6 position of adenine in 5'-GATC-3' sequences. Such 'Dam methylation' affects the transcription of certain genes when it occurs in their promoters. For example, Dam methylation in the promoter of the transposase gene of transposon Tn*10* (transposition: section 8.3) tends to inhibit synthesis of the transposase and, hence, to inhibit transposition of (chromosomally inserted) Tn*10* for most of the cell cycle; during DNA replication, however, this inhibition is transiently relieved immediately after the replication fork (Fig. 7.8) has passed the transposase gene but before Dam methylation has occurred – so that transposition of Tn*10* tends to be initiated during DNA replication (and, hence, to be linked to the cell cycle).

7.8.9 Regulation of gene expression by sigma factors

Specific sigma factors (section 7.5) are needed for the transcription of particular genes; this requirement gives the cell a mechanism for controlling the *timing* of certain events by synthesizing a given sigma factor only when necessary. Thus, e.g. a sigma factor is involved in the timing of morphological events in the life cycle of *Caulobacter* (section 4.1).

Some sigma factors are present at low levels during normal growth – higher levels induced by particular stress conditions leading to the expression of specific genes; these sigma factors include e.g. σ^H (*spo0H* gene product; Fig. 7.14), σ^{32} (*rpoH* gene product; section 7.8.2.3) and σ^s (*rpoS* gene product; a regulatory factor under various stress conditions). In some cases, stress conditions (heat shock for σ^{32}; e.g. osmotic shock for σ^s) promote the increased synthesis of a sigma factor by causing increased translation of the gene's mRNA (rather than increased transcription of the gene); the increased translation of σ^{32}, and of σ^s, appears to involve an effect of the inducing condition on a secondary structure in the mRNA.

σ^s, once associated solely with stationary-phase events, is now known to be involved in the regulation of a range of stress responses under stationary- and log-phase conditions [Hengge-Aronis (1996) MM 21 887–893] – often (perhaps always) in association with other forms of regulation. Thus, e.g. σ^s and the Fur protein are both regulatory elements in acid tolerance (section 7.8.2.6), while in osmotic up-shift (section 3.1.8), the resulting enhanced levels of K^+ and glutamate may serve to promote transcription from certain σ^s-regulated promoters [Ding et al. (1995) MM 16 649–656].

Sigma factors are also involved in phage development (see e.g. section 9.1.1).

7.8.10 Regulation of gene expression by tRNA-directed transcription antitermination

In Bacillus subtilis (and some other Gram-positive bacteria), certain genes whose products are involved in amino acid synthesis (and in the linkage of amino acids to tRNA molecules) are switched on when a shortage of the relevant amino acid gives rise to uncharged molecules of the corresponding tRNA; an uncharged tRNA molecule interacts with the leader region of the gene's mRNA and may cause a switch from a transcription terminator structure (formed when there are adequate amounts of the amino acid) to an antiterminator structure – so that transcription continues and the amino acid is subsequently synthesized.

[tRNA-directed transcription antitermination: Henkin (1994) MM 13 381–387.]

8 Molecular biology II: changing the message

DNA can change. For example, even while replicating – and despite proof-reading and repair systems (section 7.7) – about one in 10^8–10^{10} 'wrong' nucleotides are believed to be incorporated in the new (daughter) strand. Greater changes can be brought about by chemical and physical *mutagens* (section 8.1), by recombination (section 8.2), and by *transposable elements* (section 8.3). Additionally, the cell's *genome* (its 'genetic blueprint') can be supplemented by plasmids and by other pieces of 'extra' DNA via the processes of gene transfer (section 8.4).

Man-made changes in DNA (recombinant DNA technology) are considered in section 8.5.

8.1 MUTATION

In bacteria, a *mutation* is a stable, heritable change in the sequence of nucleotides in the DNA. (In some organisms – e.g. some bacteriophages (Chapter 9) – the genome consists of RNA, so that mutations in these organisms affect the RNA.) Note that a change in even a single *base-pair* (section 7.2.1) changes the sequence. Transcription of altered DNA produces altered RNA, and altered mRNA may specify a different polypeptide (Table 7.2) with different biological activity. Of course, mutations can affect control and recognition sequences as well as sequences encoding polypeptides.

Mutations occur spontaneously at low frequency without an obvious external cause; they are mainly errors in replication and repair, but chemical changes can also occur in DNA bases – for example, the deamination of cytosine to uracil (Fig. 7.3). (See also section 7.7.1.3.)

Mutation is encouraged by *mutagens*: physical agents such as ultraviolet radiation and X-rays, and chemicals such as alkylating agents, bisulphite, hydroxylamine, nitrous acid, and 'base analogues' (e.g. 5–bromouracil) which can be incorporated, in place of normal DNA bases, during DNA replication.

Mutagens work in various ways. Ultraviolet radiation can cause e.g. covalent cross-linking between adjacent thymines; correction of the resulting *thymine dimers* by 'error-prone' repair (section 7.8.2.2) can generate a variety of mutations. Cross-linking can also occur with some alkylating agents.

Bisulphites and nitrous acid can e.g. deaminate cytosine to uracil; although uracil is not a stable constituent of DNA (section 7.2), its different base-pairing specificity can cause the insertion of a different base (adenine instead of guanine) in the daughter strand at the next round of replication.

In a population of bacteria, mutations normally occur randomly, affecting different genes in different individuals. (Interestingly, in enterobacteria, mutations appear to occur more frequently at sites furthest from *oriC* [Sharp *et al.* (1989) Science *246* 808–810].) A cell in which a mutation has occurred is called a *mutant*. Mutations are often harmful – and may be lethal if the affected sequence of nucleotides encodes a vital product or function. Beneficial mutations include e.g. those which increase the cell's resistance to antibiotic(s). For example, a mutation may result in an altered ribosome such that streptomycin (section 15.4.2) no longer binds to the ribosome and (therefore) does not inhibit protein synthesis; the (mutant) cell will thus exhibit resistance to this antibiotic.

A mutation giving increased fitness for growth under existing conditions may enable the (mutant) cell to outgrow other (non-mutant: *wild type*) individuals in the population and become numerically dominant in that population; such 'natural selection' underlies the concept of *evolution*.

8.1.1 Types of mutation

Mutations occur in various ways, and they can have various effects on the genetic 'message'. Infrequently, a piece of DNA is lost, gained, inverted – or even *transposed* (section 8.3). A *point mutation* involves the loss, gain or substitution of a single nucleotide; even this, however, can have far-reaching consequences for the cell (Fig. 8.1).

8.1.1.1 Loci, genes and mutant genes: nomenclature

A given, functionally-defined location on a chromosome (= a *locus*; plural: *loci*) is designated by a group of three letters printed in italics (or underlined if handwritten); for example, *his* and *trp* are two loci concerned, respectively, with the biosynthesis of histidine and tryptophan.

When a given locus contains more than one gene, each gene is identified by a capital letter; thus, for example, the *his* locus is an operon (section 7.8.1) containing a number of genes designated *hisA*, *hisB* . . . etc. (Note that genes designated in this way do not necessarily occur in alphabetical order in the locus – see e.g. the *lac* genes in Fig. 7.11.) Genes which are *not* contiguous on the chromosome, but which have a related function, are designated in a similar way (see e.g. *ara* in section 7.8.1.1).

The above system is used when referring to a particular locus or gene – e.g. the *his* locus, the *hisA* gene.

A cell's *genotype* is its genetic make-up, a characteristic reflecting the actual sequence of nucleotides in the chromosome. To describe the genotype of a given strain we generally list the particular genes of interest, indicating whether they are wild-type or mutant. A wild-type gene is shown as e.g. *hisA+*, while the corresponding mutant gene is shown as *hisA* (or sometimes *hisA−*). Specific mutations (at particular sites) are designated by numbers: e.g. *hisA9* (a mutation at a particular site in the *hisA* gene).

A cell's *phenotype* is a set of *observable* characteristics. Phenotypic characteristics are symbolized as e.g. His⁻ (inability to grow without histidine − i.e. a histidine *auxotroph*, section 8.1.2); His+ (a histidine prototroph); Lac⁻ (inability to use lactose); Met⁻ (a methionine auxotroph); TcR (resistance to tetracyclines).

8.1.2 The isolation of mutants

A *particular* mutation occurs spontaneously only at very low frequency in a population of bacteria; for example, within a population of *E. coli*, the loss of ability to ferment galactose occurs (on average) once every 10^{10} cell division cycles, i.e. a *mutation rate* of 10^{-10}. Even in populations treated with a mutagen, cells with a *particular* mutation are still greatly outnumbered by wild-type cells and by those with other types of mutation.

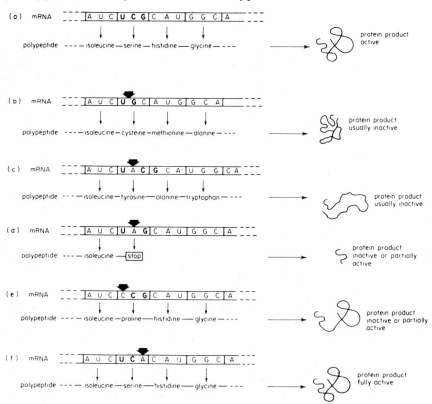

How can we isolate the one (or few) *specific* mutants from a large population of bacteria? If, for example, in a population of streptomycin-sensitive bacteria, a single cell has mutated to streptomycin resistance (section 8.1), that cell can grow on a solid medium containing streptomycin and can form a colony (section 3.3.1); all the other cells, which are inhibited by streptomycin, will not grow on such a medium. In general, this type of selective method can be used whenever the mutant can grow on a medium, or under conditions, which inhibit the growth of non-mutant cells.

However, suppose that (through mutation) a cell has lost the ability to synthesize a particular compound (e.g. amino acid) which is necessary for growth; such a metabolically dependent mutant (an *auxotroph*) can grow only if it is supplied with the appropriate compound. (The corresponding wild-type cell is called a *prototroph*.) How are auxotrophs isolated – given that any medium which allows an auxotroph to grow will also permit the growth of prototrophs? One method is to use a *minimal medium*, i.e. a medium which

---

Fig. 8.1 Point mutations: their effects on mRNA and on polypeptide synthesis. The effect of each point mutation is indicated by a heavy arrow (⬇); at the right is shown a possible effect of each of the mutations.

(a) mRNA and polypeptide synthesized from the normal (non-mutant, wild-type) gene.

(b) The *deletion* of a guanine nucleotide from DNA has resulted in the loss of cytosine (C) from the codon UCG in mRNA. The effect of this is that not only UCG but all subsequent codons are altered: compare the amino acids encoded in (a) and (b). Notice that, if one nucleotide is missing, the next nucleotide is read in its place, i.e., groups of three consecutive nucleotides continue to be read as codons. Because the genetic message is out-of-phase 'downstream' of the deletion, such a mutation is called a *phase-shift* or *frame-shift mutation*; if it occurs near the end of a gene, so that most of the polypeptide is normal, the product may have some biological activity. If a phase-shift mutation occurs in an operon (section 7.8.1) the effect will vary greatly according to the particular site affected.

(c) The *addition* of a thymine nucleotide to DNA has resulted in the addition of an adenine nucleotide to codon UCG in the mRNA; as in (b), above, this is a phase-shift mutation.

(d) In DNA, thymine has replaced guanine, so that the mRNA now contains UAG (a 'stop' codon) instead of UCG; this is a so-called *nonsense mutation*. Polypeptide synthesis stops at UAG; the polypeptide may have some biological activity if much or most of it has been translated prior to UAG.

(e) In DNA, guanine has replaced adenine, so that the mRNA now contains CCG instead of UCG – the altered codon specifying proline rather than serine; this is a so-called *mis-sense mutation*. Note that the amino acids downstream of proline are not affected. The biological activity of the polypeptide will depend on the nature and position of the incorrect amino acid.

(f) In DNA, thymine has replaced cytosine, so that the mRNA now contains UCA instead of UCG; however, the altered codon still encodes serine (Table 7.2). This is a *silent mutation*; it does not, of course, affect the biological activity of the polypeptide.

contains the minimum range of nutrients needed by prototrophs; a given auxotroph can grow on minimal medium only if the medium has been supplemented with the auxotroph's specific growth requirement(s). If, for example, we wish to isolate a histidine-requiring auxotroph, minimal medium is supplemented with a *low* concentration of histidine – allowing *limited* growth of the auxotroph; a colony formed by an auxotrophic cell will soon exhaust the histidine in its vicinity (and therefore remain small) while the colonies of prototrophs reach a normal size. Small colonies are *presumed* to be those of auxotrophs and can be tested further.

In an alternative method for isolating auxotrophs, a low-density population containing both prototrophic and auxotrophic cells is inoculated onto a *complete medium* (on which both prototrophs and auxotrophs can grow). Following incubation, normal-sized colonies are formed by all the cells, and in order to identify the colonies of auxotrophs it is necessary to inoculate each colony onto minimal medium – on which only the prototrophs will grow. This is conveniently achieved by *replica plating*. In this method a disc of sterile velvet is pressed gently onto the surface of the complete medium (the *master plate*) containing the colonies of both prototrophs and auxotrophs; cells from each colony stick to the velvet – which is then pressed lightly onto the surface of a sterile plate of minimal medium (the *replica plate*). After incubation, the positions of colonies on the replica plate are compared with those of colonies on the master plate; any colony which occurs on the master plate but not on the replica plate is presumed to be that of an auxotroph (Fig. 8.2).

8.1.3   The Ames test for carcinogens (*Salmonella*/microsome assay)

Most known carcinogens (cancer-promoting agents) are mutagens, and this test checks for potential carcinogenicity by checking for mutagenicity (the ability to cause mutations). The mutagenicity of a chemical is checked by determining its ability to reverse a previous mutation in the test organism, *Salmonella typhimurium*. (A mutation which reverses a previous mutation is called a *back mutation*.) The (mutant) test strains of *S. typhimurium* are auxotrophic for histidine, and the Ames test checks for back-mutation to prototrophy. Essentially, prototrophs are sought in an incubated mixture containing a population of the test strain, the chemical under test, and a preparation of enzymes from rat's liver; the enzymes are included because some mutagens/carcinogens need metabolic 'activation'.

8.2   RECOMBINATION

'Recombination' means the re-arrangement of one or more molecules of nucleic acid: molecules may e.g. join together, separate, or exchange strands, and a sequence within a molecule may be lost or inverted etc.

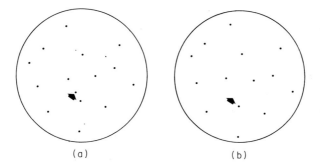

Fig. 8.2 Replica plating as used for isolating an auxotrophic mutant. (a) Master plate: complete medium with colonies of both prototrophs and auxotrophs. (b) Replica plate: minimal medium with colonies of prototrophs only. On the master plate an arrow indicates a colony of a presumed auxotroph; there is no colony at the corresponding position (arrowed) on the replica plate (auxotrophs cannot grow on minimal medium).

8.2.1   Homologous (general) recombination

Homologous recombination can involve e.g. strand exchange between two DNA duplexes. It can occur only if a long sequence of nucleotides in one duplex is very similar to a sequence in the other, i.e. the two duplexes must have an extensive region of *homology*; another requirement is the prior formation of one or more *nicks* (breaks in the sugar–phosphate backbone) allowing strand separation, or *gaps* (regions of single-stranded DNA). The RecA protein (product of the *recA* gene) has an essential role in the process. Initially, RecA binds to ssDNA to form a nucleoprotein 'filament'; the filament complexes with the other duplex – which becomes locally unwound. The filament then 'searches' the unwound duplex for a region homologous to the ssDNA of the filament; subsequent juxtaposition of homologous strands (*synapsis*) produces a *heteroduplex* (dsDNA containing one or more mismatched bases). The heteroduplex is then extended e.g. by a continuation of the strand exchange or by strand synthesis (according to different scenarios and models of recombination).

   In at least some organisms the chromosome contains certain sequences of DNA (recombinational 'hotspots') which enhance homologous recombination in their vicinity. For example, the chromosome of *E. coli* contains about 1000 copies of the so-called chi (χ) site – 5'-GCTGGTGG-3' – which enhances recombination by up to 10-fold; the influence of χ extends (with decreasing magnitude) for about 10 kb on one side of the site. Recent *in vitro* studies indicate that RecA binds preferentially to GT-rich sequences of DNA –including the *E. coli* χ sites; this has suggested that the activity of such sites may depend, at least in part, on their ability to bind RecA and thus to promote

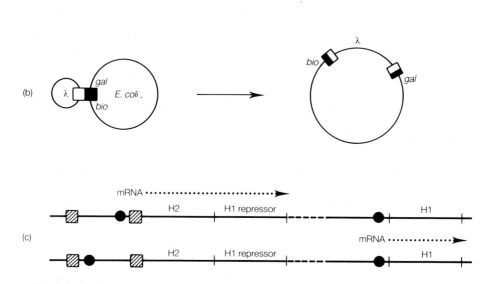

Fig. 8.3 Site-specific recombination (diagrammatic): the principle (a) and some examples (b, c). (a) A staggered break at one specific site in a DNA duplex; a staggered break, typically involving the same nucleotide sequence, is made at another site in the same duplex or in another DNA molecule. Each single-stranded region at a given breakage site can base-pair with a complementary region at the other breakage site. (The single-stranded regions are sometimes called 'sticky ends'.) The process is mediated by a protein (a *recombinase*). In the diagram, a staggered break is shown with two 4-nucleotide-long single-stranded regions. However, each single-stranded region is normally either 2 nucleotides long or 6–8 nucleotides long, the length of this 'overhang' (and the actual sequence in the sticky ends) depending on the particular recombinase involved.

(b) Integration of the (circular, double-stranded) DNA of bacteriophage λ (section 9.2.1) with the E. coli chromosome. In λ DNA, the specific recombinational site (i.e. the specific sequence of nucleotides in the duplex) is shown as a white square; in the E. coli chromosome it is shown as a black square flanked on either side by the *gal* operon (galactose utilization) and the *bio* genes (biotin synthesis). Left: the two circular DNA duplexes are shown with their recombinational sites juxtaposed. Initially, with the recombinase bound at the juxtaposed sites, a staggered break is made across each duplex; sticky ends in the chromosome then base-pair with those in the λ duplex, and the strands are ligated. Right: λ DNA incorporated in the E. coli chromosome.

(c) An example of recombinational regulation. In most strains of *Salmonella* the flagellar filament (section 2.2.14.1) can be made of either of two distinct types of protein – encoded by genes H1 and H2; normally, only one of these genes is expressed at any given time, so that the flagellar filament is made of either the H1 gene product *or* the

H2 gene product. The promoter of the H2 operon (●) is flanked by two specific sites (▨) that are recognized by a recombinase. Top: the H2 operon is transcribed; the H2 gene product forms the flagellar filament, and the H1 repressor stops transcription of the H1 gene. Bottom: site-specific recombination has occurred between the sites recognized by the recombinase, and the sequence containing the H2 promoter has been inverted; without a functional promoter, the H2 operon cannot be transcribed, but the loss of the H1 repressor permits transcription from H1 – so that the filament is now made from the H1 gene product.

Variation in the composition of the *Salmonella* flagellar filament is only one example of *phase variation*: a more general phenomenon in which the composition of certain cell surface components or structures undergoes spontaneous change. For example, in individual cells of *E. coli*, the so-called type 1 fimbriae are subject to on/off switching – resulting in spontaneous changes from a fimbriate to an afimbriate state, and vice versa; this, also, appears to be due to DNA re-arrangement.

---

the pairing of DNA strands in their vicinity [Tracy & Kowalczykowski (1996) GD *10* 1890–1903].

Heteroduplex DNA may be corrected by mismatch repair (section 7.7.1.1), either strand being used as template.

Homologous recombination occurs e.g. in some cases of plasmid–chromosome interaction.

8.2.2   Site-specific recombination

In the basic form of site-specific recombination (SSR), two *specific* sequences of duplex DNA are brought together, a protein catalyst (a *recombinase*) binding in the region of juxtaposition; cuts are made at staggered sites in each duplex, and the cut ends of one duplex are ligated (joined) to the cut ends of the other (see Fig. 8.3). [Action of site-specific recombinases: Stark, Boocock & Sherratt (1992) TIG *8* 432–439.]

SSR can control gene expression; in such *recombinational regulation*, an SSR event controls the on/off switching of gene(s) or a switch from one gene to another (Fig. 8.3c). SSR is also involved e.g. in the integration of bacteriophage λ DNA (section 9.2.1) with the *E. coli* chromosome (Fig. 8.3b), in some cases of plasmid–chromosome interaction [Campbell (1992) JB *174* 7495–7499], in the replicative form of transposition (Fig. 8.4), and in conjugative transposition (section 8.4.2.3).

8.3   TRANSPOSITION

Transposition is the transfer of a small, specialized piece of DNA – or a 'copy' of it – from one site to another in the same duplex, to a site in a different duplex in the same cell, or (see conjugative transposons, section 8.4.2.3) to a duplex in another cell. The 'small, specialized piece of DNA' is called a *transposable element* (TE; jumping gene); TEs occur e.g. in bacterial chromosomes, in the DNA of bacteriophages (Chapter 9), and in plasmids. The following account refers to 'classical' transposable elements (i.e. excluding conjugative transposons).

The two main types of TE are the *insertion sequence* (IS) and the *transposon*. Each encodes protein(s) – including a *transposase* (see later) – needed for transposition; additionally, a transposon encodes other functions – e.g. enzyme(s) which inactive particular antibiotic(s).

In all TEs, the nucleotide sequence includes at least one pair of *inverted repeats*. An example of a pair of terminal inverted repeats:

5'-CTGACTA..................TAGTCAG-3'
3'-GACTGAT..................ATCAGTC-5'

Notice that the left-hand end of the duplex has the same *polarity* as the right-hand end, i.e. the sequence is the same when read from the 5' (or 3') end in both directions – but the two ends of the duplex have opposite *orientations*. Inverted repeats seem to be necessary for transposition, probably being recognized by the appropriate transposase.

An *insertion sequence* consists of a pair of inverted repeats – each about 10–40 nucleotides long – bracketting the transposition genes.

A *class I transposon* consists of a pair of insertion sequences (which are not necessarily identical) bracketting the gene sequence; example: transposon Tn10.

A *class II transposon* consists of a pair of inverted repeats bracketting the gene sequence; example: transposon Tn3.

A TE may transfer by 'simple' transposition or by 'replicative' transposition (explained in Fig. 8.4); for some TEs the target site is highly specific, but other TEs can insert almost at random. The presence of a TE at a given site is, by convension, symbolized by a double colon – e.g. the presence of Tn3 in a phage λ genome is designated λ::Tn3.

Transposition, usually a rare event, can result in a variety of re-arrangements – including insertions, deletions and inversions. The transposition of e.g. Tn10 from a chromosomal location tends to be linked to the cell cycle (see section 7.8.8).

8.4 GENE TRANSFER

New DNA may enter a bacterium through *transformation, conjugation* or *transduction*; transduction (requiring bacteriophages) is described in Chapter 9.

8.4.1 Transformation

In transformation, a bacterium takes up from its environment a piece of DNA; one strand of this *donor* DNA is internalized and may genetically transform the recipient cell by recombining with a homologous region of the chromosome. Transformation occurs naturally in various Gram-positive and Gram-negative bacteria (though not e.g. in *E. coli* – see section 8.4.1.1); it may e.g. contribute to the spread of antibiotic resistance in pathogenic bacteria [Davies (1994) Science *264* 375–382].

Haemophilus influenzae and *Neisseria gonorrhoeae* bind donor DNA only when it includes certain *uptake-signal sequences* (which occur repeatedly in their chromosomes). This requirement seems to be lacking in *Bacillus subtilis* and *Streptococcus pneumoniae*, but these two species bind donor DNA only at a limited number of cell-surface sites.

The *competence* to bind and take up DNA is constitutive in *N. gonorrhoeae*. In *H. influenzae* it is induced by growth-inhibiting conditions – high levels of intracellular cAMP (Fig. 7.12) promoting competence. In both *B. subtilis* and *S. pneumoniae*, competence is affected by nutritional status. However, in each of these last two species, competence is also affected by the population density of the bacteria (an example of *quorum sensing* – section 10.1.2). Cell density is sensed through the levels of certain secreted peptides; in *B. subtilis*, the production of one such peptide is regulated e.g. by Spo0A~P – a protein also involved in the regulation of sporulation (Fig. 7.14). In *S. pneumoniae*, competence has also been associated with trans-membrane transport of calcium [Trombe, Rieux & Baille (1994) JB *176* 1992–1996].

[Competence in transformation: Solomon & Grossman (1996) TIG *12* 150–155.]

Transformation was first observed by Griffith in the 1920s: a live, non-pathogenic strain of *S. pneumoniae* was found to become virulent when mixed with a dead, virulent strain; it was later found that DNA, released by the dead cells, had transformed the living ones – an early indication of the role of DNA as the carrier of genetic information.

8.4.1.1   *Laboratory-induced competence in transformation*

Competence can be induced (e.g. in *E. coli*) by certain procedures which increase the permeability of the cell envelope. For example, calcium chloride solution (approx. 50 mM, 0.2 ml) containing 10^8–10^9 washed, mid-log phase *E. coli* cells, is chilled on ice, and a DNA suspension (10 μl) is added to give a final DNA concentration of approx. 0.2 μg/ml; after further chilling at 0°C (15–30 min) the suspension is heat-shocked (42°C/1.5–2.0 min) and allowed to recover – e.g. returned to ice, then incubated in Luria–Bertani broth (1 ml) at 37°C for 1 hour. (LB broth contains (per litre): 10 g tryptone, 5 g yeast extract and 10 g NaCl; pH 7.5, adjusted with NaOH.)

The acquisition of competence by *E. coli* in ice-cold calcium solutions is associated with the presence of a high concentration of poly-β-hydroxybutyrate/ calcium polyphosphate complexes in the cytoplasmic membrane. It has been suggested that these complexes may form trans-membrane channels which facilitate DNA transport, and that divalent cations may act as links between

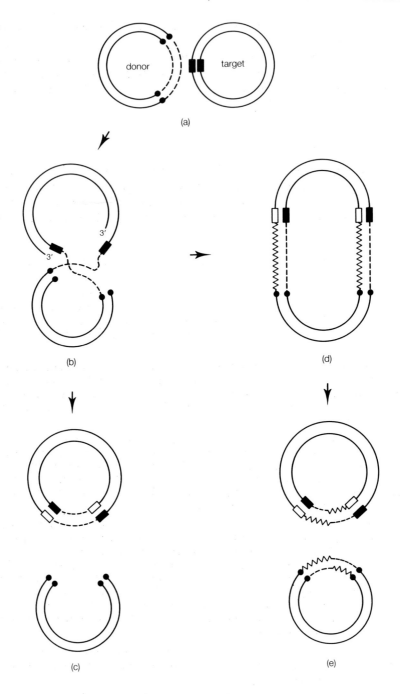

DNA and phosphate at the mouth of such channels [Huang & Reusch (1995) JB *177* 486–490].

Small, circular plasmids tend to transform more readily than do larger ones. In *E. coli*, linear dsDNA transforms poorly (if at all) as it is degraded by the RecBC enzyme; however, it can transform some *recBC* mutants (which lack the enzyme).

8.4.1.2 Electroporation

High-efficiency transformation of *E coli* with plasmids can be achieved by *electroporation*: a mixture of cells and plasmids is exposed to an electrical field of up to c. 16 kV/cm for a fraction of a second. The mechanism of the electrically induced uptake of DNA is not understood; important factors

---

Fig. 8.4 Transposition: a (diagrammatic) scheme for a transposon undergoing 'simple' and 'replicative' transposition.

(a) Two circular, double-stranded DNA molecules. The donor molecule includes a transposon (dashed lines) – either side of which is an old 'target' site (●); the donor's target site was duplicated when the transposon was originally inserted into the donor molecule (see later). The target molecule has a single target site (■) where the transposon will be inserted.

(b) An enzyme (*transposase* – not shown) mediates at least the initial stages of transposition. A staggered break has been made at the target site. A nick has been made in each strand of the transposon (at opposite ends), and the free ends have been ligated to the target molecule, as shown. In 'simple' transposition (which occurs e.g. in transposon Tn*10*) the next (and final) stage is shown at (c).

(c) The result of 'simple' transposition. DNA synthesis has occurred from each free 3' end in the target molecule, using the (single-stranded) target site as template, to form the complementary strand of each target site (▭). The remaining strand-ends of the transposon have been nicked and ligated as shown. Note that the target molecule's target site has been duplicated – compare with the donor molecule at (a). The rest of the donor molecule may be non-viable ('donor suicide').

(d) 'Replicative' transposition (which occurs e.g. in transposon Tn*3*) involves stages (b), (d) and (e). At (d) new DNA synthesis (from each 3' end in the target molecule) has continued beyond the target site, using each strand of the transposon as template; that is, the transposon has been replicated. The end of each new strand has been ligated to a free strand-end in the donor molecule. The structure shown at (d) is a *cointegrate*, the zigzag line representing newly synthesized DNA. The next stage involves a *resolvase*: an enzyme encoded by the transposon. This enzyme 'resolves' the cointegrate by promoting site-specific recombination (section 8.2.2) at a site in each transposon, forming the molecules shown at (e).

(e) Donor and target molecules each contain a copy of the transposon; notice that, in this model, each contains parts of the original transposon (dashed lines) as well as newly synthesized DNA (zigzag lines).

Modified from *Molecular Biology of the Gene*, 4th edn (1987), by J.D. Watson *et al.* (Menlo Park, California: Benjamin/Cummings Publishing Company), p. 336, with permission.

include field strength and pulse length. The frequency of transformation varies linearly with DNA concentration over a wide range of values, and the efficiency of transformation varies with cell concentration. [Electroporation in *E. coli*: Dower, Miller & Ragsdale (1988) NAR *16* 6127–6145.]

8.4.2 Conjugation

Certain (conjugative) plasmids (section 7.1) – and conjugative transposons (see later) – confer on their host cells the ability to transfer DNA to other cells by *conjugation*; in this process, a *donor* (male) cell transfers DNA to a *recipient* (female) cell while the cells are in physical contact. A recipient which has received DNA from a donor is called a *transconjugant*.

We look first at plasmid-mediated conjugation (sections 8.4.2.1, 8.4.2.2) and then at conjugative transposition (8.4.2.3).

8.4.2.1   Conjugation in Gram-positive bacteria

There are two main types of plasmid-mediated conjugation:

In strains of e.g. *Enterococcus* (formerly *Streptococcus*) *faecalis*, potential recipients secrete small amounts of a short peptide (a *pheromone*) which causes the (plasmid-containing) donor cells to synthesize an adhesive cell-surface component; plasmid DNA is subsequently transferred from the donor to the adherent recipient cell. [Review: Dunny *et al.* (1995) JB *177* 871–876.] Such matings can occur in liquid (broth) cultures. Plasmids involved in this type of conjugation typically contain genes for antibiotic resistance and/or haemolysin synthesis.

Recipient strains often secrete several pheromones, each specific for a given type of plasmid; secretion stops if the corresponding plasmid is acquired.

Some of the plasmids encode a peptide which antagonizes the action of the corresponding pheromone. The purpose of such an inhibitor may be to ensure that a mating response in the donor cell is not triggered unless the pheromone is in high enough concentration, i.e. the recipient cell is close enough for a possible random collision.

The second type of plasmid-mediated conjugation does not involve pheromones, and it requires that donor and recipient cells be present on a solid surface, e.g. a nitrocellulose filter. Such conjugation occurs e.g. in strains of *Streptococcus* and *Staphylococcus*. The plasmids encode e.g. resistance to antibiotics, and they are typically >15 kb in size. Few details of the cellular or molecular mechanism of transfer appear to be known.

8.4.2.2   Conjugation in Gram-negative bacteria

In Gram-negative donors the plasmid encodes (among other proteins) the protein subunits of the *pilus* (section 2.2.14.3); pili seem to be essential for conjugation in Gram-negative bacteria. Different types of pilus promote conjugation under different physical conditions, and they seem to function in different ways.

One well-studied type of conjugation is that involving the F *plasmid* in *E. coli* host cells. Within each donor, the F plasmid usually exists as an independent circular molecule; donors, which bear pili, are designated F^+ cells, while recipients (which lack the F plasmid) are F^- cells. On mixing F^+ and F^- cells, the tips of the pili bind to the surfaces of F^- cells (see Plate 8.1: *bottom, right*). What happens next is still unclear. American workers have claimed that DNA can pass *through* the pilus to the F^- cell [Harrington & Rogerson (1990) JB *172* 7263–7264]. However, Swiss scientists subsequently observed conjugating cells by video-enhanced light microscopy, and examined donor–recipient contacts by electron microscopy [Dürrenberger, Villiger & Bächi (1991) JSB *107* 146–156]; light microscopy showed donor and recipient cells rapidly drawn together (within a few minutes), and close wall-to-wall contact maintained for the next 80 minutes. The rapid development of wall-to-wall contact would involve pilus retraction, a feature widely accepted for some years. Specific 'conjugational junctions' at juxtaposed cell envelopes were seen by electron microscopy (Plate 8.1: *top*). From these observations, and from the kinetics of DNA transfer, the Swiss workers concluded that '. . . DNA is transferred at the state of close wall-to-wall contact rather than . . . via extended pili.'.

At some stage of contact, an (unknown) mating signal triggers DNA transfer. This starts with a nick at a specific site (*oriT*) in a specific strand of the F plasmid; it is believed that the nick is made by a plasmid-encoded protein, *helicase I* [Matson & Morton (1991) JBC *266* 16 232–16 237], which may then have a major role unwinding the duplex from the free 5' end. Details of the route and mode of DNA transfer are currently unknown. The free 5' end may enter the recipient cell – or it may remain anchored near the donor–recipient junction; if it remains anchored, the rest of the strand would be fed into the recipient cell in the form of a loop. Only the nicked strand enters the recipient – within which it acts as a template for DNA synthesis; hence a complete, circular copy of the plasmid forms in the recipient, which thus becomes F^+. Within the donor, the lost strand is replaced by DNA synthesis; such synthesis (donor conjugative DNA synthesis – DCDS) may proceed according to the rolling circle model (Fig. 8.5), although earlier claims that *primers* are needed for DCDS make this uncertain.

Infrequently, an F plasmid integrates with the host cell's chromosome. When this happens, the fusion of plasmid with chromosome somewhat resembles the integration of phage λ (Fig. 8.3) in its overall effect – though the mechanism is different. The result is an *Hfr donor*. Hfr donors form pili, and they can conjugate with F^- cells. During Hfr × F^- crosses, DNA transfer starts (as in F^+ donors) with a nick at *oriT* (see above). The transferred strand begins with plasmid DNA, but this is followed by chromosomal DNA and, finally –

if the strand doesn't break – by the remainder of the plasmid strand. (This can be understood more easily if *oriT* is imagined to be in the middle of the integrated F plasmid.) If the entire plasmid and chromosomal strands are transferred the recipient becomes an Hfr donor, but usually strand breakage occurs at some point and the recipient (which does not receive *all* of the plasmid strand) remains F⁻. However, the recipient generally receives *some* chromosomal (as well as plasmid) DNA, and if donor genes recombine with the recipient's chromosome they may alter the genetic message; the high proportion of recombinant cells resulting from Hfr × F⁻ crosses accounts for the designation 'Hfr' – i.e. 'high frequency of recombination'.

An F plasmid can also leave the chromosome, i.e. an Hfr donor can become an F⁺ donor. Sometimes, when leaving, the plasmid takes with it an adjacent piece of the chromosome; the result is an F' (F-prime) plasmid. In F' F⁻ crosses, the F' donor usually transfers donor ability to the recipient (as does an F⁺ donor) but it also transfers chromosomal DNA (as does an Hfr donor).

The F plasmid is only one of many types of plasmid, and it is not even 'typical'; for example, the ability to form Hfr donors is not common. Also, in many types of plasmid the pilus-encoding and other 'donor' genes are normally repressed, i.e. in a population of potential donors only a few cells with transiently de-repressed plasmids can actually conjugate; by contrast F⁺ cells are de-repressed, and in a population of them most or all can usually act as donors. As mentioned earlier (section 7.1), plasmids can encode – and may transfer – a range of functions.

'Universal' and 'surface-obligatory' conjugation. In Gram-negative bacteria, conjugation may be of the 'universal' or 'surface-obligatory' type, depending on the type of pilus encoded by the conjugative plasmid. Plasmids which encode long, flexible pili – as does the F plasmid (section 2.2.14.3) – promote

---

Plate 8.1   *Top.* Conjugating cells of *Escherichia coli* in wall-to-wall contact (scale: 6 cm = 1 μm). The electronmicrograph shows a 'conjugational junction': the electron-dense (dark) line (between the arrowheads) which marks the region of contact between donor and recipient. As well as conjugating, the cell at the top is about to divide; compare the conjugational junction with the site of cell division (far right). The masses of small, darkly stained 'dots' are ribosomes.

Bottom right. Conjugating cells of *E. coli*; this preparation has been stained to show an F pilus (arrowheads) going from one cell to the other. (Fragments of flagella are also visible.)

Bottom left. An apparatus used in the author's experiment to investigate 'surface-obligatory' conjugation (see text): a bundle of 80 chemically clean 73-mm glass capillary tubes (internal diameter approx. 0.8 mm) within a universal bottle. When shaken up-and-down, the mating medium forms a large number of small 'threads' of liquid in the capillary tubes, each thread having a meniscus (at both ends) beneath which donor and recipient cells may be held in close contact by surface tension.

Photographs of *E. coli* courtesy of Dr Markus B. Dürrenberger, University of Zürich, Switzerland.

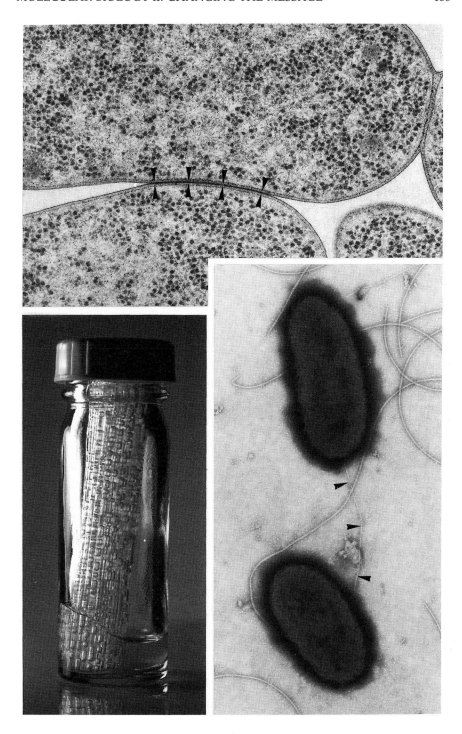

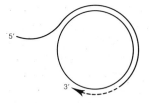

Fig. 8.5 The rolling circle model of DNA synthesis in a circular, double-stranded molecule. First, a nick is made in one strand. Then, using the un-nicked strand as template, a DNA polymerase extends the 3′ end of the nicked strand by adding nucleotides in the 5′-to-3′ direction – the 5′ end being progressively displaced. The displaced strand may itself be used as a template for the formation of Okazaki fragments (Fig. 7.8).

so-called universal conjugation; this can occur equally well in a liquid medium (e.g. a broth culture) or on a moist, *non-submerged* solid surface (e.g an agar plate). Experiments in liquid media indicate that universal conjugation can be enhanced by raising the concentration of electrolyte in the medium [Singleton (1983) FEMSML *20* 151–153].

Plasmids which encode short, rigid, nail-like pili promote so-called surface-obligatory conjugation – which occurs only on moist, *non-submerged* solid surfaces or in foams. This type of conjugation seems to require that conjugating cells be present beneath or within thin films of liquid – films which are thinner than the cells themselves; in nature, cells experience such conditions, for example, when present on soil particles that are drying by evaporation. Experiments have been carried out to see whether suitable conditions exist beneath the thin end of a liquid meniscus on chemically clean glass (Plate 8.1: *bottom left*); the results suggest that surface-obligatory conjugation requires, or is assisted by, surface tension [Singleton (1983) FEMSML *19* 179–182].

8.4.2.3   Conjugative transposition (plasmid-independent conjugation)

Conjugative transposition, which occurs primarily between Gram-positive bacteria, is a type of conjugation mediated by a *conjugative transposon*; it occurs without the involvement of a conjugative plasmid. (Conjugative transposition seems not to occur between Gram-negative bacteria; nevertheless, the DNA of a conjugative transposon has been found in some Gram-negatives (e.g. *Neisseria meningitidis*), suggesting that it might have entered these cells e.g. by transformation (section 8.4.1) or by carriage *within* a conjugative plasmid.)

Conjugative transposons, like 'classical' transposons (section 8.3), are *mobile genetic elements* which can move from one DNA duplex to another. They occur e.g. in chromosomes and plasmids, and range in size from 18 kb (Tn*916*) to > 50 kb; some of the larger ones (e.g. Tn*5253*) seem to consist of a Tn*916*-like entity inserted into another transposon (both being able to transpose independently). The conjugative transposons, each of which carries at least one antibiotic-resistance gene, are readily transferred among a broad range of host species and genera, and are often responsible for transmissible resistance to antibiotics in plasmid-less Gram-positive pathogens;

they have been identified in strains of e.g. *Enterococcus faecalis*, *E. faecium* and *Streptococcus pneumoniae*. Antibiotic-resistance genes carried by these elements include those conferring resistance to tetracycline (the *tet*(M) gene, found in all conjugative transposons from pathogens), chloramphenicol, erythromycin and kanamycin. The Tet(M) protein, unlike the TET protein (section 15.4.11), may interact directly with the machinery of protein synthesis, reducing the sensitivity to tetracyclines of ribosome–tRNA binding [Burdett (1993) JB *175* 7209–7215].

A current model for conjugative transposition is as follows. Contact between donor and recipient bacteria triggers excision of the transposon within the donor: at each end of the (integrated) transposon, a pair of staggered nicks are made such that, on excision, one strand at each end of the transposon carries with it (usually) six nucleotides of the host molecule. (These single-stranded terminal sequences are called *coupling sequences*.) Excision is mediated by a tranposon-encoded site-specific recombinase (product of the *int* gene).

The excised transposon then circularizes as a result of base-pairing (albeit mis-matched) between the coupling sequences. It seems likely that a *single* strand is transferred to the recipient, and that a complementary strand is synthesized, in the recipient, prior to insertion. Insertion of the (double-stranded) transposon into a target site in the recipient duplex is not site-specific, but neither is it random: each target site apparently contains an A-rich sequence of nucleotides and a T-rich sequence – the two sequences being separated by about six nucleotides. During insertion, the (single-stranded) terminal coupling sequences base-pair with single-stranded sequences formed in the nicked target region; mis-match is resolved at the first round of DNA replication. The integrated transposon is flanked on *one* side by a coupling sequence from the donor cell; only one end is flanked in this way because the coupling sequences (at each end of the transposon) insert into different strands of the host molecule – which separate during the first round of replication following insertion.

Conjugative transposons are characteristically resistant to restriction (section 7.4) in a new host cell. A reason for this was suggested by the finding, in Tn*916*, of a gene (*orf18*) encoding a product similar to the antirestriction proteins encoded by some plasmids.

Conjugative transposons differ from 'classical' transposons (section 8.3) in that: (i) they mediate conjugation; (ii) they do not duplicate the target site in the target molecule; (iii) the transposon is excised to form a cccDNA intermediate; (iv) the recipient duplex is characteristically in a different cell.

[Review: Scott & Churchward (1995) ARM *49* 367–397.]

8.5   GENETIC ENGINEERING/RECOMBINANT DNA TECHNOLOGY AND RELATED NUCLEIC-ACID-BASED METHODOLOGY

Nucleic acids can be manipulated and altered *in vitro* and may then be inserted into cells for replication or expression. Such technology enables us e.g. to modify cells in highly specific ways; for example, bacteria can be made to synthesize mammalian proteins such as insulin. Clearly, cells can be made to synthesize proteins of a *different species* because recombinant techniques can introduce *new* genetic material; this illustrates why such technology is superior to the 'mutation and selection' approach to innovation (which is limited to modification of *existing* genes). A further advantage of the recombinant DNA approach is that genetic change can be controlled and directed (see e.g. site-specific mutagenesis: section 8.5.5.2).

Recombinant DNA technology frequently employs bacteria and/or their enzymes, plasmids or phages. Section 8.5 explains some of the methodology and introduces a number of terms, ideas and strategies. Some of the methods described – e.g. the polymerase chain reaction (PCR) – are additionally useful in other types of work, including the detection/identification and taxonomy (classification) of bacteria and other organisms.

Note   The topics covered in section 8.5 are:

8.5.1   Cloning and associated (general) techniques – e.g. DNA restriction (cutting); screening; isolating/purifying chromosomal and plasmid DNA; centrifugation; blotting; vectors; linkers, tailing, ligation.
8.5.2   Gene fusion and fusion proteins.
8.5.3   Probes; labelling probes (including nick translation).
8.5.4   The polymerase chain reaction (PCR), including uses and variant forms.
8.5.5   Mutagenesis.
8.5.6   DNA sequencing (Sanger's chain-termination method).
8.5.7   Expressing eukaryotic genes in bacteria: limitations.
8.5.8   Functional analysis of genomic DNA.
8.5.9   Amplification of nucleic acids by the ligase chain reaction and NASBA.
8.5.10   Recombinant streptokinase: an example of recombinant DNA technology.
8.5.11   Overproduction of recombinant proteins.
8.5.12   DNA-binding proteins: some methodology (including DNase I footprinting).
8.5.13   Displaying heterologous ('foreign') proteins.

We look first at a common procedure: *cloning*. An overview (section 8.5.1) is followed by a more detailed account of various techniques associated with this procedure; *many of these techniques are also useful in other areas of recombinant DNA technology.*

8.5.1   Cloning (gene cloning, molecular cloning): an overview

Cloning is a method for obtaining many copies of a given gene (or other piece of DNA) – e.g. for sequence analysis or for making large amounts of the gene product (gp, i.e. the product encoded by the gene).

The gene is first inserted (*in vitro*) into a *vector* molecule – often a plasmid (section 7.1) or a phage (Chapter 9) – which can carry the gene into, and replicate within, a bacterium or other type of cell; the insertion of DNA into a plasmid vector is shown in Fig. 8.6. The hybrid plasmid (Fig. 8.6) can be taken up by transformation (section 8.4.1) and will continue to replicate in the host cells as they grow and divide; a bacterial population can reach very high numbers, so that we can obtain many copies of the hybrid plasmid – and, hence, many copies of the ('cloned') gene.

To harvest the gene, the cells are lysed and the hybrid plasmids are isolated e.g. by isopycnic centrifugation (Fig. 8.12) or by the QIAGEN protocol (section 8.5.1.4, Plate 8.2); using the original restriction endonuclease (Fig. 8.6), the gene can be cut from each hybrid plasmid and separated from plasmid DNA e.g. by electrophoresis.

8.5.1.1   Cloning a bacterial gene

First, the required gene must be separated from other genes in the genome. We can start by constructing a *genomic library* of the species (Fig. 8.7) and then 'screen' the library to find the required gene.

Screening is a process of selection. Suppose, for example, that the required gene encodes an enzyme needed for the synthesis of histidine. Recall that this particular gene will occur in only a small proportion of recombinant molecules in the library (Fig. 8.7); how can we isolate the particular recombinant molecules which carry this gene? First, we use transformation (section 8.4.1) to insert the whole library of recombinant molecules into a population of mutant bacteria which are defective for the given gene (i.e. which cannot synthesize histidine). These *auxotrophs* (section 8.1.2) will not grow on media which lack histidine; however, a mutant transformed with a vector molecule carrying the functional gene will be able to grow and form a colony on media lacking histidine. The transformed mutants are therefore inoculated onto a *minimal medium* lacking histidine so that any colony which develops is likely to consist of cells containing the vector and the given gene; we need to check this (see e.g. section 8.5.1.5) because a colony could also be formed by an auxotroph after back-mutation to prototrophy (section 8.1.3). A colony containing the given gene can be subcultured to produce large numbers of the cells; then, the fragments containing the gene can be isolated as indicated in section 8.5.1.

An alternative method of screening is *colony hybridization* (see later).

8.5.1.2   Cloning eukaryotic genes in bacteria

The eukaryotic genome is much larger so that, for a genomic library, it has to be cut into larger restriction fragments – otherwise there would be too many pieces for cloning; these larger pieces require a different type of vector (section 8.5.1.5).

A further problem is that a eukaryotic gene typically contains one or more nucleotide sequences (called *introns*) which do not encode any part of the gene product but which interrupt the coding sequence. Within the eukaryotic cell, any part of the mRNA which has been transcribed from an intron is later removed, leaving the mature mRNA (Fig. 8.8), i.e. the uninterrupted coding

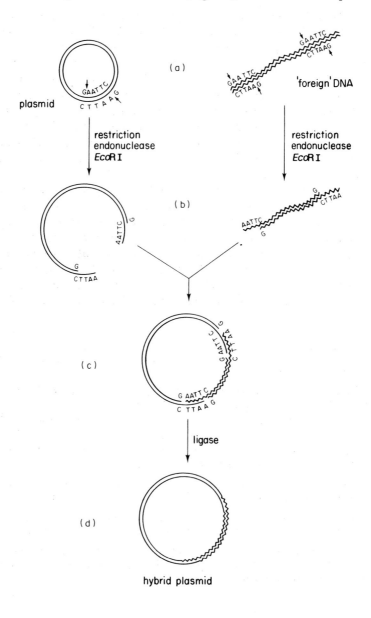

hybrid plasmid

sequence of the gene; this molecule is the one translated. Note that cloning the gene *with* introns would be fine if our purpose was subsequently to determine the gene's *in vivo* sequence, but, if expressed in bacteria, such molecules (unlike the corresponding mature mRNA) would not yield the normal gene product. However, we can make a DNA copy of the mature mRNA sequence of a given gene; this copy (= *copy DNA, complementary DNA, cDNA*) can then be cloned.

cDNA libraries. To clone a eukaryotic gene we can start by constructing a cDNA library of the given species. Initially, all the RNA ('total RNA') is isolated from cells which are expressing the gene of interest; total RNA includes the mature mRNAs of all the *active* (i.e. expressed) genes. This may seem a lot, but actually it limits the number of clones which later have to be screened: only about 1% of the cell's genes may be active at any given time.

The isolation of mRNAs from total RNA is made easier by the poly-adenylate 'tail' (A-A-A-A. . . .) which occurs at the 3' end of most *eukaryotic* mRNAs (see Fig. 8.9); this tail can bind, by base-pairing, to a synthetic oligo(dT)-cellulose, i.e. an oligonucleotide (T-T-T-T. . . .) linked to cellulose, so that mRNAs can be isolated by affinity chromotography (section 8.5.1.4).

Once isolated, the mRNAs can be converted, *in vitro*, to their corresponding cDNAs. In one method, a short synthetic oligo(dT) molecule is used as a primer; this binds to the poly(A) tail of an mRNA molecule, allowing the enzyme *reverse transcriptase* to synthesize a cDNA strand on the mRNA template. (DNA polymerase I is not used because it is relatively inefficient on an RNA template [Ricchetti & Buc (1993) EMBO Journal 12 387–396].) mRNA is then removed (e.g. with RNase H) and replaced by complementary DNA, making ds cDNA. Sticky ends can be added to each molecule (section 8.5.1.6)

Fig. 8.6 Insertion of a gene (or other piece of 'foreign' DNA) into a plasmid vector, prior to cloning (diagrammatic).

(a) A plasmid, and 'foreign' DNA. Both molecules contain the nucleotide sequence GAATTC which is recognized by the restriction endonuclease *Eco*RI (section 7.4; Table 8.1); *Eco*RI cuts between the guanosine (G) and adenosine (A) nucleotides, as shown by the arrows.

(b) As a result of *Eco*RI activity, both molecules now have 'sticky ends' (terminal, single-stranded complementary sequences of nucleotides).

(c) The 'foreign' DNA has integrated with the plasmid vector molecule through base-pairing between complementary nucleotides in the sticky ends.

(d) An enzyme, DNA ligase, has catalysed a phosphodiester bond (Fig. 7.4) between the sugar residues of each G and A nucleotide, forming a hybrid plasmid.

The hybrid plasmid can be inserted into a bacterium by transformation (section 8.4.1) for cloning (section 8.5.1). Note that, if required, the fragment can be later cut from the vector by using the same restriction enzyme.

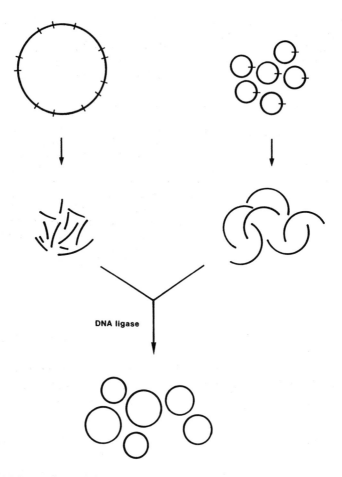

Fig. 8.7 Making a bacterial genomic library (diagrammatic). Chromosomes are first isolated from a *population* of bacteria of a given strain and are then exposed to a particular restriction endonuclease (section 7.4; Table 8.1). The enzyme cuts at specific recognition sites in each chromosome (*top, left*) forming many fragments of different sizes (*centre, left*).

A *population* of a given plasmid (*top, right*) provides vector molecules. The plasmid is one which can be cut by the same enzyme as that used to cut the chromosomes; however, note that the plasmid has only one cutting site for the given enzyme – so that cutting simply linearizes the molecule (*centre, right*).

The chromosomal fragments and linearized plasmids are then mixed in the presence of ATP and *DNA ligase*, an enzyme which catalyses phosphodiester bonds (Fig. 7.4). Some plasmids (and fragments) may simply re-circularize via their sticky ends, but concentrations are chosen such that, in many cases, plasmids and fragments will join together – randomly – via their sticky ends (Fig. 8.6) to form a large number of recombinant molecules, many consisting of a plasmid circularized with one or other of the chromosomal fragments. (Plasmid–fragment binding can be promoted by the method described under *ligation* in section 8.5.1.6.) Because the fragments are of different sizes, the recombinant plasmids will also be of different sizes (*bottom of diagram*).

Collectively, the recombinant plasmid molecules carry all the DNA in the chromosome, and this collection of molecules is therefore called a *genomic library*. By using transformation (section 8.4.1), all the recombinant plasmid molecules of the library can be taken up by a population of bacteria – *so that different cells will receive different fragments of the chromosome*. These cells can then, if required, be allowed to form individual colonies.

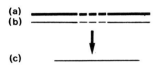

(a)
(b)
(c)

Fig. 8.8 The formation of mRNA in a eukaryotic cell (diagrammatic). (a) A gene containing a sequence (an *intron*: dashed line) which does not encode any part of the gene product. (An example of a gene containing a single intron is the human insulin gene.) (b) mRNA transcribed from the gene: the so-called primary transcript or *pre-mRNA*. (c) After removal of the intron: the final (mature) mRNA which contains the complete, uninterrupted coding sequence of the gene. The two coding sequences, initially separated by the intron, are called *exons*.

and the library of cDNA molecules is then ready for insertion into vectors prior to cloning in bacteria.

cDNA made by the above method is sometimes found to be an incomplete copy of the mRNA. The Okayama–Berg method (Fig. 8.9) produces full-length cDNA, and is particularly useful when the cloned cDNA is subsequently to be expressed.

Screening libraries for specific genes. The colonies which form from cells containing a genomic or cDNA library can be screened for a specific gene. Screening is made easier if the gene can be expressed in a bacterial host cell. Clearly, bacterial genes can be expressed (section 8.5.1.1). However, so, too, can some cDNAs if the *vector* provides certain control functions – e.g. a promoter recognizable by the bacterial RNA polymerase (so that mRNA, and then protein, can be synthesized). Vectors which provide such functions are called *expression vectors* (see later). A *cDNA expression library* is a library of cDNA molecules inserted into expression vectors; when cells containing these vectors form colonies, it is sometimes possible to identify a colony expressing a given cDNA by the method shown in Fig. 8.10.

If a cloned gene or cDNA is *not* expressed in the bacterial host we can screen by other methods. *Colony hybridization* starts (as in Fig. 8.10) with a replica of colonies on a nitrocellulose filter; the cells on the filter are lysed, their DNA is denatured (i.e. the strands separated), and the single-stranded DNA is bound to the filter. A labelled *probe* (section 8.5.3), which can bind to a specific sequence in the required gene or cDNA, is then added to the filter; when unbound probe is washed away, the label on the probe identifies the required

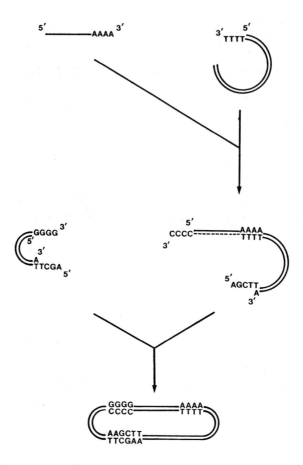

Fig. 8.9 Making cDNA from a mature mRNA template by the Okayama–Berg method (diagrammatic). A mature, eukaryotic mRNA (*top, left*) is shown with the typical poly-adenylate 'tail' at the 3′ end. A specific plasmid (*top, right*) has been cut (i.e. linearized), and an oligo(dT) tail has been added (*in vitro*) to each 3′ end (see *tailing*, section 8.5.1.6); only one tail is shown. mRNA and plasmid interact (*centre, right*) by base-pairing between the A and T tails. The T tail acts as a primer, allowing synthesis, by the enzyme *reverse transcriptase*, of a DNA strand (*dashed line*) on the mRNA template. An oligo(dC) tail is added to the 3′ end of the new DNA strand; because only a *completed* new strand can be extended in this way, this tail allows subsequent selection for complete, full-length molecules of cDNA. Next, the plasmid is cut by *Hind*III, leaving one 'sticky end' – the overhang 5′-AGCT. A specially constructed fragment (*centre, left*), together with a ligase, can now circularize the plasmid; an enzyme, *RNase H*, then removes the mRNA strand, and DNA polymerase synthesizes a DNA strand in its place – thus (with ligase action) completing the dsDNA recombinant plasmid (*bottom*) ready for insertion into a bacterial host for cloning.

Note that this method ensures that the cDNA is inserted into the plasmid in a known orientation (compare Fig. 8.6, where DNA can insert in either direction); this is important when the recombinant plasmid encodes control functions for the *expression* of the gene (see vectors, section 8.5.1.5).

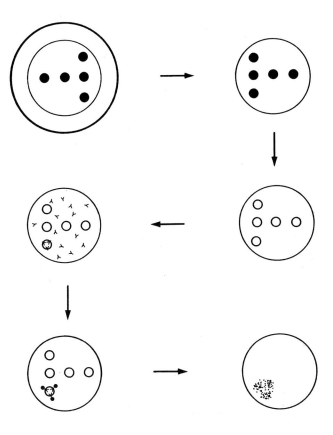

Fig. 8.10 Screening a cDNA expression library (diagrammatic). The cDNA library, within expression vectors (section 8.5.1.2), has been inserted into a population of bacteria, and the latter have formed individual colonies on a plate. The object now is to find the colony whose cDNA encodes (and expresses) a particular protein. cDNA-containing colonies (*top, left*) are lightly overlaid with a nitrocellulose filter; when lifted off, the filter contains a (mirror image) replica of the colonies (*top, right*). The cells on the filter are lysed, and the proteins from these cells (including any encoded by cDNA) are bound to the filter (*centre, right*). The filter is then treated with *antibodies* (ʌ) (section 11.4.2.1) specific to the required protein (*centre, left*); antibodies have bound to the given protein (present in *one* of the colonies). Unbound antibodies are washed away, and the filter is treated with *protein A* – a protein (derived from *Staphylococcus aureus*) which binds to antibodies; unbound protein A is washed away, leaving some protein A (*small dots*) bound to the specific antibody–protein complex (*bottom, left*). Before use, the protein A is radioactively labelled so that its presence can be detected by *autoradiography*; this involves exposing the nitrocellulose filter to a photographic film and then developing the film. When the film is developed (*bottom, right*) it identifies that colony on the original plate which contains the cDNA of interest. This colony can then be used as an inoculum to grow more of the cells.

DNA – and, hence, the corresponding colony on the original plate. A radioactive label, for example, can be detected as in Fig. 8.10.

How do we choose the probe's sequence of nucleotides? If the required protein has a known amino acid sequence we can 'work backwards' – deducing possible sequences for the corresponding mature mRNA (and, hence, DNA); we would choose a short sequence containing an uncommon amino acid (e.g. methionine) and then, using the genetic code (Table 7.2), work out some corresponding mRNA sequences. More than one mRNA/DNA molecule is possible for a given amino acid sequence because most amino acids are encoded by more than one codon (Table 7.2); for example:

Amino acid sequence: L y s -T r p-Me t -L y s -G l u-Hi s -Ph e

Possible mRNA: AAA-UGG-AUG-AAA-GAG-CAU-UUC
Possible DNA: T T C-ACC-T AC-T T T-C T C-GT A-AAG

Possible mRNA: AAG-UGG-AUG-AAG-GAA-CAC-UUU
Possible DNA: T T C-ACC-T A C-T T C-C T T-GT G-AAA

Each of the possible DNA sequences can be used, in separate experiments, as a probe.

8.5.1.3 Cutting DNA precisely

DNA can be cut precisely and predictably by type II restriction endonucleases (REs) (section 7.4); these are enzymes which cut within (or very close to) specific sequences of nucleotides (Table 8.1).

Many type II REs make staggered cuts, forming 'sticky ends', but some (e.g. *Hpa*I) cut straight across both strands, forming 'blunt' ends (Fig. 8.11).

The 8-nucleotide recognition sequence of *Not*I, being uncommon, results in infrequent cutting; this is useful for making large fragments (e.g. >1 million bp long) for eukaryotic genomic libraries (section 8.5.1.2).

The activity of REs is affected e.g. by pH and electrolyte, and it may also be affected by the type of DNA: linear or supercoiled. The *specificity* of cutting of some REs can be upset by factors such as high enzyme concentration, the use of different buffers, and glycerol concentrations >5% w/v. So-called *star activity* refers to a change or reduction in specificity which is shown by some restriction endonucleases (e.g. *Bam*HI, *Eco*RI, *Sal*I) under non-optimal conditions.

Different REs (from different species) which recognize the same sequence are called *isoschizomers*.

8.5.1.4 Isolating/purifying nucleic acids

For *in vitro* work, nucleic acids are obtained e.g. by lysing cells and separating the different forms of DNA and RNA. The starting point: colonies, or a pellet

Table 8.1 Restriction endonucleases: some common examples[1]

Restriction endonuclease (source)		Recognition sequence, cutting site [2,3]
*Aat*II	(*Acetobacter aceti*)	GACGT/C
*Alu*I	(*Arthrobacter luteus*)	AG/CT
*Bam*HI	(*Bacillus amyloliquefaciens*)	G/GATCC
*Bcl*I	(*Bacillus caldolyticus*)	T/GATCA
*Bgl*II	(*Bacillus globigii*)	A/GATCT
*Dpn*I	See section 7.4	
*Eco*RI	(*Escherichia coli*)	G/AATTC
*Hind*III	(*Haemophilus influenzae*)	A/AGCTT
*Hpa*I	(*Haemophilus parainfluenzae*)	GTT/AAC[4]
*Kpn*I	(*Klebsiella pneumoniae*)	GGTAC/C
*Not*I	(*Nocardia otitidis*)	GC/GGCCGC[5]
*Pal*I	(*Providencia alcalifaciens*)	GG/CC
*Pst*I	(*Providencia stuartii*)	CTGCA/G
*Sal*I	(*Streptomyces albus*)	G/TCGAC
*Sau*3AI	(*Staphylococcus aureus*)	/GATC
*Sma*I	(*Serratia marcescens*)	CCC/GCC[4]
*Srf*I	(*Streptomyces* sp)	GCCC/GGGC[5]
*Xba*I	(*Xanthomonas campestris* var *badrii*)	T/CTAGA
*Xho*I	(*Xanthomonas campestris* var *holcicola*)	C/TCGAG
*Bae*I	(*Bacillus sphaericus*)	/(10N)ACNNNNGTANC(12N)/[6]

[1] For further information see section 8.5.1.3.
[2] The sequence is written in the 5'-to-3' direction, and only one strand is shown. For each enzyme, compare the recognition sequence in the table with its *complementary* sequence (also read in the 5'-to-3' direction).
[3] The solidus (/) indicates the cutting site.
[4] Produces 'blunt-ended' cuts; note that the cut is in the *centre* of the recognition sequence.
[5] The 8-nucleotide recognition sequence of a 'rare-cutting' enzyme which is used e.g. for making large chromosomal fragments for eukaryotic genomic libraries.
[6] A newly discovered RE [Sears *et al.* (1996) NAR 24 3590–3592] (see section 7.4 and Fig. 8.11); in this particular strand the recognition sequence is flanked by a 10-nucleotide sequence at the 5' end and a 12–nucleotide sequence at the 3' end, the cutting sites being immediately beyond the flanking sequences.

Fig. 8.11 Cutting DNA by restriction endonucleases. A staggered cut (*top, left*) by *Hind*III (Table 8.1) has produced 'sticky ends' – i.e. the two 5'-AGCT overhangs; these (complementary) overhangs may bind together (thus re-forming the duplex) or each may base-pair with a complementary overhang on another molecule. (Note that some REs, e.g. *Pst*I, form 3' overhangs, while other REs form 5' overhangs.) 'Blunt' ends (*top, right*) are formed e.g. by *Hpa*I (Table 8.1); blunt ends can be fitted with *linkers* or *homopolymer tails* (section 8.5.1.6). *Below:* the cutting sites of a newly discovered RE, *Bae*I [Sears *et al.* (1996) NAR 24 3590–3592] – one of a family of REs which cut on *both* sides of the recognition sequence (see section 7.4); the recognition sequence of *Bae*I is underlined in the top strand (N = nucleotide), and the arrows indicate the cutting sites.

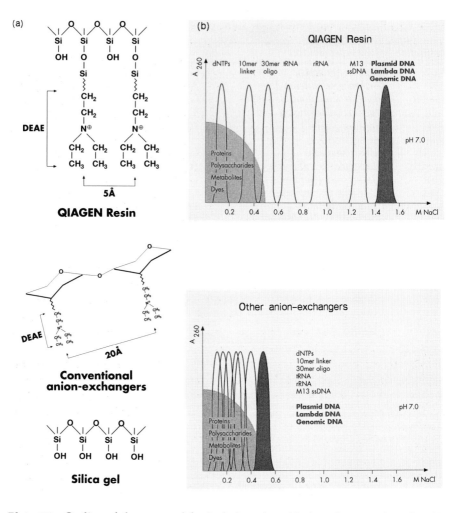

Plate 8.2 Outline of the protocol for isolating plasmids from bacteria (e.g. *E. coli*) using the QIAGEN® plasmid mini kit; up to 20 μg of plasmid DNA can be isolated with this kit (larger kits can yield up to 10 mg).

The bacterial pellet is resuspended in a Tris–EDTA buffer containing RNase A; in Gram-negative bacteria Tris–EDTA disrupts the outer membrane. This is followed by controlled alkaline lysis with NaOH and the detergent sodium dodecyl sulphate (SDS). SDS disrupts the cytoplasmic membrane, releasing soluble cell contents (e.g. proteins, RNA, plasmids). NaOH denatures plasmid (and chromosomal) DNA and proteins. RNase A digests contaminating RNA. Lysis time is optimized (i) to allow maximum release of plasmids without release of chromosomal DNA, and (ii) to avoid irreversible denaturation of plasmids (denatured plasmids are resistant to restriction enzymes). The lysate is neutralized by a pre-chilled high-salt buffer (pH 5.5) which precipitates SDS; complexes of salt and SDS entrap proteins etc., but plasmids renature and stay in solution. Gentle mixing is used to ensure precipitation of all SDS; vigorous agitation is avoided as it could shear the chromosome – contaminating the supernatant with chromosomal fragments. Centrifugation (at e.g. 16 000 g) is used to obtain a cleared lysate (containing the plasmids).

The cleared lysate is filtered through QIAGEN anion-exchange resin (a). This resin was developed specifically for isolating nucleic acids; it has a high density of (+ve) anion-exchange groups (distance between adjacent N^+ only 5Å) which is ideal for binding nucleic acids. Moreover, different types of molecule elute at *widely* differing concentrations of salt – see elution profiles in (b); this permits step-wise elution and efficient separation of molecules by the use of simple salt buffers. Given the (low) electrolyte concentration and pH of the cleared lysate, only plasmid DNA binds to the resin – degraded RNA and proteins etc. are not retained. The resin is washed with buffer (which includes 1 M NaCl, pH 7) to eliminate contaminants (and also to remove any DNA-binding proteins). Plasmid DNA is then eluted with buffer containing 1.25 M NaCl, pH 8.5. The DNA is precipitated (at room temperature) with isopropanol, the tube centrifuged, and the supernatant discarded; after a wash with 70% ethanol, the DNA is air-dried (5 min) and suspended in buffer.

Line drawings courtesy of QIAGEN GmbH, Hilden, Germany.

---

from a centrifuged broth culture; sedimentation of the cells in a broth culture requires centrifugation of about 5000 g for 20 minutes.

In Gram-positive bacteria, the cell wall peptidoglycan may be disrupted by buffered *lysozyme* (see Fig. 2.7). (For *Staphylococcus aureus* the enzyme *lysostaphin* may be used; this enzyme cleaves the oligopeptide links between the backbone chains of peptidoglycan.) In Gram-negative bacteria the outer membrane (section 2.2.9.2) is commonly disrupted by chelating its stabilizing cations with Tris-buffered EDTA; lysozyme can then reach the peptidoglycan. (EDTA also inhibits nucleases in the cell lysate, thus protecting the nucleic acids.) Sucrose (e.g. 0.3 M) may be used to prevent violent osmotic lysis (section 2.2.8.1).

We often need specifically to isolate *plasmids* from bacteria – e.g. in cloning (section 8.5.1). Of the various methods available, a rapid and simple procedure – which yields highly pure plasmid DNA – is described in Plate 8.2; this involves gentle, controlled lysis in which the cytoplasmic membrane is disrupted by the detergent SDS.

Isopycnic centrifugation (at appropriate g values) can separate macromolecules of different densities. The sample is layered on top of a solution whose concentration, and hence density, increases with depth (i.e. a *density gradient column*); centrifugation to equilibrium then brings each molecule in the sample to that part of the density gradient which corresponds to its own density. Thus, e.g. 'total RNA' (section 8.5.1.2) can be separated from DNA and from protein in a gradient of caesium trifluoroacetate; this reagent also inhibits RNase (section 7.6.1), present in cell lysate, thus helping to protect isolated RNA. Different forms of DNA can be separated in a caesium chloride gradient (see Fig. 8.12).

Chromatography. Dissimilar molecules in a sample can be separated by passing the sample over or through an appropriate stationary medium which

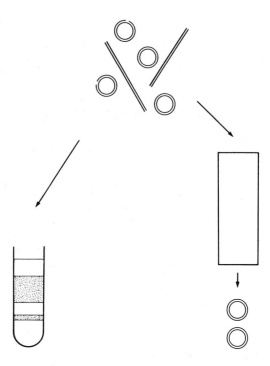

Fig. 8.12   Two methods for separating supercoiled DNA from linear and 'nicked' circular DNA (diagrammatic). DNA isolated from lysed bacteria may contain (*top*): linear fragments of chromosomes; circular plasmids in which one strand has been 'nicked' (i.e. circular DNA in which a phosphodiester bond (Fig. 7.4) has been broken); and circular, supercoiled plasmids.

If DNA is treated with the dye *ethidium bromide*, molecules of dye tend to insert between adjacent base pairs – thus tending to elongate the sugar–phosphate backbone of the molecule; this distortion decreases the *density* of DNA. Linear and nicked circular DNA take up more dye than do supercoiled molecules, so that their density is decreased by a greater amount; hence, if dye-treated DNA is subjected to isopycnic centrifugation (section 8.5.1.4) in a caesium chloride gradient, supercoiled DNA forms a separate band below the less-dense linear and nicked circular molecules (*left*).

Another method, involving *spun column chromatography* (section 8.5.1.4), is quicker and easier. Alkali causes strand separation (denaturation) in DNA; on return to neutral pH, re-naturation (to the double-stranded state) is much more efficient in supercoiled molecules than in linear or nicked circular molecules – so that a column which strongly binds single-stranded DNA will tend to allow only supercoiled molecules to emerge with the effluent (*right*).

Both methods can be used to isolate supercoiled DNA in milligram quantities.

retains or transmits molecules according to their physical properties. *Affinity chromatography* exploits the specificity of binding between certain types of molecule – e.g. the A-A-A-A. . . . tail of a eukaryotic mRNA (in the sample) and the T-T-T-T. . . . of oligo(dT)-cellulose in a stationary matrix (see e.g.

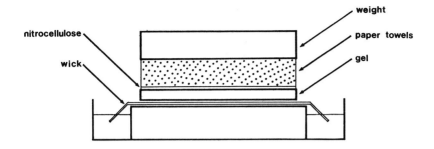

Fig. 8.13 Southern blotting: the transfer of DNA fragments from a gel strip to a sheet of nitrocellulose by capillary action. Within the gel, the fragments are distributed in discrete zones, according to size, having been separated earlier by gel electrophoresis (section 8.5.1.4). The fragments are first denatured (made single-stranded) by exposing the gel to alkali; the gel is then exposed to neutral buffer and arranged as shown in the diagram. Driven by capillary action, the neutral, saline solution in the dish rises, via the wick, into and through the gel, through the (permeable) nitrocellulose, and into the stack of paper towels; this upward stream of liquid carries the DNA fragments from the gel to the sheet of nitrocellulose. (The zones of fragments on the sheet have the same relative positions as they had in the gel.) The sheet is removed and baked at 70 °C under vacuum to bind the DNA. The fragments can then be examined e.g. by exposing the nitrocellulose to a specific labelled *probe* (section 8.5.3); probe–fragment binding (*Southern hybridization*), detected by a probe's label, identifies a particular sequence in a given fragment.

One alternative to nitrocellulose is a paper which incorporates 2-aminophenylthioether (*APT paper*); before use, APT is chemically modified to the reactive diazo derivative, so that single-stranded nucleic acids are bound *covalently*, i.e. no baking is required. This (more robust) preparation allows removal of probes and the subsequent use of different probes on the same sheet.

Northern blotting refers to the transfer of RNA from gel to matrix.

Western blotting refers to the transfer of proteins from gel to matrix.

Southwestern blotting: see section 8.5.12.1.

Electroblotting is a faster method which uses an electric field (instead of capillary action) for effecting transfer.

Immunoblotting refers to the transfer of proteins to a matrix – followed by exposure of the matrix to labelled antibodies; the binding of antibodies to a particular protein may indicate the corresponding antigen.

section 8.5.1.2). *Spun column chromatography* uses centrifugal force (e.g. 300–400 *g*) to hasten the transit of a liquid sample through e.g. a tightly packed column of fine particles.

Electrophoresis can separate mixed, charged molecules. In *gel electrophoresis* the sample is placed in a well at one end of a gel strip; a voltage applied between the two ends of the strip causes each molecule to move, through the gel, towards the anode (+ve) or cathode (−ve), depending on the net charge (−ve or +ve, respectively) on the molecule. Small and/or highly charged molecules usually move faster than do larger and/or weakly charged molecules.

Polyacrylamide gels, which have small mesh sizes, are used for separating small fragments of DNA, or RNAs; agarose gels are used for larger pieces of DNA.

Within the gel, molecules of nucleic acid are separated into discrete zones (according to size) which may be stained *in situ* (revealing spots or bands) and/or transferred for later analysis to a paper-like matrix by *Southern blotting* (Fig. 8.13) or by electroblotting.

Pulsed-field gel electrophoresis (PFGE). Standard electrophoresis can separate DNA fragments up to about 20 000 bp long. PFGE can separate molecules of over 10 million bp in length; it uses two electrical fields, applied alternately from different angles, so that the net migration of a given molecule depends on the speed with which it can change its direction of movement. [A useful source of information is *Pulsed Field Gel Electrophoresis* (Birren, B) Academic Press, 1993; ISBN 0 12 101290 5.]

8.5.1.5   Vectors

Vectors carry DNA into cells (section 8.5.1). Subsequently, it's often helpful if the vector can signal its presence in the cell. For example, when a population of bacteria is transformed with a library (section 8.5.1.1) some of the cells may not receive a vector; moreover, some vectors may lack an insert – the vector having re-circularized without incorporating a fragment (see Fig. 8.7). We therefore need ways of confirming the uptake of vector and insert in order to avoid the kind of uncertainty described in section 8.5.1.1. How can we identify those colonies which contain both vector *and* insert?

Some vectors contain antibiotic-resistance genes which can signal their presence (and that of the insert) by expression (or by lack of expression) in the host cell: see, for example, Fig. 8.14. Other vectors may contain the gene for β-galactosidase: an enzyme which cleaves lactose (section 7.8.1.1), and which can also form a blue-green product by cleaving *Xgal* (5-bromo-4-chloro-3-indolyl-β-D-galactoside); when constructing a library (Fig. 8.7) fragments can be made to insert into (and thus inactivate) this gene simply by choosing a cutting (restriction) site within the gene itself. Subsequently, colonies known to contain the vector (e.g. through expression of antibiotic resistance) can signal the presence or absence of an insert: on an Xgal-containing medium, white colonies are formed when vectors contain an insert (β-galactosidase inactivated by insert), while blue-green colonies are formed when vectors lack an insert (β-galactosidase functional). Of course, this indicator system is useful only for cells which are not themselves forming β-galactosidase.

Cloning large fragments. Large fragments of DNA (section 8.5.1.2) in plasmid vectors make large recombinant molecules that would not transform efficiently (section 8.4.1.1). However, we can use certain phage vectors (phages: Chapter 9) which *inject* their nucleic acid into cells; fragments for cloning are inserted into the phage genome and are replicated when the genome replicates inside

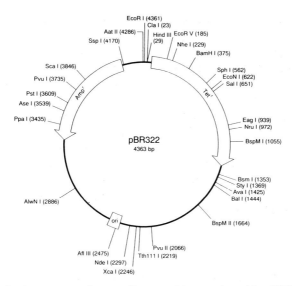

Fig. 8.14 A cloning vector: the small, recombinant plasmid pBR322 which was constructed *in vitro* from other plasmids; it is 4363 bp long. Plasmid vectors are generally small (often 3–5 kb) because (i) transformation is more efficient with small plasmids, and (ii) small plasmids are more resistant to damage during isolation from cells.

Various restriction sites are shown (with their distances, in bp, clockwise from the top); each RE cuts pBR322 at one site only (linearization). The variety of restriction sites allows flexibility in the construction of recombinant molecules from various sources of DNA.

The presence of genes encoding resistance to ampicillin and tetracycline (transcribed in opposite directions, as shown by the arrows) enables us to test for (i) the presence of the plasmid in a cell, and (ii) the presence of an insert in the plasmid. Thus, if pBR322 and *Bam*HI are used when making a library (Fig. 8.7), fragments will insert into – and thus inactivate – the tetracycline-resistance gene; the ampicillin-resistance gene will remain functional. Hence, any bacterium containing the plasmid can form a colony on media containing ampicillin; if these colonies are then replica plated (section 8.1.2) onto a medium containing tetracycline, cells containing vector *and* insert should fail to grow – owing to inactivation of the tetracycline-resistance gene. (Cells containing re-circularized plasmids grow on both media.)

A cloning vector must, of course, contain an *origin* of replication (section 7.3). In pBR322, the origin (*ori*) is derived from a plasmid which replicates under relaxed control (section 7.3.1.1). Thus, pBR322, also relaxed, can continue to replicate in non-growing (chloramphenicol-inhibited) bacteria; this effect is used to obtain an *amplification* of up to several thousand copies of the vector per cell.

Line drawing of pBR322 courtesy of Pharmacia Biotech Inc., Molecular Biology Reagents Division, Milwaukee, Wisconsin, USA.

a cell. A *replacement vector* is made by removing a sequence of non-essential DNA (the 'stuffer') from the phage genome and replacing it with the fragment to be cloned; thus, a fragment of about 20 kb can be cloned in phage λ (compared with < 10 kb in plasmid pBR322). Even bigger fragments (up to about 40 kb) can be cloned in *cosmids* (Fig. 8.15).

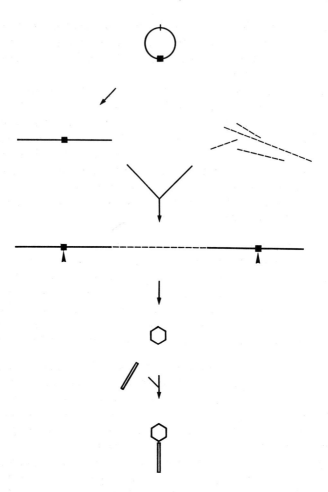

Fig. 8.15 A cosmid cloning vector (diagrammatic). The cosmid (*top*) is a small plasmid into which has been inserted the *cos* site (*solid black square*) of phage λ (see Fig. 9.2); *cos* enables the plasmid, with an insert, to be packaged (*in vitro*) in a phage λ head.

Cosmids are cut at a single restriction site; the linearized molecules are then mixed, in the presence of ligase, with DNA fragments (*dashed lines*) cut by the same enzyme. Cosmids and fragments bind randomly, and in some cases a fragment will be flanked by two cosmids (*centre*); if, in such a molecule, the distance *cos* to *cos* is about 40–50 kb long, then the *cos–cos* sequence is the right size for packaging in a phage λ head. Enzymic cleavage at the *cos* sites (*arrowheads*) creates sticky ends, and packaging of the recombinant molecules occurs in the presence of phage components. After addition of the tail, the particle can inject its recombinant DNA into a suitable bacterium (just like an ordinary phage λ). The injected DNA circularizes, via the *cos* sites, and replicates as a plasmid; its presence in the cell can be detected e.g. by the expression of plasmid-encoded antibiotic-resistance genes.

Screening for specific sequences is carried out in a way similar to that used for plasmid vectors.

Fragments of up to 300 kb can be cloned in *bacterial artificial chromosomes* (BACs): large, circular vector molecules, based on the F plasmid, which replicate in *E. coli*; the uptake of such large molecules by host cells is achieved by electroporation (section 8.4.1.2).

Yeast artificial chromosomes (YACs) can carry inserts of up to 2000 kb; these linear vector molecules, constructed *in vitro*, contain those parts of the yeast (*Saccharomyces cerevisiae*) chromosome necessary for replication and segregation in yeast cells. YACs provide a useful alternative to bacterial systems for the cloning or expression of eukaryotic genes; they can carry large eukaryotic genes, together with promoter(s) and control sequences, and are therefore suitable for studies of gene function. YACs have been used for mapping in the human genome project. [Artificial chromosomes (review): Monaco & Larin (1994) TIBTECH *12* 280–286.]

Expression vectors encode functions for transcription/translation of the insert (Fig. 8.16); not shown in the diagram are the vector's origin of replication or its 'marker' genes (e.g. antibiotic-resistance genes). Expression can be controlled in various ways – e.g. in some systems the gene can be switched on by a specific change in temperature. If required, a gene can be cloned in bacteria and then expressed in eukaryotic cells, a single vector carrying a prokaryotic origin of replication and control functions for expression in eukaryotes.

Shuttle vectors can replicate in different types of organism. They can, for example, contain both prokaryotic and eukaryotic *ori* sequences, allowing replication in both prokaryotic and eukaryotic cells; such vectors have prokaryotic markers (e.g. antibiotic-resistance genes) and eukaryotic markers (e.g. a gene for leucine synthesis, useful in auxotrophic host cells). Eukaryotic components are from the yeast *Saccharomyces cerevisiae*.

To isolate a given yeast gene, a library of yeast genome fragments in shuttle vectors is inserted into a population of mutant yeast cells (defective for the given gene); then, a procedure analogous to that described earlier (section 8.5.1.1) may be used to screen for the vector carrying the required fragment. This vector can then be amplified in *E. coli*.

A *gapped shuttle vector* (= *retrieval vector*) can copy e.g. a specific mutant yeast gene, *in vivo*, for subsequent analysis. The corresponding wild-type gene, together with its flanking chromosomal sequences, is inserted in the vector; the gene is then cut out (with restriction enzymes), its flanking sequences remaining in the vector. This 'gapped vector' is inserted into cells containing the mutant gene; the flanking sequences in the vector bind to complementary sequences flanking the mutant gene, and the mutant gene is copied (Fig. 8.17). The end product is a complete (circular) vector carrying a copy of the mutant gene; it can be subsequently amplified in *E. coli*.

A *phagemid* is a flexible vector (constructed *in vitro*) which includes (i) a plasmid's origin of replication; (ii) a multiple cloning site (MCS) – see Fig.

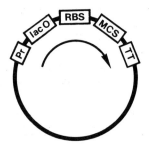

Fig. 8.16 A generalized plasmid expression vector used in *Escherichia coli* (one of many possible *in vitro* constructions).

The promoter (Pr) is commonly a strong, hybrid promoter – e.g. the *tac* promoter (constructed, *in vitro*, from the promoters of the *trp* and *lac* operons of *E. coli*); the arrow indicates the direction of transcription. Promoter activity (in this case) is controlled by the host cell's *lac* regulator protein (see Fig. 7.11) which binds to the operator site (lacO) on the plasmid, inhibiting transcription. Activity can be switched on by adding isopropyl-β-thiogalactoside (IPTG) to the medium; IPTG binds to – and inactivates – the regular protein, allowing transcription to proceed. For cells which do not synthesize the *lac* regulator protein we can use a vector which incorporates a *lacI* gene (Fig. 7.11).

The ribosome binding site (RBS) ensures that, following transcription, the mRNA product will contain the Shine–Dalgarno sequence (section 7.6) – needed for binding to the ribosome. The RBS is a short sequence of purine nucleotides.

MCS is a *multiple cloning site* (also called a *polylinker*): a sequence of nucleotides containing a number of different restriction sites (which often overlap); part of an MCS might be as follows:

....GGATCCCGGGAATTC....	containing the sites
GGATCC	*Bam*HI
CCCGGG	*Sma*I
GAATTC	*Eco*RI

DNA to be expressed is inserted in the MCS; the range of restriction sites allows flexibility in the preparation of the insert.

Three important points. (i) The insert must be in the correct orientation relative to the direction of transcription. (ii) The insert's equivalent of the initiator codon (transcribed as AUG in mRNA – section 7.6) must be the correct distance from the RBS so that AUG can align with the 'P' site on the ribosome (see Fig. 7.9); this distance has been reported to be between 5 and 13 nucleotides – in a given case the optimal distance being necessary for maximum efficiency of translation. (iii) If the insert lacks an initiator sequence, the vector must provide one, upstream of the insert; in this case, care must be taken to ensure that, following transcription, the insert's mRNA is in the correct *reading frame*, i.e. its codons are in-phase with the initiator codon – otherwise the effect will be analogous to that of a phase-shift mutation (Fig. 8.1b).

One way of correctly orientating the insert is shown in Fig. 8.9. Alternatively, the gene to be expressed can be isolated on a fragment which has been created by cutting with two different restriction enzymes – and which has non-complementary sticky ends; each sticky end can bind to its complementary sequence in a vector molecule which has been cut by the same two enzymes. (This procedure also prevents re-circularization of the vector, and of the insert, thus encouraging vector–insert binding.)

The mRNA is in the correct reading frame when the number of nucleotides separating the initiator codon from the first base of the first complete codon is 0, 3, or a multiple of 3; if not, this can be adjusted by inserting a linker (section 8.5.1.6) of the right length between insert and initiator sequence.

TT is a sequence specifying a rho-independent transcription terminator (section 7.5); it is often included to prevent unwanted *readthrough* (continued transcription) which may otherwise occur.

Fig. 8.17 A gapped shuttle vector copying a mutant yeast gene (diagrammatic) (see text). In the yeast chromosome, strand separation has occurred in the region of the mutant gene (▅▅), and strands in the vector's free ends have base-paired with complementary sequences on the chromosome. The strand binding to the lower (5′-to-3′) chromosomal strand acts as a *primer* (Fig. 7.8) for DNA synthesis (*dashed line*). Similarly, synthesis occurs from the opposite end of the other strand (*dashed line*) – thus, after ligation, completing the (circular) plasmid; note that the newly synthesized DNA includes a copy of the mutant gene.

8.16; (iii) a marker gene (e.g. for kanamycin resistance); and (iv) the origin of replication of a filamentous phage (e.g. f1 – Table 9.1). Sample DNA can be inserted into the MCS and the phagemid then inserted into a bacterium by transformation.

A phagemid can be used for making double-stranded copies of the insert, i.e. conventional cloning. Alternatively, if we infect the phagemid-carrying bacteria with a suitable 'helper phage' (e.g. a strain of M13), replication will be initiated from the phage origin in the phagemid, thus producing single-stranded copies of the phagemid (for mechanism see section 9.2.2); these single-stranded copies will be packaged in phage coat proteins (encoded by the helper phage) and exported into the medium. Single-stranded copies of the insert (when freed from the phage coats) can be used e.g. for sequencing (section 8.5.6).

Many phagemids contain a promoter on either side of the MCS (on opposite strands); this permits transcription of the cloned fragment, if required.

8.5.1.6 *Linkers, adaptors, tailing, ligation*

To join molecules of DNA we can often use the method shown in Fig. 8.6; this method allows the molecules to be separated again, if required, by using the same restriction enzyme. However, molecules may lack restriction sites or may have incompatible ones; in such cases we can alter one or both molecules, as follows. Sticky ends (if present) are first changed to 'blunt' ends (Fig. 8.11) either by removing single-stranded overhangs (e.g. with endonuclease S1) or by synthesizing a complementary strand on each 5′ overhang (thus converting overhangs to dsDNA). To each (blunt) end is now ligated (i.e. covalently bound) a *linker*: a short, synthetic dsDNA molecule containing a restriction site – for example:

5′-GGAATTCC-3′
3′-CCTTAAGG-5′

containing the *Eco*RI site (Table 8.1). Because a linker is a *palindromic* sequence (i.e. the 5′-to-3′ sequence is the same in both strands) either end can be ligated. After ligation, linkers are cleaved with the given restriction enzyme to form sticky ends. *Before* ligation, it's necessary to ensure that the enzyme's restriction sites (if any) in the insert/fragment itself are methylated (section 7.4) so as to avoid subsequent cleavage.

Linkers are also used for reading-frame adjustments (Fig. 8.16).

Adaptors are similar to linkers but contain more than one type of restriction site, and may be designed to have pre-existing sticky ends. They are used e.g. to insert one or more restriction sites into a vector; for example, an adaptor with *Eco*RI sticky ends and an internal *Sma*I site can be inserted into a vector's *Eco*RI site, thus introducing the *Sma*I site.

Tailing involves adding nucleotides to the 3′ end of DNA, using the enzyme *terminal deoxynucleotidyl transferase*; blunt-ended DNA can be tailed, though less efficiently than ssDNA or the 3′ overhangs of sticky ends. Commonly, nucleotides of only one type are added – forming a *homopolymer tail* (e.g. a deoxythymidine (dT) tail – see Fig. 8.9). Homopolymer tailing can be used e.g. to join blunt-ended DNA molecules; for example, if one molecule is tailed with deoxycytidine (dC) the other would be tailed with deoxyguanosine (dG). One advantage of this is that base-pairing can occur only between different molecules – i.e. a given molecule cannot circularize because both ends have the same tail; however, joining molecules in this way does not usually result in the creation of restriction sites, so that it can be difficult subsequently to separate the molecules (if required).

Ligation means covalently joining free, juxtaposed 5′-phosphoryl and 3′-hydroxyl ends of DNA (to form a phosphodiester bond – Fig. 7.4); the 5′ end must be phosphorylated, and energy is required. The enzyme commonly used is an ATP-dependent ligase encoded by phage T4.

Ligation shown in Fig. 8.6 is single-strand ligation in different parts of the molecule. Blunt-ended ligation (the joining of two blunt ends) requires a higher concentration of enzyme, a lower temperature, and a longer time.

When making a library (Fig. 8.7) we can prevent re-circularization of vector molecules (and thus maximize vector–fragment binding) by de-phosphorylating the 5′ ends of the vectors with the enzyme alkaline phosphatase; phosphoryl groups on the *fragments* allow some covalent binding, and this stabilizes the recombinant molecules during transformation. Subsequently, the cell's DNA repair enzymes can join the remaining strands.

8.5.2 Gene fusion and fusion proteins

An expression vector (section 8.5.1.5; Fig. 8.16) can carry sequences from two different genes with their reading frames in phase; both sequences can be transcribed from a single promoter to yield a single transcript – translation of which gives rise to a hybrid or *fusion protein*. Of what use is this technique?

When expressed, some cloned genes yield a protein product which is difficult to detect or isolate. In such cases, a sequence from the given gene can be fused with one from a second gene whose product can be easily isolated; for example, the second gene may be that encoding the enzyme glutathione S-transferase (GST) – the resulting fusion protein being readily isolated by affinity chromatography (section 8.5.1.4) using glutathione in the stationary matrix. (In this example GST is the *affinity tail* on the protein of interest.) A fusion protein may be cleaved enzymatically (*in vitro*), by a site-specific protease, to yield a separate product for each of the two genes. Cleavage of fusion proteins may also be achieved by certain chemical agents – e.g. cyanogen bromide (which cleaves at methionine residues) or hydroxylamine (which cleaves between asparagine and glycine residues); problems with chemical cleavage include (i) cleavage sites *within* the required protein, and (ii) unwanted chemical modification of the required protein.

The fusion technique is also useful for producing proteins from genes which, for various reasons, may not be highly expressed. In such cases the required gene can be fused downstream of a highly expressed gene (the 'carrier' or 'partner') within the expression vector. (The resulting fusion protein may also benefit from the presence of the carrier protein e.g. by increased solubility and/or stability.) In a so-called '*lacZ* fusion', the carrier is the *lacZ* gene (which encodes β-galactosidase).

In some cases, the gene of interest can be fused to a gene whose product is normally secreted so that the fusion protein will be secreted into the medium – see e.g. section 8.5.11.4(b); this can facilitate isolation/purification of the required gene product.

Fusion can also be used to study promoter activity, the promoter of interest being inserted upstream of a promoter-less gene. For example, when the expression of a gene cannot be monitored (owing e.g. to difficulty in assaying the gene product) the activity of the gene's promoter can be studied separately by fusing it to a so-called *reporter gene* whose product can be readily assayed; reporter genes include those encoding β-galactosidase, chloramphenicol acetyltransferase and galactokinase. Although this approach usually works well, certain reporter genes have been found to influence the activity of some promoters [Forsberg, Pavitt & Higgins (1994) JB *176* 2128–2132]; clearly, we have to bear in mind the possibility of unpredictable interactions between the components of recombinant DNA.

8.5.3 Probes

Suppose that we wish to find out whether a given plasmid or genome etc.

contains a specific short sequence of nucleotides. One way is to use a *probe*: a piece of single-stranded DNA (or RNA) whose sequence is complementary to the sequence of interest and which has been 'labelled' in some way; by taking advantage of the specificity of base-pairing, the given sequence can be located (if present) by the binding (base-pairing) of the probe – binding being detected by the probe's label. In practice, the target DNA is typically examined in denatured (single-stranded) form attached to a support such as a nitrocellulose filter. The target DNA is first exposed to many molecules of the probe, and probe–target binding occurs (by base-pairing) if the specific sequence is present. When free (non-bound) probes are washed away, any remaining probes – detected by their label – indicate the presence of the given sequence. Probes labelled with ^{32}P, for example, can be detected by autoradiography (see e.g. Fig. 8.10).

Non-radioactive (= non-isotopic) labelling of probes is common. In *direct* non-isotopic labelling, the label – bound covalently to the probe – may be an enzyme (e.g. alkaline phosphatase or horseradish peroxidase) or a fluorescent molecule (e.g. fluorescein or rhodamine). An enzymic label is detected e.g. by adding a suitable substrate which the enzyme can cleave to coloured end-products (assayed by colorimetry). Fluorescein labels are detected by ultraviolet radiation.

In *indirect* non-isotopic labelling, a compound such as biotin or digoxigenin is bound covalently to the probe. Digoxigenin is detected by first adding the corresponding *antibody* (section 11.4.2.1) – antidigoxigenin – to which a label (e.g. alkaline phosphatase) has previously been covalently linked; the antibody–label complex binds to digoxigenin and is detected by adding a suitable substrate for the enzyme.

An enzymic label can also be detected by *chemiluminescence*. For example, a probe labelled with alkaline phosphatase can be detected by adding a substrate which, when cleaved by this enzyme, emits light – the light being recorded either by photographic film or by an instrument. Chemiluminescent substrates include the 1,2-dioxetanes CSPD and AMPPD (both trade designations of Tropix, Bedford, Massachusetts, USA). In this method, alkaline phosphatase can be detected either as a direct or indirect label. Such labelling can be used not only for probes but also e.g. for the primers in DNA sequencing by the dideoxy method (section 8.5.6).

Probes can be made by cloning the particular sequence (section 8.5.1; Fig. 8.6). Before excision from the vector, the cloned probe can be labelled by so-called *nick translation*, as follows. An endonuclease makes random single-strand nicks in each hybrid plasmid; then, another enzyme (e.g. *E. coli* DNA polymerase I) removes nucleotides from the nicked strands (using the nicks as starting points) and replaces them with labelled nucleotides which are present, in excess, in the reaction mixture. After ligation, each probe – labelled as part of the hybrid plasmid – is cut out by restriction endonucleases.

Probes can also be made by PCR (section 8.5.4) or synthesized directly according to any required nucleotide sequence.

8.5.4 The polymerase chain reaction (PCR)

PCR is a method for copying DNA in which repeated replication of a given sequence (sometimes called the *amplicon*; usually <2 kb long) forms millions of copies within hours; the method depends on the ability of a DNA polymerase to extend a primer on a template strand (section 7.3; Fig. 7.8). PCR can be used e.g. to copy a particular sequence for insertion into expression vectors and it also has many other uses in bacteriology and molecular biology (see later).

The basic method is outlined in Fig. 8.18. Note that a high temperature (e.g. 72°C) is used to keep the strands of sample dsDNA separated during primer extension (a condition necessary for synthesis – section 7.3); hence, it is necessary to use a polymerase which can work at such temperatures, so that the thermostable polymerase is the key component in PCR. The *Taq* polymerase, which has been used in PCR, is from *Thermus aquaticus*, a bacterium which lives in hot springs and which has an optimum growth temperature of about 66–75°C; however, the enzyme lacks 3′-to-5′ exonuclease activity (i.e. it cannot cleave nucleotides from the 3′ end of a strand) – an activity needed for proof-reading (section 7.7). Polymerases from *Pyrococcus furiosus* and *Thermococcus littoralis* have been reported to be even more thermostable than that of *T. aquaticus* and to have proof-reading ability.

For some purposes (e.g. AP-PCR: section 16.2.2.4; Fig. 16.4c) a primer is required to bind to a target sequence with which it is not fully complementary. In a given case, primer–target binding may occur only under *low-stringency* conditions, i.e. particular values of e.g. temperature, pH and electrolyte concentration which optimize binding by helping to overcome or minimize the effects of the mismatch. *High-stringency* conditions are those which allow binding to occur only when a primer and target sequence are exactly complementary to one another, or very nearly so. The greater the stringency of the conditions (e.g. the higher the temperature) the greater will be the degree of matching required between primer and target for successful binding to occur.

In general, successful PCR requires close attention to detail, e.g. temperature and enzyme concentration, and it is essential to safeguard against contamination with extraneous DNA. Extraneous DNA may, by chance, contain (i) site(s) which are able to bind the primers being used, and/or (ii) sequences which are able to prime unwanted sites in the sample DNA; in either case, unwanted sequences (as well as the required sequence) would be amplified, so that PCR would give rise to a troublesome mixture of products and the efficiency of amplification of the required amplicon would be diminished. One approach to solving problem (i) (above) is outlined in section 8.5.4.3.

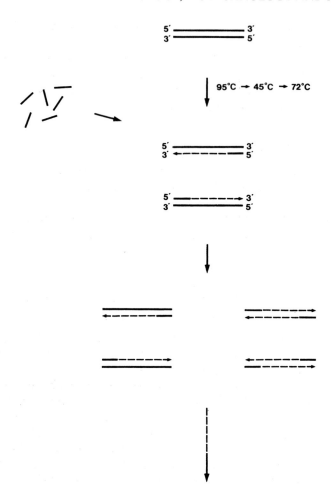

Fig. 8.18   The polymerase chain reaction (PCR) – the basic method (diagrammatic). The reaction mixture includes (i) the sample dsDNA; (ii) small pieces of ssDNA (the primers, each about 20–30 nucleotides long) of two types: one type complementary to the 3′ end of one strand of sample DNA, the other type complementary to the 3′ end of the other strand; (iii) deoxyribonucleoside triphosphates of all four types; (iv) a thermostable DNA polymerase; (v) appropriate buffer etc.

The mixture is heated to about 95 °C so that the sample dsDNA (*top*) is denatured to the single-stranded state; transient cooling to about 45°C allows primers (*top, left*) to bind at the 3′ end of each strand, and subsequent heating to 72°C permits DNA synthesis (*dashed line*) from each primer. When the temperature cycle is repeated, the newly synthesized strands (as well as the original strands) act as templates; the number of copies made by n cycles of PCR is theoretically 2^n, but after about 20–25 cycles the rate of synthesis decreases due e.g. to the increased number of template strands competing for the finite quantity of enzyme.

Inhibitors of PCR can occur in some samples – e.g. haemin in blood (see section 11.7.1.1).

In some cases it may be difficult to design a suitable primer owing to inadequate data on the sequence in target DNA. Surprisingly, we can use a so-called 'universal' nucleotide in place of one or more unknown nucleotides [Nichols *et al.* (1994) Nature *369* 492–493].

8.5.4.1 The uses of PCR

PCR is used in fields as diverse as forensic medicine and palaeobiology. Copies of a given sequence of DNA may be needed e.g. for analysis, but sometimes PCR is used simply to indicate whether a given sequence (delimited by a specific pair of primers) is present or absent in a sample; the assumption is that, if present, the sequence will be amplified and can be detected e.g. by gel electrophoresis. This method of detecting a particular sequence may be useful e.g. for detecting the presence of a given pathogen in a clinical sample.

Specific uses of PCR in bacteriology and molecular biology include:

1. Making probes (section 8.5.3).
2. Classification/taxonomy (see e.g. sections 16.2.2.4 and 16.2.2.5).
3. Detecting a specific point mutation in a DNA molecule whose wild-type sequence is known (see Fig. 8.19a).
4. Detecting specific sequence(s) of DNA of a given pathogen in clinical material – particularly DNA from pathogens which cannot be cultured (i.e. which do not grow *in vitro*) or which are slow-growing in culture. Hence, PCR is useful for the diagnosis of certain diseases and for epidemiological studies. It is particularly useful, for example, for the rapid diagnosis of tuberculous meningitis [Kaneko *et al.* (1990) Neurology *40* 1617–1618] since treatment of this disease must be started quickly; PCR can confirm the diagnosis in hours (rather than weeks by usual methods).

   For diagnostic purposes, the primer(s) must clearly be highly specific to sequence(s) found *only* in the given pathogen – e.g. a sequence in a gene encoding a unique toxin. Conditions must be arranged so that extension of primers by PCR occurs only if the specific target sequence is present in the sample. Note, however, that amplification of a given sequence does not necessarily indicate the presence of the *living* pathogen.

[The future of PCR technology: White (1996) TIBTECH *14* 478–483.]

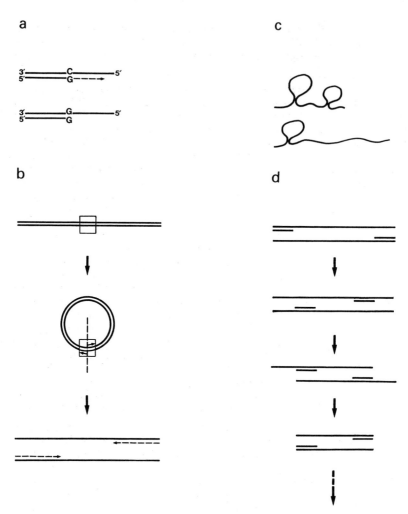

Fig. 8.19   PCR: variant forms/applications (diagrammatic).

(a) Amplification-refractory mutation system (ARMS). This procedure can detect a specific point mutation in DNA whose wild-type sequence is known. In the diagram, a G→C (guanine to cytosine) mutation is shown as 'C' in one strand of the molecule (*top*). To detect this mutation, a primer is designed with a 3'-terminal guanosine nucleotide that will align with the potentially mutant base; when hybridized to the mutant strand, this primer can be extended by PCR – thus indicating the presence of the mutation. With a wild-type strand (*bottom*), the primer would not be extended owing to the absence of base-pairing between the two guanosine nucleotides.

(b) Inverse PCR. This procedure copies DNA which *flanks* a known sequence (rather than the sequence itself); it can be used e.g. to investigate the site of insertion of a known sequence in a chromosome or plasmid. The diagram shows a linear fragment (*top*) in which the known sequence occurs within the square. The fragment is first circularized (*centre*). PCR primers are designed such that, when bound to the fragment, their 3' ends will be extended in the directions shown by the arrows. A cut at a restriction site (*dashed line*) is made at a position between the 5' ends of the two primers; such linearization gives the primer–template relationship in the standard form of PCR (*bottom*).

(c) Single-strand conformation polymorphism (SSCP) analysis. SSCP analysis compares different (but related) samples of single-stranded DNA, detecting differences in their sequences by revealing differences in their electrophoretic speeds in polyacrylamide gel; strands which differ by even one base may be distinguished. Although not strictly a form of PCR, PCR can be used to produce the samples of single-stranded DNA.

Related strands of DNA are likely to differ electrophoretically when difference(s) in their nucleotide sequences cause different levels of intra-strand base-pairing; intra-strand base-pairing confers a specific conformation which influences electrophoretic speed. The figure shows two related strands of DNA; in the lower strand a change in nucleotide sequence has resulted in a loss of local base-pairing – so that the two strands now have different conformations and are therefore likely to have different electrophoretic speeds. The gel used in SSCP analysis is a *non*-denaturing gel so as to avoid disrupting any existing intra-strand base-pairing.

A related method – denaturing gradient gel electrophoresis (DGGE) – compares samples of related PCR-generated (or restriction-generated) *double*-stranded DNA by two-dimensional electrophoresis. In this method [see e.g. Vijg (1995) Biotechnology 13 137–139], the fragments are initially separated (by size) in the first phase of electrophoresis. Then, in the same gel, the fragments are moved electrophoretically at right-angles to their original path – this time through an increasing gradient of DNA-denaturing agents (e.g. urea + formamide); at given levels in the gradient, localized *sequence-dependent* 'melting' (DNA strand separation) occurs within part(s) of the different fragments (base-pairing being stronger in GC-rich regions) – this affecting electrophoretic speeds and allowing separation of fragments in the gel. DGGE has been used e.g. to characterize organisms by comparing PCR-amplified sequences from their 16S rRNA genes [Ferris, Muyzer & Ward (1996) AEM 62 340–346].

(d) Nested PCR (nPCR). Initially, standard PCR is carried out for about 25 cycles (*top*). Then, using part of the product as template, a further 25 cycles of PCR are carried out using a pair of primers complementary to *sub-terminal* sequences in the template. The final product is thus many copies of a sequence contained within the original sequence (*bottom*). nPCR has been reported to increase both the sensitivity and specificity of PCR; it has been used e.g. when only small amounts of target DNA are available or when the material is not in good condition.

8.5.4.2   Variant forms of PCR

Amplification-refractory mutation system (ARMS). See Fig. 8.19a.

Arbitrarily-primed PCR (AP-PCR). See section 16.2.2.4.

Asymmetric PCR. In this procedure, one of the two types of primer is used at a much lower concentration (e.g. 1:50), and this primer will be used up after a few cycles of PCR; however, a plentiful supply of the other primer ensures that *one* strand of the sample dsDNA will be copied a significant number of times. This method can be used e.g. for making single-stranded DNA for sequencing (section 8.5.6) or for use as probes (section 8.5.3).

Hot-start PCR involves withholding (or blocking) an essential component of the reaction mixture until the mixture has been initially heated to above the primer-binding temperature. In one commercial process, the polymerase is initially inactivated by its antibody during heating up to 70°C; such inactivation is lost as the temperature continues to rise for the first denaturing step, and PCR then proceeds normally. Hot-start PCR enhances specificity and sensitivity by avoiding extension of primers from non-complementary sites. [See also Birch *et al.* (1996) Nature *381* 445–446.]

Inverse PCR. See Fig. 8.19b.

Multiplex PCR. Several different pairs of primers are used, simultaneously, allowing more than one sequence in the sample to be examined.

Nested PCR. See Fig. 8.19d.

REP-PCR. See section 16.2.2.5.

Reverse transcriptase PCR (rtPCR). Initially, the enzyme reverse transcriptase forms a cDNA copy of the sample RNA (section 8.5.1.2); PCR is then carried out in the usual way. With appropriate primers, this technique can be used e.g. to detect a specific mRNA molecule or to detect the presence, in a bacterium, of the RNA genome of certain phages (Table 9.1).

Single-strand conformation polymorphism (SSCP). See Fig. 8.19c.

8.5.4.3   The problem of extraneous DNA: inactivation by the UNG method

It was mentioned earlier that contamination with extraneous DNA can jeopardize the success of PCR. The apparatus to be used must therefore be de-contaminated carefully. One approach to this important problem is to carry out a preliminary procedure, resembling PCR, in which deoxyuridine-triphosphate (dUTP) is used instead of deoxythymidinetriphosphate (dTTP). In this way, any sequence (in the contaminating DNA) which is amplifyable from the given primers will give rise to products which incorporate U instead of T; such products can be subsequently degraded by the enzyme uracil-*N*-

glycosylase (UNG), which removes the uracil from deoxyuridine nucleotides in uracil-containing DNA (section 7.7.1.3). After an appropriate number of cycles, the product strands will greatly outnumber the template strands of (contaminating) DNA, so that any template strand is likely to base-pair with a (uracil-containing) product strand. Following degradation by UNG, the (double-stranded) DNA can be cleaved by appropriate treatment (e.g. 95°C heating), thus destroying the contaminating templates.

8.5.4.4   Cloning PCR products

A simple and convenient method for cloning PCR products – e.g. prior to sequencing (section 8.5.6) – involves the use of a 'rare-cutting' restriction endonuclease (section 8.5.1.3), SrfI [Schlötterer & Wolff (1996) TIG 12 286–287]. SrfI makes a blunt-ended cut (section 8.5.1.3) at its recognition sequence; because such sequences are rare, it can be assumed that, in general, PCR products will not contain an SrfI site.

A cloning vector was constructed by inserting an SrfI site into the genome of phage M13 (see section 9.2.2). To insert PCR products into the (double-stranded, circular) vector molecules, the products and vectors are incubated together in a mixture containing both SrfI and a ligase. A given SrfI site will undergo cutting (thus linearizing the vector molecule), ligation, cutting, re-ligation etc. until a cut end becomes ligated to a PCR fragment (by blunt-ended ligation – section 8.5.1.6); when this happens the vector is no longer susceptible to SrfI, and ligation can occur between the free ends of vector and fragment to form a circularized, fragment-containing vector. Such vectors can be inserted into bacteria by transformation.

SrfI cuts (/) at the following recognition sequence

5'-GCCC/GGGC-3'
3'-CGGG/CCCG-5'

and is manufactured by Stratagene.

8.5.4.5   Preparation of single-stranded DNA from PCR products

Purified single-stranded DNA is needed e.g. for sequencing (section 8.5.6) and for use as probes; purified ssDNA can be obtained from the products of PCR by the rapid and simple method described by Pagratis [(1996) NAR 24 3645–3646]. Prior to PCR, one of the two types of PCR primer is labelled with biotin. Following PCR, streptavidin (a protein obtained from Streptomyces avidinii) is added to the reaction mixture; streptavidin binds tightly to biotin, i.e. it binds to one of the two types of ssDNA product. The mixture is subjected to gel electrophoresis in a denaturing gel (one containing e.g. concentrated urea) which inhibits base-pairing between complementary strands; the (complementary) ssDNA products of PCR therefore cannot anneal (base-pair, hybridize). Because *one* of the PCR products is bound to protein (streptavidin), its electrophoretic mobility (i.e. its speed during electrophoresis) is greatly reduced; hence the *other* product will form a well-separated band in the gel. If a *particular* strand is required, the primer of the other strand is biotinylated prior to PCR.

8.5.5 Mutagenesis

We have seen that nucleic acids can be altered and manipulated by methods such as restriction and ligation, gene fusion etc. Mutagenesis (involving the deliberate creation of mutations) is a further method which can be approached in various ways.

When bacteria are exposed to mutagens, mutations develop in different genes in different individuals (section 8.1); by using appropriate *selective* conditions we can isolate cells which have the required type of mutation and which display a particular altered characteristic or function.

8.5.5.1 *Mutagenesis without mutagens in a cloned gene: mutator strains*

An effective way of introducing random mutations into a cloned gene is to insert the vector (carrying the gene) into a *mutator strain* of bacteria, i.e. a strain which has mutations in gene(s) encoding DNA repair systems (section 7.7); such a strain is defective in repairing mutations which arise spontaneously either in the bacterial genome or in a plasmid or cloning vector. Growth of the vector-containing mutator strain therefore permits spontaneous mutations to accumulate in the cloned gene. One such mutator strain is Epicurian Coli® XL1–Red (marketed by Stratagene); in this strain the mutation rate is several thousand times higher than it is in the corresponding wild-type strain, so that mutations can be introduced efficiently without the use of mutagens.

8.5.5.2 *Making specific mutations: site-specific mutagenesis*

This *in vitro* procedure allows us to make mutation(s) in a specific sequence of nucleotides; it is used for studying gene expression and also for modifying gene products – e.g. improving the efficiency of an industrial enzyme by re-coding particular nucleotide(s) in its gene. Point mutations (section 8.1.1) and even more extensive changes can be created at pre-determined sites. The basic idea behind current methods of site-specific mutagenesis (also known as *oligonucleotide-directed mutagenesis*) is shown in Fig. 8.20.

Recently, a simpler method for site-specific mutagenesis – QuikChange™ – has been marketed by Stratagene. The method does not require single-stranded templates – it can use e.g. double-stranded insert-containing circular vector molecules prepared in 'miniprep' quantities. The sample DNA is first mixed with two types of mutagenic oligonucleotide primer – analogous to the

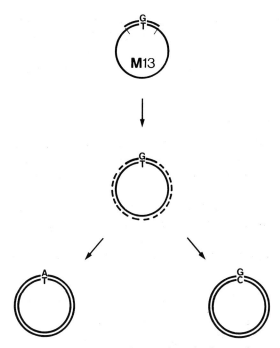

Fig. 8.20 Site-specific mutagenesis (= oligonucleotide-directed mutagenesis) using a phage M13 cloning vector: creating a point mutation at a known site in a given gene (diagrammatic) (see section 8.5.5.2).

The purpose of M13 is to make single-stranded copies of the given gene. Initially, the (double-stranded) gene is inserted into a circular, double-stranded form of M13 (see section 9.2.2); replication of M13 (in a bacterium) produces a single-stranded copy of the gene *incorporated in each of the ccc ssDNA genomes of M13*. One such genome, carrying the target gene, is shown in the diagram (*top*).

A 15–20–nucleotide sequence of the target gene (its size exaggerated for clarity) is shown between two bars (*top*); within this small sequence we wish to replace deoxythymidine (T) with deoxycytidine (C). To do this we initially synthesize copies of a *complementary* sequence (an oligonucleotide, 15–20 nucleotides long) containing one mismatch at the required site: deoxyguanosine (G) opposite deoxythymidine (T); each oligonucleotide is then bound to a copy of the target gene (*top*) and used to prime *in vitro* DNA synthesis (*dashed line*) (*centre*). The resulting dsDNA molecules (each with the single mismatch) are inserted into bacteria (by transformation). Within the cells, DNA repair mechanisms (section 7.7) correct the mismatch – sometimes replacing T with C, sometimes replacing G with A (deoxyadenosine). Even without repair, replication of the mismatched duplex will produce some molecules with the mutant sequence (*bottom, right*). Cells containing the mutant sequence can be identified e.g. by colony hybridization (section 8.5.1.2), using labelled oligonucleotides as probes.

Simpler methods, using double-stranded vectors, are now available – see section 8.5.5.2.

primer in Fig. 8.20; the two types of primer are complementary to sequences on different strands of the insert, and both primers include the required mutation. The vector molecules are thermally denatured, the primers allowed to bind to their respective sites in the insert, and a thermostable DNA polymerase, *Pfu* (from *Pyrococcus furiosus*), extends the primers. Temperature cycling produces nicked, circular copies of both strands of the parent vector carrying an insert which incorporates the mutant sequence; the nicks in the copied strands are in staggered positions, so that complementary strands of the copied vector can hybridize to form a nicked, circular molecule. Next, the non-mutant 'parent' vectors are eliminated by treatment with the restriction endonuclease *Dpn*I. (Note that the parent vector molecules, which had replicated in *E. coli*, will have undergone *dam* methylation in the sequence 5'-GATC-3' – see e.g. section 7.8.8; hence, parent vector molecules will be restricted by *Dpn*I – see section 7.4. The vectors carrying the mutant insert, having been synthesized *in vitro*, will not be methylated and will not be affected by *Dpn*I.) The vectors containing the mutant insert are then inserted into *E. coli* by transfomation, the nicks being subsequently repaired by the bacterial DNA repair systems.

8.5.6   DNA sequencing

In DNA, the primary source of information is the *sequence* in which the nucleotides occur (section 7.1). From a known sequence we can, for example, deduce the composition of a protein – identifying first the DNA equivalents of initiator and stop codons (section 7.6). We can also study promoters and other control sequences, and identify features such as REP sequences (section 16.2.2.5). In eukaryotic genes, a comparison of gene sequence with corresponding cDNA can indicate the number, positions and sequences of introns (section 8.5.1.2).

One method for sequencing is explained in Fig. 8.21. In another method, single-stranded templates for sequencing are made by asymmetric PCR (section 8.5.4.2) – rather than by cloning in M13 – the primers being complementary to known sequences either side of an unknown sequence. Single-stranded templates can also be made in phagemid vectors (section 8.5.1.5). (See also section 8.5.4.5.)

Sequencing for a maximum of about 500 nucleotides can be carried out from a single primer-binding site. Further sequencing of a given fragment can be achieved by removing nucleotides from the double-stranded form of the fragment, in the direction of sequencing, and sequencing the new single-stranded fragment from a new primer; repetition of this process (*nested deletions*) can be carried out until the entire fragment has been sequenced.

Once a large genome/chromosome has been fully sequenced, comparison of variant forms of it may be facilitated by probe-based methods such as those described by Chee *et al.* [(1996) *Science* **274** 610–614].

8.5.7 Expressing eukaryotic genes in bacteria: limitations

Most eukaryotic genes would not be expressed in bacteria: bacteria cannot (for example) deal with their introns – hence the use of cDNA (section 8.5.1.2). Additionally, many eukaryotic proteins undergo post-translational modification – for example, enzymic cleavage, or the addition of oligosaccharides (*glycosylation*) at particular sites; these modifications, which are carried out in eukaryotic cells and which are typically necessary for normal biological activity, are not carried out by bacteria. Hence, gene expression must often be studied in the eukaryotes themselves, but this can be difficult owing to their complex control systems. In an attempt to overcome this problem, certain well-characterized bacterial gene-regulatory systems have been used for gene control in eukaryotic cells [Gossen, Bonin & Bujard (1994) TIBTECH *12* 58–62].

As an alternative to the use of bacterial systems, studies on the expression of eukaryotic genes can be carried out with YACs (section 8.5.1.5); for example, wild-type genes, cloned in YACs, can be introduced into mutant eukaryotic cells to study *complementation* (i.e. the ability of the gene to compensate for the defect(s) of the corresponding mutant gene when present in the same cell).

8.5.8 Functional analysis of genomic DNA

Sequencing of genomic DNA from various species has revealed many genes of unknown function. A systematic approach to the functional analysis of genomic DNA from *Rhodobacter capsulatus* has been described by Kumar, Fonstein & Haselkorn [(1996) Nature *381* 653–654]. This method involves a set of 192 *cosmids* (Fig. 8.15) whose overlapping inserts, collectively, cover the entire (3.7 megabase) chromosome of *R. capsulatus*. Each cosmid was digested with the restriction enzyme *Eco*RV and the resulting (gene- or operon-sized) fragments subjected to Southern blotting (Fig. 8.13); this produced a set of 192 blots. The complete set of blots thus displays the entire chromosome in such a way that any given fragment can be related to a particular cosmid-sized location on the chromosome.

Collectively, the 192 blots were used as a 'high-resolution hybridization template' (HRHT) to detect altered transcription patterns under different physiological conditions. For example, when heat-shocked (section 7.8.2.3), the pattern of transcription differs from that under normal conditions; 'total RNA' (section 8.5.1.2) isolated from such cells can be suitably labelled and used as *probes* (section 8.5.3) to identify various parts of the chromosome containing sequences which are active under these conditions. In this way, specific locations on the chromosome can be associated with particular physiological state(s) of the cell.

The HRHT can also be used e.g. to examine the transcription patterns of deletion mutants of *R. capsulatus*.

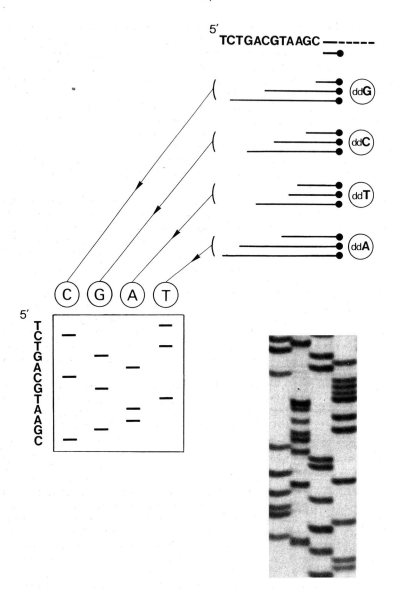

Fig. 8.21 Determining the nucleotide sequence of a cloned DNA fragment (i.e. 'sequencing'): Sanger's chain-termination method (the dideoxy method) (diagrammatic).

The fragment to be sequenced is first obtained in single-stranded form – e.g. by cloning in phage M13 (see section 9.2.2 and Fig. 8.20).

In this example, the unknown sequence is TCT....AGC (*top*). If cloning has been carried out in M13, the unknown sequence will be integrated with the M13 genome and will therefore be flanked on either side by (single-stranded) phage DNA. Hence, the DNA following AGC in the 5'-to-3' direction (*dashed line*) will consist of a *known* sequence of nucleotides; this allows us to construct a primer, carrying a label (*black disc*), that will bind next to the unknown sequence (*top*) such that the first nucleotide to be added to the primer will pair with the first 3' nucleotide of the unknown sequence.

In the laboratory, *in vitro* DNA synthesis is usually carried out with a reaction mixture which includes: (i) templates; (ii) primers; (iii) the four types of deoxyribonucleoside triphosphate (dNTP: dATP, dCTP, dGTP and dTTP – Table 7.1); and (iv) DNA polymerase. When bound to the template, the primer is extended in the 5'-to-3' direction by the addition of nucleotides as dictated by the template.

For sequencing, there are four separate reaction mixtures (G, C, T, A) – each containing all the constituents mentioned above (including millions of copies of the unknown template sequence, and of the primer). In addition, each mixture contains a given *di*deoxyribonucleoside triphosphate (ddNTP) (Fig. 7.2); thus, the G mixture contains dideoxyguanosine triphosphate (ddG), the C mixture ddC, the T mixture ddT, and the A mixture ddA.

When a dideoxyribonucleotide is added to a growing strand of DNA it prevents the addition of the *next* nucleotide; this is because dideoxyribonucleotides lack the 3'-OH group necessary for making the next phosphodiester bond (Fig. 7.4). Hence, extension of a primer will stop (= chain termination) at any position where a dideoxyribonucleotide has been incorporated. In a given reaction mixture, the concentration of the ddNTP is such that, in most growing strands, synthesis will be stopped – at some stage – by the incorporation of a dideoxyribonucleotide; because a ddNTP may pair, randomly, with any one complementary base in the template, chain termination occurs in different places on different copies of the template strand – so that fragments of different sizes are produced. For example, with ddG (see diagram), the three fragments are of different sizes because, in each case, ddG has paired with a different cytosine residue in the template; thus, in this case, the *length* of each fragment is related to the position of a particular cytosine residue in the unknown sequence. Analogous comments apply to other ddNTPs.

At the end of the reaction, new strands are separated from templates by formamide. Each of the four reaction mixtures is then subjected to electrophoresis in a separate lane of a polyacrylamide gel. (The gel contains urea to stop base-pairing between new strands and their templates.) During electrophoresis, small fragments travel further than larger ones, in a given time; moreover, fragments which differ in length by only one nucleotide can be distinguished – the shorter fragment moving just a little further. After electrophoresis the gel is dried. If the primers' label is ^{32}P, bands of fragments can be detected in the gel by autoradiography: exposing photographic film to the gel and then developing etc. The locations of the bands (*bottom, left*) indicate the relative lengths of the fragments – the shorter fragments having moved further down the gel (from top to bottom in the diagram). Note that the first unknown 3' nucleotide (C) is identified by (i) the shortest fragment (which has moved the furthest), and (ii) the fact that this fragment came from the G mixture, indicating a base which pairs with G, i.e. C. Similarly, the next unknown (G) is the next shortest fragment – which came from the C mixture. The whole unknown sequence can be deduced this way.

The autoradiograph (*bottom, right*) was derived from part of a sequencing gel (courtesy of Joop Gaken, Molecular Medicine Unit, King's College, London).

———— 3' A T T C G G C A G A T G T T A C 5' ————

5' | T A A G C C G T | C T A C A A T G | 3'

Fig. 8.22   The ligase chain reaction (section 8.5.9.1) – binding of oligomers to the target sequence. The strand containing the DNA target sequence is shown at the top: 3'-AT.......AC-5'. The two oligomers, each boxed, base-pair correctly with the target sequence and will undergo ligation. Note that the 5' nucleotide of the oligomer on the right must be phosphorylated so that a phosphodiester bond (Fig. 7.4) can be formed.

8.5.9   Methods – other than PCR – for amplifying nucleic acids

8.5.9.1   Ligase chain reaction (LCR)

Like PCR, LCR involves thermal cycling in which the heating phase is needed to separate the template strands; while the strands are separated, two oligomers bind to the amplicon on *each* strand in such a way that those on a given strand cover the entire amplicon (Fig. 8.22). The oligomers of a given pair are then joined by a heat-stable *ligase* (see ligation, section 8.5.1.6). The cycle is repeated e.g. 25 times. Note that, for ligation to occur (i) both juxtaposed bases in a given pair of oligomers must base-pair correctly with nucleotides in the template strand, and (ii) the 5' end of one oligomer in each pair must be phosphorylated. Ligation, and amplification, argue for the presence of the target sequence in the sample.

8.5.9.2   Nucleic acid sequence-based amplification (NASBA)

This method is intended primarily for the amplification of specific sequences in RNA, and this use is outlined in Fig. 8.23. NASBA is more complicated than PCR, but it works at 37°C – i.e. it does not need thermal cycling. The product consists of (i) copies of RNA which are *complementary* to the original amplicon, and (ii) copies of cDNA.

This type of procedure is used e.g. in an assay system for *Mycobacterium*; in this assay (Gen-Probe), NASBA amplifies a particular sequence in rRNA, and this sequence is then detected by means of chemiluminescent DNA probes.

8.5.10   Recombinant streptokinase: an example of recombinant DNA technology

Streptokinase is a protein which, in nature, is produced by certain species of *Streptococcus*. It has the ability to convert human plasminogen to plasmin (fibrinolysin), an enzyme which breaks down fibrin clots; thus, during infection, streptokinase may act as an *aggressin* (section 11.3.2) – helping to spread infection by promoting the lysis of fibrin barriers which may enclose (and hence localize) streptococcal lesions.

Interest in streptokinase as a therapeutic agent relates to its ability to promote solubilization of the fibrin component of blood clots, and, hence, its value in the treatment of e.g. thrombosis. As a commercial source of streptokinase, streptococci give low yields; expressing the streptokinase gene in *Escherichia coli* (see Fig. 8.24) increases the yield by more than 10-fold [Estrada *et al.* (1992) Biotechnology *10* 1138–1142].

8.5.11 Overproduction of recombinant proteins in *Escherichia coli*: strategies for optimization of gene expression

'Overproduction' refers to the synthesis of abnormally high concentrations of specific proteins – usually heterologous ('foreign') proteins – under experimental conditions; this procedure has been used e.g. for the production of therapeutic agents such as streptokinase (section 8.5.10).

Although *E. coli* is widely used for overproduction, it cannot be used for the expression of every gene; for example, in some cases a eukaryotic gene product requires post-translational modification which is not carried out in a prokaryote (see section 8.5.7). For genes which *can* be expressed, efficient, high-level expression requires attention to a range of factors which can greatly affect the yield of gene product; some of these factors are considered below. The gene of interest is assumed to be inserted in a suitable *expression vector* (Fig. 8.16).

8.5.11.1 *Optimization of transcription*

The promoter of the given gene should be *strong* (section 7.8). It should also be 'tightly regulated' – i.e. the gene should not be expressed prior to its induction or derepression. Why? Typically, cells are grown to high density prior to expression of the given gene in order to maximize the yield of gene product; gene expression before the appropriate time may e.g. depress growth rate and lower the yield of recombinant protein. Tight regulation is even more important if the gene product is toxic for *E. coli*.

The transcription regulator must also be appropriate for the purpose; thus, the inducer IPTG (Fig. 7.11), which is toxic for man, is not an optimal regulator in the overproduction of therapeutic proteins.

8.5.11.2 *Optimization of translation*

The efficiency of translation is affected e.g. by the precise sequence of the Shine–Dalgarno (SD) sequence (section 7.6) and the number of nucleotides between the SD sequence and the start codon. Moreover, some genes encode a 'translational enhancer' called the *downstream box* – a specific sequence of nucleotides downstream of, and close to, the start codon. These factors are relevant to the construction of an effective expression vector.

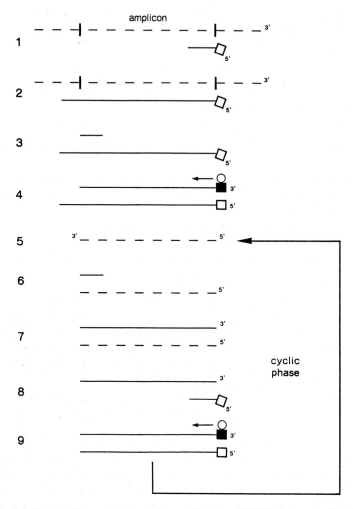

Fig. 8.23 Nucleic acid sequence-based amplification (NASBA) (section 8.5.9.2) used for the amplification of an RNA sequence (diagrammatic). The dashed lines are strands of RNA, the solid lines strands of DNA. The following gives an outline of the stages involved.

1. A strand of RNA showing the target sequence (amplicon) delimited by the two, short vertical bars. A primer (primer 1) has bound to the 3' end of the amplicon. The 5' end of this primer is tagged with a short sequence (☐) containing the *promoter* (section 7.5) of an RNA polymerase.

2. The enzyme reverse transcriptase (section 8.5.1.2) has extended the primer to form a strand of cDNA.

3. The enzyme RNase H has degraded (removed) the RNA strand, and a different primer (primer 2) has bound to the amplicon sequence in cDNA.

4. Reverse transcriptase (which can also synthesize DNA on a DNA template) has

extended primer 2 to form double-stranded cDNA – so that a functional (double-stranded) promoter is now present. An RNA polymerase (○) has bound to the promoter and will synthesize an RNA strand in the direction of the arrow, i.e. in a 5'-to-3' direction.

5. The newly synthesized RNA strand. Note the polarity of the strand: it is *complementary* to the amplicon in the sample RNA (compare 5 with 1).

6. The RNA strand has bound primer 2.

7. Reverse transcriptase has synthesized cDNA on the RNA strand.

8. RNase H has removed the RNA strand, and primer 1 has bound to the amplicon sequence in cDNA.

9. Reverse transcriptase has synthesized a complementary strand of cDNA; RNA polymerase has bound to the promoter and will synthesize an RNA strand identical to the one shown at stage 5.

Operation of the cyclic phase gives rise to many copies of the amplicon in the form of (i) *complementary* RNA, and (ii) cDNA.

8.5.11.3   The problem of inclusion bodies

Inclusion bodies (in this context) are insoluble aggregates of unfolded/ incorrectly folded proteins which often form in the cytoplasm during overproduction. Provided that such proteins can later be correctly folded *in vitro*, inclusion bodies can be advantageous in that they facilitate purification of the protein product (e.g. by centrifugation), and they may help to prevent degradation of the product by intracellular proteases.

The reasons for inclusion body formation are not fully understood so that, in some cases, *in vitro* folding may result in little or no product with biological activity. Strategies for tackling this problem include gene fusion (section 8.5.2) to increase solubility, co-expression of chaperones (section 7.6) to promote folding, modification of pH and reduction in growth temperature. However, such approaches may be largely empirical; thus, e.g. not all fusion partners will increase the solubility of a given protein, and the particular chaperone(s) required to promote the folding of a given protein may need to be ascertained by trial and error.

8.5.11.4   Avoiding proteolysis

The *E. coli* cytoplasm contains many proteases (protein-degrading enzymes), and there are also some proteases in the periplasmic region; the gene product should clearly be protected from these enzymes. Strategies for avoiding – or minimizing – proteolysis include the following.

(a) The gene product may be targeted to the periplasm (fewer proteases) by incorporating a signal sequence (section 7.6) in the gene.

(b) The gene product may be made secretable by fusing its gene to another

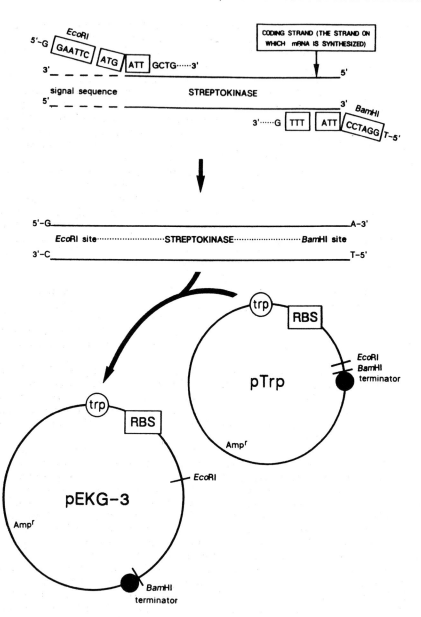

Fig. 8.24 Expression of the streptokinase gene in *E. coli* (section 8.5.10): an outline of the recombinant DNA technology [Estrada *et al.* (1992) Biotechnology *10* 1138–1142].

Cells of *Streptococcus equisimilis* are lysed, the genomic DNA separated, and the streptokinase gene (without its signal sequence) copied by PCR (section 8.5.4); earlier studies had suggested that the signal sequence of this gene may cause problems with gene expression in *E. coli*. The two strands of the streptokinase gene are shown at the top of the figure. The PCR primers are 5'-GGAATTCATGATTGCTG····3' (primer SK2) and 5'-TGGATCCTTATTTG···3' (primer SK3). Note that the template on which SK2 is extended is the coding strand ('sense strand') of the streptokinase gene, i.e. when fully extended, SK2 will form a strand *complementary* to the coding strand. After several cycles of PCR, an increasing number of PCR-derived strands will act as templates, and the major product of PCR will be as shown in the centre of the figure: the streptokinase gene flanked on either side by a specific restriction site (Table 8.1).

Note that ATG in primer SK2, when copied in the *coding* strand, will appear as 3'····TAC····5' (see base-pairing, section 7.2.1); during transcription of the gene (section 7.5), the corresponding mRNA will be 5'-AUG-3' – i.e. the *initiator* codon (section 7.6). (See also footnote 3 in Table 7.2.) It also follows that ATT (in primer SK2) corresponds to the amino acid isoleucine. Primer SK3, when fully extended by PCR, forms the coding strand; note that 5'-TTA-3' (in SK3) corresponds, in mRNA, to the codon UAA – the *ochre* codon (a stop codon – Table 7.2).

The PCR-amplified fragment (centre of figure) is inserted into plasmid pTrp, between the *Bam*Hl and *Eco*Rl restriction sites, to form plasmid pEKG-3 (both plasmids shown single-stranded for simplicity); trp is the promoter of the *E. coli* tryptophan operon, RBS is a ribosome binding site (Fig. 8.16), terminator is a transcription terminator (section 7.5), and Amp^r is a gene encoding resistance to ampicillin.

A strain of *E. coli* is transformed (section 8.4.1) with plasmid pEKG-3, and the transformed cells are plated on an ampicillin-containing medium to select those cells which have taken up pEKG-3. Several colonies are chosen, and a check is carried out on the nucleotide sequence of the streptokinase gene in cells from each colony. The gene can then be cloned by culturing cells from those colonies which are known to contain the correct sequence.

In pEKG-3, trp, the promoter of the *E. coli* tryptophan operon, is used to control transcription of the streptokinase gene. In *E. coli*, regulation of the *trp* promoter involves a repressor protein (TrpR, encoded by gene *trpR*); in the presence of tryptophan, TrpR represses transcription from the promoter. Some (mutant) strains of *E. coli* do not form TrpR; the strain chosen for expression of the streptokinase gene (*E. coli* W3110) is one which is *known* to form TrpR, i.e. a strain which is 'wild-type' for gene *trpR*. In strain W3110, transcription from the *trp* promoter can be induced ('switched on'), as required, by using an analogue of tryptophan, 3–β–indole acrylic acid, as inducer (compare with allolactose in the *lac* operon – Fig. 7.11).

During the growth of W3110, maximal expression of the streptokinase gene was obtained at a plasmid density of 420 copies per cell. When synthesized, the streptokinase remained *intra*cellular (reminder: the signal sequence was omitted); the yield of streptokinase was 25% of the total cell protein. After cell lysis, the recombinant streptokinase was purified by a procedure involving affinity chromatography (section 8.5.1.4), the stationary matrix containing human plasminogen.

gene (e.g. that of α-haemolysin – section 5.4.1.2) whose product is exported to the cell's exterior (see gene fusion: section 8.5.2). Fusion with the gene of the periplasmic protein DsbA targets the required gene product to the periplasm.

(c) Particular proteolytic site(s) in the protein may be eliminated by re-coding the gene in a way which does not affect the required function of the gene product.

(d) Use can be made of E. coli strains which are mutant in the rpoH gene. The accumulation of abnormal or heterologous proteins in the cytoplasm (as in overproduction) triggers the heat-shock response (section 7.8.2.3) – resulting in the production of the Lon protease (which degrades abnormal proteins). RpoH 'switches on' synthesis of the Lon protease, and rpoH mutants, defective in Lon, have been found to give greatly increased yields of heterologous proteins in E. coli.

Proteins whose N-terminal amino acid is arginine, leucine, lysine, phenylalanine, tryptophan or tyrosine tend to be inherently unstable (i.e. they have short half-lives); this may suggest the re-coding of a gene. Note that the penultimate N-terminal amino acid in a bacterial protein is also important as it may potentiate removal of the terminal N-formylmethionine (section 7.6) and thus itself become the terminal amino acid.

A detailed, well-referenced and comprehensive account of the optimization of high-level gene expression in E. coli has been given by Makrides [MR (1996) 60 512–538].

8.5.11.5 The cell's response to overproduction

In the example of streptokinase (section 8.5.10), up to 25% of the protein synthesized by E. coli was – from the cell's point of view – gratuitous (i.e. non-functional). Commercially, this may represent a good yield, but how do the cells themselves respond to such situations? As noted above, the accumulation of abnormal proteins triggers the heat-shock response – an indication of stress. In fact, at very high levels of gratuitous protein bacteria have been found to degrade their own ribosomes, and rRNA, thus bringing about an inhibition of translation which can lead to cell death [Kurland & Dong (1996) MM 21 1–4].

8.5.12 DNA-binding proteins: some methodology

Protein–DNA interactions are important e.g. in transcription (section 7.5.1) and in the initiation of DNA replication. Information about such interactions may be useful e.g. for engineering specific changes in the DNA in order to manipulate the binding affinity of protein(s) in studies on the regulation of gene expression etc. Outlined below are some methods for (i) detecting such interactions, and (ii) determining the binding site in the DNA.

8.5.12.1   Southwestern blotting

This technique can detect DNA-binding proteins in a cell lysate etc. Essentially, the sample is subjected to gel electrophoresis, and the *proteins* thus separated are electroblotted onto a nitrocellulose filter. The affinity of any of the proteins for a specific DNA sequence may then be determined by using labelled DNA sequences as probes and detecting the label of any probe which has bound to a given protein.

8.5.12.2   Electrophoretic mobility-shift assay (EMSA)

EMSA can determine e.g. whether any protein (in a lysate etc.) can bind DNA of a particular sequence. Labelled DNA of the given sequence is incubated with the lysate to permit protein–DNA complexing. Electrophoresis then separates unbound from (any) bound DNA – the mobility of the latter (protein–DNA complex) being different from that of free DNA.

8.5.12.3   Footprinting

In footprinting, the DNA sequence to which a protein binds can be determined by identifying that region of the DNA which is protected from enzymic (or chemical) cleavage by the shielding effect of the bound protein. Essentially, two sets of homologous end-labelled DNA fragments (one set with bound protein) are subjected to enzymic (or chemical) cleavage under conditions in which (ideally) each fragment is cut at only one of a number of potential cleavage sites. The fragments yield labelled subfragments in a range of different lengths (produced by cleavage at different sites). If, in the *protein-bound* fragments, the protein obscures one or more cleavage sites, then none of the subfragments will end at these sites; hence, when the two sets of subfragments are compared by gel electrophoresis, the absence of certain size(s) of subfragment from the protein-bound set will appear as a gap in the gel. This gap, which reflects the shielding effect of the protein, is the protein's *footprint*; its location in the gel, in relation to the bands of subfragments, indicates the site of the binding sequence in the DNA.

DNase I footprinting (using enzymic cleavage with DNase I – see Plate 8.3) tends to give a large, clear footprint because DNase I, being quite a large protein, does not cut at sites which are very close to a DNA-bound protein. However, better resolution of the binding site may be obtained by using *in vitro*-generated hydroxyl radical which causes sequence-independent cleavage of DNA; all unprotected phosphodiester bonds are susceptible to cleavage by this agent, so that the electrophoretic pattern contains a band for each unshielded position in the sequence.

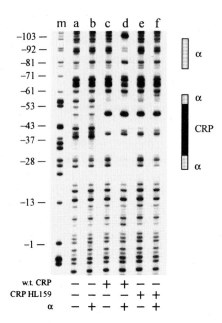

Plate 8.3. DNase I footprinting: a technique used for examining the interactions between DNA and proteins (see section 8.5.12.3 for rationale).

Here, the sample DNA is the regulatory region of the *E. coli* galactose (*gal*) operon – which includes the transcription *start site* (section 7.5), the promoter P1, and a binding site for the CRP (= CAP) protein which is involved in catabolite repression (section 7.8.2.1) of the *gal* operon; the sample DNA was labelled, at one end, with ^{32}P. This DNA was examined for interactions with (i) the α-subunit of RNA polymerase, and (ii) CRP (= CAP). Specifically, the object of the work was to study the role of the α-subunit during the activation of the *gal* P1 promoter by CRP. Note that the α-subunit and CRP are only *part* of the assembly of proteins normally found at the promoter site at the start of transcription *in vivo*; the assembly normally includes other subunits of RNA polymerase and a sigma factor. Note also that CRP binds to DNA as a *dimer*, i.e. a unit consisting of two molecules of the CRP protein.

The photograph shows the gel electrophoresis pattern from six footprinting experiments involving interactions between sample DNA and various combinations of (i) the α-subunit of RNA polymerase, (ii) wild-type (i.e. normal) CRP, and (iii) CRP HL159, a mutant (altered) form of CRP. In each experiment, sample DNA was pre-incubated with protein(s), subjected to cleavage by DNase I, and examined by gel elecrophoresis.

Lane m is a form of calibration: a *Maxam–Gilbert G sequence ladder*. To prepare such a ladder, many copies of the end-labelled sample DNA are subjected to chemical action (dimethyl sulphate followed by piperidine) which cleaves phosphodiester bonds (Fig. 7.4) *specifically* between a guanine (G) base and the adjacent base. Under appropriate conditions, each of the (many) copies of sample DNA will be cleaved at only *one* of its guanine bases; the resulting fragments are in a range of sizes because cleavage occurs at different guanines in different copies of sample DNA. On electrophoresis, these fragments (of different sizes) move at different speeds, thus forming the 'ladder' seen in lane m. To the left of lane m are numbers which indicate the locations of some of the guanine bases in the sample DNA, and this provides a useful scale; -1 refers to the first base upstream of the start site (see section 7.5). The CRP-binding site is centred between bases -41 and -42. As the nucleotide sequence of the entire *gal* regulatory region is already known, and because the exact locations of the promoters and CRP-binding site are also known, this calibration ladder can be used to indicate the *location* of DNA–protein interactions detected in the footprinting experiments.

Lane a shows bands of fragments resulting from cleavage of sample DNA, by DNase I, in the absence of bound proteins. The fragments are of different sizes because DNase I can cut at any of a number of sites in the sample DNA. Only fragments carrying the end label (^{32}P) will be detectable; that is, a copy of sample DNA which is cleaved into two pieces will yield only one piece (that carrying the label) detectable in the gel.

Lane b shows the result of DNase I action on sample DNA in the presence of the α-subunit of RNA polymerase. No footprint is seen, indicating a lack of interaction between sample DNA and the subunit in the absence of other proteins.

Lane c shows the 'footprint' of CRP (between -29 and -54) (compare c with a). Not all of this region of DNA has been protected from DNase I by the presence of CRP: as shown, cleavage has occurred at e.g. -40; this region of DNA is accessible to DNase I because it presumably passes over a cleft between the two molecules of the CRP *dimer*.

Lane d shows the action of DNase I on sample DNA in the presence of both CRP *and* α-subunits. The presence of α-subunits extends the CRP footprint both upstream and downstream. This suggests that the α-subunit can bind to a site on each of the upstream and downstream components of the CRP dimer; however, while upstream binding is believed to occur – and to be important in transcription – the downstream binding was interpreted as an artefact due to the absence of the constraining effect of other polymerase subunits. The reason for the footprint in the -80 to -90 region is not known.

Lanes e and f show the action of DNase I on sample DNA in the presence of a mutant form of CRP, with and without α-subunits. The results indicate that this particular mutation in CRP suppresses binding between CRP and the subunit – because the effects of the subunit (seen in lane d) have been suppressed. As the mutation in CRP seems to inhibit binding of the α-subunit, it appears that binding between normal (non-mutant) CRP and the subunit occurs at that site in CRP which has been altered by the mutation.

Photograph reprinted from Attey *et al.* (1994) *Nucleic Acids Research* 22 4375–4380 with permission from Oxford University Press, and by courtesy of Professor Stephen Busby, University of Birmingham, UK.

8.5.13 Displaying heterologous ('foreign') proteins

'Foreign' proteins can be *displayed* on bacterial or phage surfaces. Thus, e.g. we can display the product of an unknown gene and characterize it by affinity chromatography (section 8.5.1.4) against known, immobilized ligands (e.g. antibodies); to do this, the unknown gene is fused (section 8.5.2) to the gene of a bacterial or phage *surface* protein. [Examples: Lu *et al.* (1995) Biotechnology 13 366–372; Jespers *et al.* (1995) Biotechnology 13 378–382; Maurer *et al.* (1997) JB *179* 794–804.]

9   Bacteriophages

Most or all bacteria can be infected by specialized viruses (*bacteriophages*, usually abbreviated to *phages*). A *virus* is an organism which does not have a cell-type structure and which cannot, by itself, metabolize or reproduce. However, when inside a suitable living cell the genome of a virus may 'take over' the synthesizing machinery and direct it to make copies of the virus; the newly-formed viruses are released and can then infect other cells.

Many phages consist simply of nucleic acid enclosed within a protein *capsid* ('coat'); depending on phage, the genome may be dsDNA, ssDNA, dsRNA or ssRNA. Some phages are polyhedral, others filamentous or pleomorphic, and some have a 'tail' with which they attach to the host cell (Table 9.1, Fig. 9.1, Plate 9.1). In many cases a given phage can infect the cells of only one genus, species or strain.

The effect of phage infection depends on the particular phage and host cell – and, to some extent, on conditions. *Virulent* phages multiply within and *lyse* their host cells, i.e. the host cells die and break open, releasing phage progeny. *Temperate* phages can establish a stable, non-lytic relationship (*lysogeny*) with their host cells. Still other phages can multiply within their hosts without destroying them, phages being released from the living cells.

9.1   VIRULENT PHAGES: THE LYTIC CYCLE

The lytic cycle begins when a virulent phage adsorbs to a susceptible host cell; it ends with cell lysis and the release of phage progeny.

9.1.1   The lytic cycle of phage T4 in *Escherichia coli*

Initially, a phage attaches by the tips of its tail fibres (Fig. 9.1) to specific sites on the outer membrane. The base-plate then binds, and contraction of the tail sheath causes the inner core to penetrate the cell's outer membrane so that phage DNA can pass, via the central duct, into the periplasmic region. Uptake of DNA across the cytoplasmic membrane appears to require a pmf.

Within 5 minutes of infection the host cell stops making its own DNA, RNA and protein. 'Early' genes of T4 are transcribed and translated, and T4-encoded nucleases degrade the host's chromosome – releasing nucleotides used later for the synthesis of T4 DNA.

Middle and then late genes are subsequently transcribed. Throughout,

Table 9.1 Bacteriophages: some examples

Name	Genome[1]	Morphology[2]; size[3]	Main hosts(s)
λ	dsDNA (linear)	Isometric head, long non-contractile tail; ca. 200 nm	Escherichia coli
Mu	dsDNA (linear)	Isometric head, long contractile tail; ca. 150 nm	enterobacteria
MV-L3	dsDNA (linear)	Isometric head, short tail; ca. 80 nm	Acholeplasma laidlawii
T4	dsDNA (linear)	Elongated head, long contractile tail; ca. 200 nm	Escherichia coli
PM2	dsDNA (ccc)	Icosahedral, with internal lipid membrane; ca. 60 nm	Alteromonas espejiana
f1	ssDNA (ccc)	Filamentous; >750 nm × 6 nm	enterobacteria (only conjugative donor cells can be infected)
M13	ssDNA (ccc)	Filamentous; >750 nm × 6 nm	as for f1
φ29	dsDNA (linear)	Complex (see Plate 9.1, item 5)	Bacillus spp
φX174	ssDNA (ccc)	Icosahedral; ca. 30 nm	Escherichia coli
M12	ssRNA (linear)	Icosahedral; ca. 25 nm	enterobacteria (only conjugative donor cells)

[1] The form in which the nucleic acid exists within the virus.
[2] Isometric means approximately spherical, commonly icosahedral; 'icosahedral' means resembling an *icosahedron*: a solid figure bounded by 20 plane faces, all the faces being equilateral triangles of the same size.
[3] Excludes tail fibres (where present).

transcription seems to involve the host cell's RNA polymerase, though the enzyme undergoes several phage-induced changes which enable it to recognize different promoters; one such change is *ADP-ribosylation*: the transfer, from NAD (Fig. 5.1) to the polymerase, of an ADP-ribosyl group. Additionally, a phage-encoded sigma factor (section 7.5) is used for transcription of the late T4 genes.

About 5 minutes after infection, T4 DNA begins to replicate. Leading strand synthesis (Fig. 7.8) may be primed by the cell's RNA polymerase, but Okazaki fragments seem to be primed by phage-encoded proteins. New phage DNA undergoes modification (sections 7.4 and 9.6).

The T4 genome is *linear*, double-stranded DNA, about 170 kb, which contains hydroxymethylcytosine instead of cytosine; synthesis of the hydroxymethylcytosine requires phage-encoded enzymes. The DNA is *circularly permuted* and *terminally redundant* – terms which can be illustrated as follows:

5'-ABCD|EFGH....ABCDEFGH|ABCD-3'

The above shows (alphabetically for clarity) the order of nucleotides in one

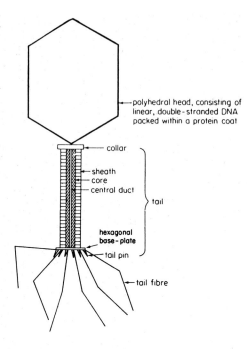

Fig. 9.1 Bacteriophage T4, a complex, tailed phage which infects *Escherichia coli* (diagrammatic). The purely structural parts are made of protein; the genome (dsDNA) is enclosed within the head. Extended fibres are needed for infection. Extension of tail fibres is inhibited e.g by low pH, low temperature and low ionic strength. For size see Plate 9.1.

Plate 9.1. Some bacteriophages (see text and Table 9.1). **1.** Bacteriophage lambda (*λ*). **2, 3, 3a.** Bacteriophages T6, T4 and T2, respectively (all to the same scale). These phages are morphologically very similar; they differ e.g. in their receptor (binding) sites on *E. coli*. **4.** Bacteriophage fd, a filamentous phage containing ss cccDNA; the bar is 50 nm. Below, higher magnification shows clearly that the two ends of fd are different. **5.** Bacteriophage *φ*29. The elongated head (approx. 35–40 nm), which contains linear dsDNA, bears protein fibres and is connected to a short, non-contractile tail. **6.** Bacteriophage Q*β*, a small icosahedral phage (approx. 25 nm in diameter) which contains ssRNA. **7, 8.** Bacteriophages T3 and T7, respectively, both to the same scale (bar = 50 nm); both phages contain linear dsDNA, and each has a short, non-contractile tail. **9.** Bacteriophage T1 (bar = 50 nm); the head contains dsDNA, and the long tail is non-contractile.

Courtesy of Dr Michel Wurtz, University of Basel, Switzerland, and reprinted, with permission, from selected micrographs in 'Bacteriophage structure' by Dr M. Wurtz, *Electron Microscopy Reviews* vol. 5(2) (1992), Pergamon Press plc.

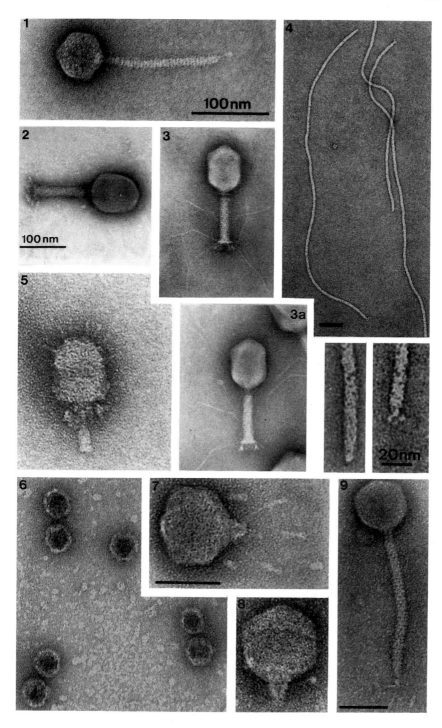

strand of the DNA in a given T4 genome. In another T4 genome, the sequence may be e.g.:

5'-EFGH|ABCDEFGH....ABCD|EFGH-3'

Notice that, in circularly permuted genomes, the *sequence* is the same in all molecules, but the starting points are different. 'Terminally redundant' means that the same sequence occurs at both ends of a given molecule (as shown above).

Replication of T4 DNA creates molecules with single-stranded 3' ends because the last Okazaki fragment of the lagging strand (see Fig. 7.8) is not synthesized on the (linear) template strand. The single-stranded ends 'invade' complementary sequences in other phage genomes within the cell – e.g. by displacing a short homologous sequence in another duplex; such recombinational activity leads to further DNA replication (the 3' ends acting as primers) and the formation of a complex, branched network of replicating phage genomes. The result is a number of *concatemers*, each consisting of a number of phage genomes joined end-to-end. (Concatemers are also formed e.g. by phage λ – see Fig. 9.2 – but by a different mechanism.)

The phage head, tail etc. are encoded by late genes. Head construction, which involves about 20 genes, starts with the assembly of a *prohead* on the inner surface of the cell's cytoplasmic membrane; the prohead is built around a core of *scaffolding proteins* (which determine the shape and size of the prohead). The core is later removed, and the prohead detaches from the membrane and undergoes further conformational changes. DNA is packaged in the head by the *headful mechanism*: one end of a concatemer is inserted, and DNA is fed into the head until it is full; the DNA is then cleaved. The tail is polymerized (core first) on the base plate, and the completed tail joins spontaneously to the DNA-filled head; the fibres are then added.

Phage progeny are released by osmotic lysis: the cell envelope is weakened by T4-encoded lysozyme (Fig. 2.7) assisted by a T4–encoded *lysis protein* (section 9.1.2).

9.1.2   Lytic cycles of other phages

Lytic cycles can differ markedly from that of phage T4 – for example: (i) not all phages bind initially to the cell wall; some bind e.g. to specific sites on the pilus: M12, for example, binds to the side (not the tip) of an F or F-type pilus. (ii) In some cases the host's chromosome survives for some time so that certain host genes can be exploited. (iii) In small ssRNA phages (such as M12, MS2 and Qβ) the phage genome itself acts as mRNA, i.e. only translation is required. For replication, RNA phages must themselves encode at least part of the replicase system because no bacterium encodes the machinery for synthesizing RNA on an RNA template. (iv) Concatemer formation in phage λ involves the rolling circle mechanism (Fig. 9.2).

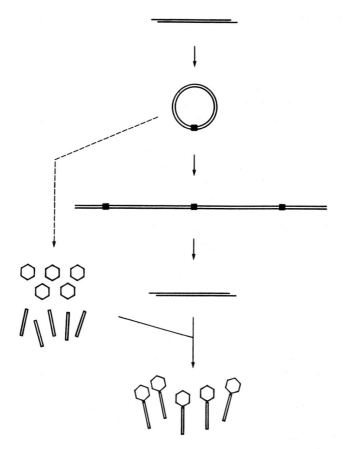

Fig. 9.2 Bacteriophage λ: outline of the lytic cycle (diagrammatic). DNA which the phage injects (through its tail) into a cell is a linear, double-stranded molecule, about 49 kb long, with complementary sticky ends, each 12 bases long (*top*). Within the cell the DNA quickly circularizes via the sticky ends; the double-stranded region formed by the base-paired sticky ends is called the *cos* site (*solid black square*).

Within the host cell, transcription of the phage's 'early' genes yields regulatory proteins whose activities are mutually antagonistic; the outcome, which is influenced by conditions in the host cell (and in the cell's environment), is either lysogeny (section 9.2) or lysis, as follows.

The DNA initially replicates by a Cairns-type mechanism (Fig. 7.8), i.e. the circular DNA is replicated. Later, replication follows the rolling circle pattern (Fig. 8.5) which gives rise to a continuous chain of (double-stranded) phage genomes, joined end-to-end, in which the *cos* site is regularly repeated (*centre*); such a tandemly repeated molecule is called a *concatemer*.

Phage head and tail proteins are encoded by the late genes – transcribed from the circular DNA. In the concatemer, a *cos* site is cleaved enzymatically, and a *cos* sticky end locates in a phage head; DNA is fed into the head until the next *cos* site is reached – and cleaved. Thus, the DNA between two consecutive *cos* sites is packaged, so that the head contains linear DNA with terminal *cos* sticky ends. A tail is added to the head, and the phage – released by cell lysis – is ready to infect a new host cell.

Many of the phages of Gram-negative and Gram-positive bacteria encode an enzyme which degrades peptidoglycan and which is thus able to lyse the host cell and allow release of phage progeny at the appropriate time; such enzymes, which include e.g. lysozyme and endopeptidase, have been referred to collectively as *endolysins*. The endolysins lack a 'signal sequence' (section 7.6); how, then, do they pass through the cytoplasmic membrane to reach the peptidoglycan? In addition to the endolysin, some of these phages have been shown to encode a specific type of *lysis protein* (also called a *holin*) which localizes in the cell's cytoplasmic membrane – forming a 'pore' through which the endolysin can gain access to peptidoglycan. In phage λ, for example, the holin is the product of the *S* gene; in phage T4, the *t* gene product appears to have the same role.

Clearly, a holin must be active at a specific time in the phage-infected cell: it must promote lysis only after normal phage assembly has been completed; if lysis occurred earlier, fewer complete phages would be released when the cell burst (i.e. the 'burst size' would be smaller). Such early lysis has been found to occur with a strain of phage λ containing a mutation in the *S* gene [Johnson-Boaz, Chang & Young (1994) MM *13* 495–504].

Interestingly, in some cases the sequence encoding the holin includes *two* start codons which are separated by one or two codons – i.e. two polypeptide chains (of slightly different length) are encoded; the shorter polypeptide is the holin, while the longer product appears to be an inhibitor of holin activity. The way in which these gene product(s) may influence the timing of cell lysis is discussed by Bläsi & Young [(1996) MM *21* 675–682].

9.1.3 The effect of virulent phages on bacterial cultures

If virulent phages are added to a broth culture of susceptible bacteria, most or all of the cells will subsequently lyse; this can cause a dense, cloudy culture to become clear.

If a *small* number of virulent phages is added to a *confluent* layer of susceptible bacteria (section 3.3.1), each individual phage will infect and lyse a single cell, releasing many phage progeny; progeny phages can then infect and lyse neighbouring cells, and so on. In this way, a visible, usually circular, clearing – called a *plaque* – develops in the opaque layer of confluent growth at each site where one of the original phages infected a cell.

9.2 TEMPERATE PHAGES: LYSOGENY

A *temperate* phage can enter into a stable, non-lytic relationship with a bacterium; the relationship is called *lysogeny*, and the host bacterium is said to be *lysogenic*. A lysogenic bacterium is immune from attack by other phages of similar type (*superinfection immunity*). One mechanism for superinfection immunity is illustrated by the lysogenic infection of *Salmonella typhimurium* with phage P22; this phage brings about the glycosylation of the host cell's lipopolysaccharides (section 2.2.9.2) – thus effectively destroying the cell-surface receptor sites for P22. In most cases of lysogeny the phage's genome (i.e. its nucleic acid, called the *prophage*) integrates with the host's chromosome; in a few cases (e.g. phage P1) the prophage does not integrate with the chromosome but exists within the bacterium as a circular molecule ('plasmid'). Either way, replication of the prophage and host cell is co-ordinated so that, at cell division, each daughter cell receives a prophage.

Maintenance of the lysogenic state involves synthesis of a phage-encoded repressor protein. Loss of active repressor protein results in *induction* of the lytic cycle – i.e. a prophage normally retains the potential for virulence. In a population of lysogenic bacteria induction may occur spontaneously in a small number of cells.

Lysogeny seems to be common in nature.

9.2.1   Lysogeny of phage λ in *Escherichia coli*

DNA enters the bacterium as linear dsDNA and immediately circularizes (Fig. 9.2). While the 'early' phage genes are being transcribed either lysogeny or lysis may follow. The lysis/lysogeny decision depends on the 'success' of one or other of two early gene products: the cI protein (which represses transcription from certain operators, and which promotes lysogeny) and the cro protein, which inhibits transcription of the *cI* gene. The cell's internal state can affect this decision. For example, lysogeny is favoured by starvation; one explanation of this is that starvation increases levels of cAMP (Fig. 7.12), a factor believed to favour synthesis of the cI protein.

In lysogeny, the λ genome integrates with the host's chromosome by site-specific recombination (Fig. 8.3b); a phage-encoded protein (product of the *int* gene) acts as the specific recombinase.

Induction (the switch from lysogeny to virulence) occurs spontaneously in a few cells in a lysogenic population. Induction in most or all of the cells can be brought about e.g. by DNA-damaging agents such as ultraviolet radiation or the antibiotic mitomycin C; under these conditions, when the SOS system is operative (section 7.8.2.2), the activated RecA protein cleaves the phage repressor protein (cI protein), allowing expression of the lytic cycle. This requires *excision* of the prophage – the reverse of integration; excision is mediated by the phage-encoded xis protein. The genome is then replicated, phage components are synthesized, and the assembled phages are released on cell lysis.

9.2.2   DNA replication in phage M13

M13 is a filamentous phage (genome: ccc ssDNA) of the same group as phages f1 (Table 9.1) and fd (Plate 9.1); modified forms of M13 are used as cloning vectors for producing single-stranded copies of DNA for sequencing (section 8.5.6) or for site-specific mutagenesis (section 8.5.5).

Following infection of a bacterium, the host cell's enzymes synthesize a complementary (*c* or 'minus') strand on the single-stranded phage genome (the *v* or 'plus' strand), forming the ccc dsDNA *replicative form* (RF); this molecule is then supercoiled by the cell's gyrase (the supercoiled molecule is called RFI) and is transcribed from the *c* strand. The *v* strand is later nicked (by a phage-encoded enzyme), and replication occurs by the rolling circle mechanism (Fig. 8.5). Initially, genome-length pieces of *v* strand ssDNA are cleaved, circularized, and converted to daughter RFs; subsequently, a particular phage-encoded protein accumulates in the cell and coats the newly formed *v* strands, blocking their conversion to RFs. When this happens, the newly cleaved and circularized ssDNA genomes are packaged in phage coat proteins and extruded through the cell envelope of the (living) host.

9.3 ANDROPHAGES

Androphages infect only certain bacteria which contain a conjugative plasmid. For example, f1 and fd are filamentous, ssDNA phages (Table 9.1) which adsorb specifically to the *tips* of certain types of pili; penetration of the host cell may involve pilus retraction. The progeny phages of f1 and fd are released through the cell envelope; host cells remain viable, but they grow more slowly than do uninfected cells.

M12, MS2, f2 and Qβ are small, icosahedral ssRNA phages which adsorb to the *sides* of certain types of pili.

9.4 PHAGE CONVERSION

Bacteria infected with phage may have certain characteristics not shown by uninfected cells; such *phage conversion* (= bacteriophage conversion) may be due e.g. to expression of phage genes by the cells, or to inactivation of chromosomal genes through integration of the prophage. For example, those strains of *Corynebacterium diphtheriae* which cause diphtheria are lysogenized by a certain type of phage which encodes and expresses a potent toxin; strains which lack the phage cannot form the toxin and do not cause diphtheria. (See also cholera toxin in section 11.3.1.1.) In *Staphylococcus aureus*, integration of the genome of phage L54a causes loss of lipsase activity due to inactivation of the relevant chromosomal gene. (See also *Salmonella* in the Appendix.)

9.4.1 Bacteriophage Mu

The DNA of phage Mu (Table 9.1) inserts into the bacterial chromosome by simple transposition (Fig. 8.4, a–c), an event requiring the phage *A* gene product (a transposase which binds to the ends of the phage DNA). Insertion occurs regardless of whether lysis or lysogeny is to follow, and it occurs at near-random sites in the chromosome; insertion of Mu DNA produces a 5 bp duplication in the target site DNA. Because phage Mu DNA often inserts into a bacterial gene, inactivating it, detectable mutations in a Mu-lysogenized bacterial population may be found in e.g. 2–3% of the cells – hence the name 'Mu' (mutator) phage.

Lysogeny involves the activity of a repressor protein (the *c* gene product) which e.g. regulates synthesis of the transposase.

Lytic infection involves about 100 *replicative* transpositions between one part of the chromosome and another, the phage genome behaving as a giant transposable element; the resulting deletions, inversions etc. in chromosomal DNA are sufficient to cause the death of the host cell. Phage genomes are excised and then packaged by the 'headful' mechanism, each genome carrying with it some bacterial DNA at each end.

9.5 TRANSDUCTION

The transfer of chromosomal (or plasmid) DNA from one cell to another via a phage is called *transduction*.

9.5.1 Generalized transduction

In this process, any of a variety of genes may be transferred from one cell to another. In a population of phage-infected bacteria it occasionally happens that, during phage assembly, chromosomal or plasmid DNA is incorporated in place of phage DNA; such abnormal phages can, once released, attach to other cells and donate DNA but (as it is not phage DNA) neither lysogeny nor lysis will result.

In the recipient cell (*transductant*) the transduced DNA may (i) be degraded by restriction endonucleases (section 7.4); (ii) undergo recombination with the chromosome (or plasmid) so that some donor genes can be stably inherited (*complete transduction*); (iii) persist as a stable but non-replicating molecule (*abortive transduction*). (If an abortive transductant gives rise to a colony, only one cell in the colony will contain donor DNA.)

Donor genes with functional promoters (whether in a complete or abortive transductant) may be expressed in the transductant.

In some cases, any given host gene has a similar chance of being transduced. However, in the *Salmonella*/phage P22 system, a certain region of the cell's chromosome has a greater chance of being transduced; this region resembles a sequence of the phage's genome concerned with packaging of DNA into the phage head.

The transfer of any given gene by generalized transduction is a rare event. If two or more genes are transduced together (*co-transduction*), this is taken as evidence of closeness on the chromosome; such information has been useful for the detailed mapping of donor chromosomes and plasmids – distances between genes being estimated from co-transduction frequencies.

9.5.2   Specialized (restricted) transduction

Specialized transduction can be brought about only by a temperate phage in which there is a phase of integration with the host's chromosome (see e.g. section 9.2.1). On excision, a prophage will occasionally take with it some of the adjacent chromosome – an event similar (in principle) to the formation of an F' plasmid (section 8.4.2.2). In the *E. coli*/phage λ system, the prophage is normally flanked on either side by the host's *gal* and *bio* genes (Fig. 8.3b); hence, in one of the rare 'aberrant' excisions, the prophage may take with it *gal* or *bio* gene(s) – often leaving behind certain phage genes from the opposite end of the prophage. Such a recombinant phage genome may be subsequently packaged in a phage head and the rest of the phage may be assembled normally. Such a phage (a specialized transducing particle, or STP) may lack the ability to replicate (i.e. it may be *defective*); nevertheless, an STP can inject its DNA into a recipient cell and (hence) transfer specific donor genes (*gal* or *bio*). In other bacterium/phage systems, the genes flanking the prophage are those which can be specifically transduced.

9.6   HOW DOES PHAGE DNA ESCAPE RESTRICTION IN THE HOST BACTERIUM?

In bacteria, the main purpose of restriction (section 7.4) seems to be to protect against 'foreign' DNA – particularly phage DNA. How, then, do phages manage to replicate in bacteria? They do so by means of various antirestriction mechanisms [Bickle & Krüger (1993) MR 57 434–450]. For example, in the DNA of some phages (e.g. phage T7 of *E. coli*) there are few (or no) cutting sites for the host cell's endonucleases, such sites having been limited or lost through the process of selection (*counterselection*).

Some phages encode specific inhibitors of restriction enzymes, and some encode methylases with the host cell's own specificity.

The T-even phages (T2, T4 and T6 of *E. coli*) are resistant to restriction because their DNA, being glycosylated, is not susceptible to *E. coli* endonucleases. Even so, on infection with T-even phages some strains of *E. coli* produce an enzyme which cleaves their own tRNA^{lys} (tRNA which carries lysine) – thus inhibiting protein synthesis in the host cell and (hence) inhibiting phage replication; however, phage T4 encodes enzymes (including an RNA ligase) which can repair the host's self-inflicted damage.

10 Bacteria in the living world

Bacteria are often thought of as pests to be destroyed, or as convenient 'bags of enzymes' – useful for experimental purposes. However, bacteria have a life of their own outside the laboratory, and many of their activities are important not only to man but also to the whole balance of nature. This aspect of bacteriology has many facets, and only a brief outline of some of them can be given.

10.1 MICROBIAL COMMUNITIES

Most bacteria are *free-living*, i.e. they do not necessarily form specific associations with other organisms; nevertheless, they are part of the web of life, and in nature they can rarely grow without affecting – or being affected by – other organisms. Bacteria normally occur as members of mixed communities which may include fungi, algae, protozoa and other organisms. Such communities can be found in a wide variety of natural habitats – e.g. in water, in soil, on the surfaces of plants, and on and within the bodies of man and other animals. Those microorganisms which are normally present in a particular habitat are referred to, collectively, as the *microflora* of that habitat.

Microorganisms which colonize a given habitat may affect each other in various ways; for example, they may have to compete for scarce nutrients, for oxygen, or for space etc., and those organisms which cannot compete effectively are likely to be eliminated from the habitat. In some cases an organism can actively discourage at least some of its competitors by producing substances which are toxic to them – a phenomenon termed *antagonism*; a microorganism which produces antibiotics (section 15.4), for example, may have a competitive advantage. There may also be relationships in which one or both organisms benefit and neither organism is harmed; for example, an acid-producing organism can help to create favourable conditions for another organism whose growth depends on a low pH.

If a habitat remains undisturbed there will eventually develop a stable community of organisms in which the various beneficial and antagonistic interactions have reached a delicate state of balance. An alien microorganism will often have difficulty in establishing itself in such a community – unless a disturbance in the environment upsets the balance in the community. For example, in the intestine of an animal, the natural microflora can often discourage the establishment of a pathogen because (i) they occupy space (thereby hindering access), and (ii) they are well adapted to the intestine, and, for that reason, can usually outgrow a pathogen; however, any disturbance to the microflora – due e.g. to antibiotic therapy – may enable a pathogen to become established and cause disease.

10.1.1 Transient communities

In contrast to the stable, mixed communities of microorganisms in many habitats, there are occasions when one, or a few, species transiently predominate.

In cholera (section 11.3.1.1), the patient's intestine becomes a living incubator for the causal organism, *Vibrio cholerae*, and the so-called 'rice-water stools' may contain up to 10^9 cells/ml of *V. cholerae*.

In lakes, reservoirs and other bodies of water, certain conditions can encourage prolific growth of particular organisms. The result is a so-called *bloom*: a visible (often conspicuous) layer of organisms at or near the surface; the organisms include (or may consist mainly of) certain cyanobacteria – particularly those which form gas vacuoles (section 2.2.5) – and/or certain eukaryotic microorganisms. Blooms can be encouraged e.g. by an excess of nutrients (such as nitrogen leached from agricultural fertilizer) and/or by thermal stratification in the water. The death/decomposition of the bloom-forming organisms (due e.g. to cessation of favourable factors) can cause a severe depletion of oxygen in the water, sometimes resulting in the asphyxiation of fish and other aquatic animals.

In some cases bloom formation can be discouraged by pumping (circulating) the water to avoid stratification and/or by using anti-cyanobacterial chemicals such as *dichlone* (dichloronaphthoquinone).

In reservoirs, a substance (*geosmin*) may be produced by *Anabaena* and some other bloom-formers, and this can impart an 'earthy' or 'musty' taste to the water (and to fish living in the water).

Some bloom-forming cyanobacteria (e.g. species of *Anabaena*, *Microcystis*, *Nodularia*) produce toxins which can be lethal to fish and/or other animals.

10.1.2 Quorum sensing

In some cases, cells in a high-density population exhibit characteristics which are absent in the *same* cells in low-density populations ('density' referring to the number of cells per unit volume). For example, *Photobacterium fischeri* (= *Vibrio fischeri*) can be either a free-living organism (occurring at low densities) or a *symbiont* (section 10.2.4) occurring at high densities in the light-emitting organ of certain fishes; in the latter role, *P. fischeri* produces blue-green light (= *bioluminescence*: light from a living organism) whereas, when free-living (low densities), this bacterium produces little or no light. (The fish use bacterial bioluminescence to signal one another; the bacteria benefit from a stable habitat in the fish.)

How do individual cells *sense* the density of the population, i.e. in the example above, how do they know *when* to 'switch on the light'? The signal actually develops as a direct consequence of growth in population. Thus, all the cells secrete signal molecules of a special kind, and if cell density reaches a certain minimum (*quorum*) these molecules accumulate to a level which activates certain genes within the cells; in this case the *lux* genes are induced, and the bacteria produce light.

The (low-molecular-weight) signalling molecule is called an *autoinducer* (because the cells themselves produce it). Different bacteria may produce different autoinducers which regulate different characteristics; in some cases different species produce the same autoinducer – but for regulating different genes. Some bacteria produce a range of autoinducers for controlling the expression of various properties. In many cases (including the *P. fischeri* example) the autoinducer is one of a group of N-acyl-L-homoserine lactones (AHLs) [AHL-based quorum sensing: Swift *et al.* (1996) TIBS *21* 214–219].

Other examples of quorum sensing include e.g. transformation (section 8.4.1) in *Bacillus subtilis*.

10.2 SAPROTROPHS, PREDATORS, PARASITES, SYMBIONTS

10.2.1 Saprotrophs

Organisms which obtain nutrients from 'dead' organic matter are called *saprotrophs*. Some saprotrophs use only soluble compounds, but others can degrade cellulose (section 6.2) and other polymers, outside the cell, and assimilate soluble products. Complex substrates may be degraded in a stepwise manner, each of several species of saprotroph carrying out one (or a few) steps in the process; co-operation of this sort is important in the breakdown, and hence re-cycling, of organic matter in nature. Indeed, there are very few biological compounds which cannot be readily broken down by a community of saprotrophs working as a team. Without the (heterotrophic) saprotrophs the carbon cycle (Fig. 10.1) – and, hence, the other cycles of matter – would stop.

10.2.2 Predators

Bacteria of the order Myxobacterales are Gram-negative rods which live in soil and on dung and decaying vegetation. Most species prey on other microorganisms: they release enzymes which lyse other bacteria and fungi, and live on the soluble products. Of course, the myxobacteria are not typical predators: a predator usually ingests its prey *before* digesting it.

Another unusual predator, *Bdellovibrio* (see Appendix), attacks and lives within certain other bacteria; the name '*Bdellovibrio*' derives partly from the

Greek word for 'leech'. Interestingly, *Bdelovibrio* can penetrate cells of *E. coli* even when the latter have a thick capsule [Koval & Bayer (1997) Microbiology *143* 749–753].

10.2.3 Parasites

A *parasite* is an organism which lives on or within another living organism (the *host*) and which benefits (in some way) at the expense of the host; in almost all cases a parasite obtains nutrients from its host. The host may suffer varying degrees of damage – ranging from slight inconvenience to death.

Parasitism may be adopted as an alternative way of life by certain free-living bacteria, but in some bacteria it is obligatory. Bacteria such as *Mycobacterium leprae* (which causes leprosy) can grow only within particular types of (eukaryotic) host cell; obligate parasites such as this depend heavily on their host's metabolism, and often they cannot be grown in the laboratory except in specialized preparations of living cells.

A parasite which affects its host severely enough to cause disease, or death, is called a *pathogen*. Not all parasites are pathogens, and not all pathogens are parasites; an example of a non-parasitic pathogen is *Clostridium botulinum* (section 11.3.1.2).

10.2.4 Symbionts

Originally, *symbiosis* meant any stable, physical association between different organisms (the symbionts) – regardless of the nature of their relationship. Later, the meaning of the term was restricted to cover only those instances in which the relationship was one of mutual benefit. However (despite the potential for confusion) there is now a general tendency to move back to the original meaning. Hence, in the common understanding of symbiosis, the possible relationships between symbionts include both mutual benefit (*mutualism*) and parasitism. Using the same terminology, a *commensal* is a symbiont which gains benefit from another symbiont such that the latter derives neither benefit nor harm from the association.

10.2.4.1 Mutualistic symbioses

Mutually beneficial relationships between bacteria and other organisms are quite common. For example, ruminants (e.g. sheep, cows etc.) cannot produce enzymes to digest the cellulose in their diet of plant material; however, in these animals the gut includes a specialized compartment (the *rumen*) containing vast numbers of microorganisms (including bacteria such as *Ruminococcus*) which convert cellulose to simple products that the animal can absorb. In return, the microbes benefit from a warm, stable environment and the abundance of nutrients which the animal swallows.

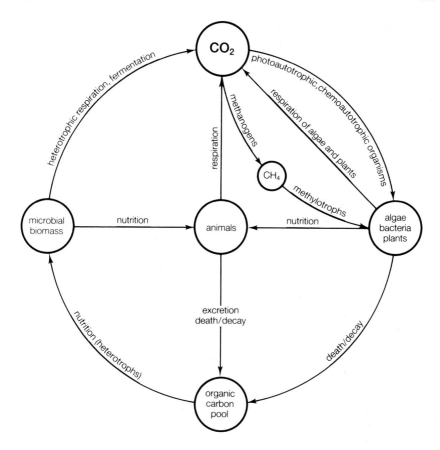

Fig. 10.1 The biological carbon cycle: a simplified scheme showing some major interconversions of carbon in nature. Bacteria have significant roles both as autotrophs and heterotrophs (Chapter 6), and members of the Archaea have unique roles as methanogens (section 5.1.2.2); the methylotrophs (section 6.4) include methane-utilizing bacteria. Microorganisms (including bacteria) are responsible for the essential conversion of 'dead' organic carbon to biomass and CO_2; without this process the cycle would stop. Microbial biomass, as such, is used e.g. by filter-feeders (oysters etc.) and, via food chains, by fish and other animals. The role of *elemental* carbon seems to be minimal (compare with nitrogen and sulphur in their respective cycles).

In certain insects, specialized cells (*mycetocytes*) in the gut lining contain (intracellular) bacteria which, in at least some cases, supply essential nutrients to the host. A *mycetome*, a distinct organelle composed of a group of mycetocytes, may be associated with the gut.

In leguminous plants (peas, beans, clover etc.) the roots have small swellings (*nodules*) containing bacteria of the genus *Rhizobium*; in this arrangement, the plant provides nutrients and protection, while *Rhizobium* supplies the plant with 'fixed' nitrogen from the atmosphere (section 10.3.2). Root nodules enable these plants to thrive in nitrogen-poor soils.

Nitrogen-fixing bacteria also form associations with non-leguminous plants. For example, nodules in the roots of the alder tree (*Alnus*) contain bacteria of the genus *Frankia*, and the small floating fern *Azolla* contains *Anabaena azollae* within specialized cavities.

10.3   BACTERIA AND THE CYCLES OF MATTER

The elements which make up living organisms occur on Earth in finite amounts; accordingly, for life to continue, the components of dead organisms must be re-used (re-cycled). Bacteria, together with other microorganisms, play a vital role in this process.

10.3.1   The carbon cycle

In all living organisms the major structural element is carbon (Chapter 6), so that the re-cycling of carbon is of fundamental importance. In the biological carbon cycle (Fig. 10.1) the chief contribution of bacteria is the use and degradation of 'dead' organic matter by the (heterotrophic) saprotrophs. In terrestrial systems, dead organic matter includes the remains of plants and animals.

For *aquatic* systems, dissolved organic carbon derives at least partly from *phytoplankton*, i.e. microscopic, photosynthetic (CO_2-fixing) organisms. (The growth of phytoplankton, seaweeds and terrestrial green plants – all photosynthetic organisms – is collectively referred to as *primary production*.) In lakes, carbon from phytoplankton is likely to be augmented by organic carbon of terrestrial origin (which arrives in run-off from the land). In the open ocean almost all carbon is derived from phytoplankton. Although much phytoplankton is used as food by aquatic animals, some provides carbon for bacterial growth (i.e. biomass production) and respiration. Recent work indicates that the proportion of dissolved carbon used for bacterial *respiration* is greater than previously supposed; moreover, it appears that when and where aquatic primary product is low, the production of CO_2 from bacterial respiration can actually exceed the rate of CO_2 fixation by phytoplankton. This means that, for at least certain periods of time, the cycling of carbon is unbalanced in parts of the ocean; this has suggested e.g. the possibility that periods of net autotrophy (when CO_2 fixation exceeds respiratory CO_2 output) may occur at different times of the year to compensate for the net heterotrophy [del Giorgio, Cole & Cimbleris (1997) Nature *385* 148–151].

A knowledge of the carbon cycle is relevant to an understanding of the 'greenhouse effect' (section 10.6), and the evolution of the cycle is of interest in itself [past and present cycle of carbon on our planet (review): Schlegel (1992) FEMSMR *103* 347–354].

10.3.1.1   Xenobiotics

Many *xenobiotics* (environmental pollutants such as pesticides, other agro-chemicals, and detergents) tend to resist biodegradation owing to the presence of ether bond(s) (C–O–C) in their molecules; in this respect they resemble the natural compound *lignin* – which tends to persist in the environment and which eventually gives rise to e.g. coal. Despite this in-built resistance to biodegradation, there are nevertheless communities of bacteria which have developed the ability to degrade xenobiotics; this is fortunate because such activity helps to limit the undesirable build-up of these compounds in the environment. The bacterial cleavage of ether bonds has been discussed by White, Russell & Tidswell (1996) MR *60* 216–232].

10.3.2   The nitrogen cycle

Nitrogen is a component of proteins and nucleic acids – and is therefore essential to all organisms. Gaseous nitrogen (dinitrogen) makes up over three-quarters of the Earth's atmosphere, but most organisms cannot use this form of nitrogen; however, some can: see section 10.3.2.1.

Ammonia assimilation. Many bacteria assimilate nitrogen as ammonia, primarily by incorporating an amino group (from ammonia) into either 2-oxoglutarate or glutamate – forming glutamate or glutamine, respectively; the glutamate or glutamine, in turn, act as nitrogen donors in various transamination reactions in the synthesis of other nitrogenous compounds. Thus, glutamate provides nitrogen for the synthesis of e.g. L-alanine and L-aspartate (Fig. 6.3) while glutamine is a nitrogen donor in the synthesis of purines and pyrimidines (bases in nucleic acids) and the amino acids histidine and tryptophan.

In *E. coli*, the pathway of ammonia assimilation depends on the concentration of ammonia; NADPH is used in both pathways.

In high concentrations of ammonia, the enzyme glutamate dehydrogenase catalyses the reductive amination of 2-oxoglutarate to glutamate.

In low concentrations of ammonia (e.g. $<1\,mM$) there is a two-stage reaction. First, glutamine synthetase (GS) catalyses the formation of glutamine from glutamate and ammonia; in the second stage, glutamine: 2-oxoglutarate aminotransferase (GOGAT; glutamate synthase) catalyses a reaction in which glutamine (1 molecule) and 2-oxoglutarate (1 molecule) from glutamate (2 molecules).

Nitrate assimilation. Some bacteria can assimilate nitrogen as nitrate, though the nitrate is first reduced to ammonia by *assimilatory nitrate reduction* (Fig. 10.2); this differs from *nitrate respiration* (section 5.1.1.2; Fig. 10.2) e.g. in that it does not yield energy.

Nitrate, nitrite and ammonia in energy metabolism. Nitrate or nitrite can be used by some bacteria as electron acceptors in respiratory metabolism; according to species, either organic or inorganic substrates can be used as electron donors in this type of metabolism. (A test for nitrate reduction is described in section 16.1.2.12.) Several pathways are known.

Denitrification is a respiratory process (section 5.1.1.2; Fig. 10.2) in which nitrate, or nitrite, is reduced to gaseous products (mainly dinitrogen and/or nitrous oxide); the process can be important in agriculture as it causes the loss of biologically useful nitrogen from the soil (section 10.3.2.2).

Dissimilatory reduction of nitrate to ammonia (DRNA) is a respiratory process carried out by certain bacteria (e.g. species of *Enterobacter* and *Vibrio*) and which has been reported to occur in habitats such as marine sediments. DRNA appears to occur more readily with low concentrations of nitrate – higher concentrations being reduced to nitrite and correspondingly smaller amounts of ammonia [Bonin (1996) FEMSME *19* 27–38].

Nitrification (Fig. 10.2) is an oxygen-dependent, energy-yielding process in which ammonia is oxidized to nitrite, and nitrite is oxidized to nitrate, by certain chemolithotrophic bacteria (section 5.1.2).

10.3.2.1 Nitrogen fixation

The 'fixation' of atmospheric nitrogen (reduction of nitrogen to ammonia) is apparently carried out only by certain types of bacteria and by some members of the Archaea; these *diazotrophs* include some cyanobacteria (e.g. *Anabaena, Nostoc*), some species of *Bacillus* and *Clostridium, Klebsiella pneumoniae* (some strains), and members of the families Azotobacteriaceae and Rhizobiaceae and of the order Rhodospirillales.

Fixation is catalysed by the enzyme complex *nitrogenase*. Electrons from a source such as hydrogen, or NADPH, are transferred e.g. to a ferredoxin (section 5.1.1.2) and thence to nitrogenase; the reduction of nitrogen requires much energy – about 12–16 molecules of ATP for each molecule of nitrogen fixed.

The ammonia produced by nitrogen fixation may be assimilated e.g. by the amination of glutamate to glutamine.

Nitrogenase is highly sensitive to oxygen. Many diazotrophs (e.g. clostridia) fix nitrogen anaerobically or microaerobically, and in some cyanobacteria the process is carried out within *heterocysts* (section 4.4.2). Aerobic nitrogen-fixers have special mechanisms for protecting their nitrogenase. In unicellular, aerobic, nitrogen-fixing cyanobacteria of the genus *Cyanothece*, nitrogen fixation occurs only in the dark period during alternating 12-hour light/dark cycles, but is continuous during 24-hour illumination; nitrogen fixation during the dark period would avoid exposure to oxygen evolved during photosynthesis (section 5.2.1.1), but continuous nitrogenase activity during continuous illumination requires additional explanation – and possibly

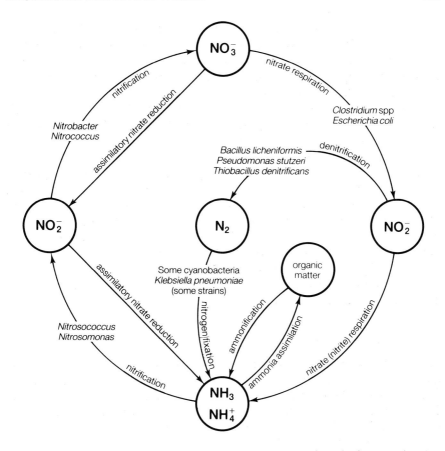

Fig. 10.2 The nitrogen cycle: some interconversions carried out by bacteria (section 10.3.2). Nitrification is an aerobic process, while nitrate respiration/denitrification and nitrogen fixation are typically associated with anaerobic or microaerobic conditions (but see section 5.1.1.2). In some cyanobacteria nitrogen fixation occurs in heterocysts (section 4.4.2). Nitrate, or ammonia, can be assimilated (i.e. used as a source of nitrogen) by many types of bacteria – nitrate being reduced, intracellularly, to ammonia (*assimilatory nitrate reduction*). Ammonification is part of the process of mineralization (section 10.3.4).

involves the 'inclusion granules' which develop under nitrogen-fixing conditions [Reddy *et al.* (1993) JB *175* 1284–1292]. [Nitrogen fixation by non-heterocystous cyanobacteria: Bergman *et al.* (1997) FEMSMR *19* 139–185.]

Diazotrophs occur as free-living organisms in soil and water, and some are involved in symbioses (section 10.2.4.1).

Finally, an intriguing question: why does nitrogen fixation occur only in certain *prokaryotes* – why not, for example, in plants? During evolution, plants are believed to have incorporated prokaryotes as organelles (e.g. mitochondria), but the ability of prokaryotes to fix nitrogen may not have developed until later, in free-living organisms; this, together with an insufficiency in selection pressure, is believed to be the reason [Postgate (1992) PTRSLB *338* 409–416].

10.3.2.2   Agriculture and the nitrogen cycle

Our knowledge of the roles of bacteria in the nitrogen cycle can be put to good use in improving agricultural food production. Food crops can be limited in yield by a shortage of available nitrogen in the soil; hence, by knowing how nitrogen is lost – and by exploiting nitrogen fixation – we can often take appropriate measures to increase crop yields.

Nitrogen is taken from the soil when crops are grown and harvested, and it may also be lost by denitrification and nitrification. Denitrification typically occurs maximally under anaerobic or microaerobic conditions in the presence of nitrate and organic nutrients – conditions found e.g. in waterlogged farm soils; thus, denitrification can often be reduced by improving soil structure and drainage so as to minimize the development of anaerobic conditions.

The harmful effect of denitrification is obvious, but why does *nitrification* lead to a loss of nitrogen? The answer is that, although nitrate and ammonia are both soluble, ammonium ions adsorb readily to soil particles (clay particles typically bear a net negative charge) whereas nitrate ions do not; for this reason, nitrate is much more readily washed (*leached*) from the soil by rain or flooding. Hence, if nitrogen fertilizers are required there is an advantage in choosing ammonium compounds rather than nitrates. Nitrification can often be prevented by adding a 'nitrification inhibitor' to the fertilizer; such compounds primarily block the oxidation of ammonia, and one of them, *etridiazole*, is additionally useful as a fungicide – being used e.g. for soil and seed treatment against certain 'damping off' diseases.

Nitrogenous fertilizers can replace lost nitrogen, but they are expensive and can be afforded least by countries in greatest need of them. However, there are alternatives. For example, 'nitrogen-fixing plants' such as clover and lucerne can be included in crop rotation schemes, while plants such as *Azolla* (section 10.2.4.1) can be used as 'green manure' – a practice common in South-East Asia; in rice paddies, fertility can be increased by encouraging the growth of free-living nitrogen-fixing cyanobacteria. Even better would be the creation, by genetic manipulation, of plants capable of fixing nitrogen without help from prokaryotes.

Nodulation of non-leguminous plants. Attempts to express the bacterial *nif* (nitrogen fixation) genes in plants have so far been unsuccessful, but the finding of *Rhizobium*-containing nitrogen-fixing nodules (section 10.2.4.1) in the *non*-legume *Parasponia* (a sub-tropical shrub) suggested an alternative approach. Thus, efforts have been made to promote the development of nitrogen-fixing nodules in certain non-legumes – particularly cereals such as rice and wheat – which apparently do not contain such nodules in nature; the object of this work is to reduce the need for nitrogenous fertilizer in agriculture in order to e.g. (i) lower the cost of cereal production and (ii) help minimize the use of (environmentally unfriendly) fixed-nitrogen fertilizers (such as nitrates).

In certain (tropical) legumes, 'invasion' of the plant by nitrogen-fixing bacteria occurs when the plant produces lateral rootlets; at this stage of growth the bacteria enter at sites where emergent rootlets penetrate the root cortex. This mode of bacterial invasion is called *crack entry*. The rootlets subsequently develop as nitrogen-fixing nodules. Experiments have been carried out to determine whether wheat can be 'infected' with bacteria of the nitrogen-fixing species *Azorhizobium caulinodans* (which nodulates the tropical legume *Sesbania rostrata*). *A. caulinodans* became established in the xylem and root meristem (i.e. *endophytically*), and the wheat showed significantly increased dry weight and nitrogen content compared with uninoculated control plants [Sabry *et al.* (1997) PRS 264 341–346].

10.3.3 The sulphur cycle

Sulphur is a component e.g. of the amino acids cysteine and methionine, of ferredoxins (section 5.1.1.2), and of cofactors such as coenzyme A. Green plants (and many bacteria) can assimilate sulphur in the form of sulphate, a substance commonly available in adequate amounts under natural conditions. Before incorporation, sulphate must be reduced to sulphide by *assimilatory sulphate reduction* (Fig. 10.3); this process differs from *sulphate respiration* (section 5.1.1.2; Fig. 10.3) in much the same way as assimilatory nitrate reduction differs from nitrate respiration (section 10.3.2). Some bacteria can assimilate sulphide direct from the environment.

In some habitats (e.g. stagnant anaerobic ponds) much sulphide is formed by sulphate respiration; this sulphide may, in turn, be used as electron donor (i.e. oxidized to sulphite or sulphate) by anaerobic photosynthetic bacteria (section 5.2.1.2).

Elemental sulphur can be used e.g. by the archaen *Sulfolobus*, and by *Thiobacillus thiooxidans* and *T. ferrooxidans* (see Appendix). Some species of *Thiobacillus* have been shown to form a filamentous matrix with which they adhere to particles of sulphur in sewage sludge, and it has been suggested that such a matrix may play an important role in the colonization and oxidation of sulphur in the natural environment [Blais *et al.* (1994) PB 29 475–482].

Thiosulphate and polythionates appear to be important intermediates in the bacterial oxidation of *pyrite* (metal sulphides), a process involved in the (environmentally harmful) formation of acid mine drainage. In this process, ferric (Fe III) ions are produced by bacterial oxidation of ferrous (Fe II) ions from the pyrite. It seems that ferric ions oxidize the sulphur in pyrite – initially to thiosulphate and then to tetrathionate; thiosulphate is regenerated, via trithionate, in a proposed cyclical reaction. A continuous supply of ferric ions is produced by the ongoing oxidative metabolism of lithotrophic bacteria such as *Thiobacillus ferrooxidans*. [Sulphur chemistry of bacterial leaching of pyrite: Schippers, Jozsa & Sand (1996) AEM 62 3424–3431.]

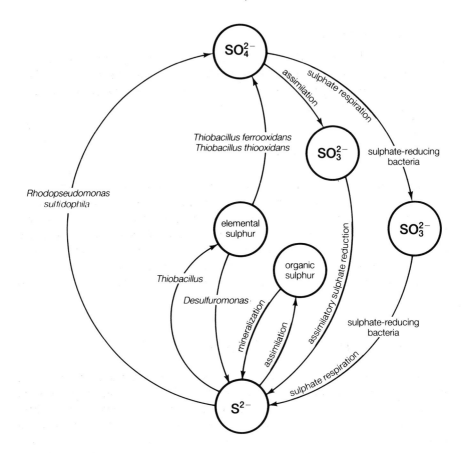

Fig. 10.3   The sulphur cycle: some interconversions carried out by bacteria. Many species can use sulphate (SO_4^{2-}) as a source of sulphur (needed e.g. for the synthesis of certain amino acids); this is shown in the figure as assimilatory sulphate reduction. In this process, sulphate is reduced, intracellularly, to sulphide, the sulphide being incorporated in different ways by different organisms; in e.g. *Escherichia coli*, sulphide is incorporated into *O*-acetylserine to form cysteine. Sulphate respiration (= dissimilatory sulphate reduction) is carried out e.g. by the 'sulphate-reducing bacteria' – organisms which use sulphate (or, e.g. sulphite) as a terminal electron acceptor in anaerobic respiration (section 5.1.1.2). *Desulfuromonas* uses elemental sulphur as a terminal electron acceptor in anaerobic respiration. *Thiobacillus* spp typically carry out aerobic respiration in which they oxidize e.g. sulphide (S^{2-}) and/or elemental sulphur. *Rhodopseudomonas sulfidophila* is one of a number of species which use sulphide as an electron donor in anaerobic phototrophic metabolism (section 5.2.1.2).

10.3.4 Mineralization

Figures 10.1–10.3 show that, in the cycles of matter, complex organic substances are broken down to simple inorganic materials such as carbon dioxide, sulphide, sulphate, ammonia and nitrate; this conversion of organic to inorganic matter is called *mineralization*.

10.4 ICE-NUCLEATION BACTERIA

At temperatures just below 0 °C, some bacteria promote water-to-ice transition by acting as nuclei around which ice crystals can form. Some of these 'ice-nucleation bacteria' are commonly found on the surfaces of plants, and they have been implicated as contributory factors in frost damage in various agricultural crops; they include strains of *Erwinia*, *Pseudomonas* and *Xanthomonas*.

The ability to promote ice-crystal formation can be transferred e.g. to *E. coli* by transferring a particular fragment of DNA; this fragment appears to encode one of the proteins of the outer membrane.

10.5 BACTERIOLOGY *in situ* – FACT OR FICTION?

Ideally, every organism should be studied *in situ*, i.e. in its natural environment; biology is, after all, about the real, living world. Of course, some aspects – e.g. intracellular structures – have to be studied *in vitro* (in the laboratory), but, where possible, a cell's normal *behaviour* is best observed under conditions which most closely resemble its normal habitat.

Clearly, any meaningful *in situ* study demands an understanding of the particular environment since, without this, the design of the experiment can be faulty; unfortunately, many studies which claim to be '*in situ*' involve obvious (sometimes extreme) distortions of nature, so that they are of little or no scientific value.

10.5.1 Membrane-filter chambers

To test for survival in rivers, suspensions of bacteria have been enclosed in 'membrane-filter chambers' – each essentially a wide plastic tube which is sealed at both ends by a membrane filter (pore size 0.22–0.45 μm); once a chamber had been immersed in a river, the bacteria inside it were considered to be essentially in 'natural' conditions, i.e. the sample was believed to be *in* the environment simply because the *chamber* had been immersed. However, the sample in such a chamber has limited contact with the environment: it generally receives less light, and it is shielded from natural turbulence (and hence from fluctuations in temperature etc.). Moreover, access or exit of molecules (including waste products) can occur only by diffusion through the minute pores of the filter; the rate of diffusion is necessarily extremely low in those membrane filters whose pores are small enough to retain the test organisms.

10.5.2   Conjugation *in situ?*

In an attempt to demonstrate conjugation (section 8.4.2) in a Welsh river and canal, suspensions of donors and recipients were mixed and then filtered through a membrane filter; the filter, with its layer of bacteria, was placed (face-down) on a flat stone and held in place by a glass-fibre filter which was secured to the stone by rubber bands. After 24 hours immersion in the river or canal the whole was tested for transconjugants. Transconjugants were detected, and the overall conclusion of the experiment was that "plasmid transfer is possible between bacteria in the river epilithon" [Bale, Fry & Day (1987) JGM *133* 3099–3107]. Clearly – in order to draw this conclusion – the layer of bacteria on the filter must have been taken to represent *epilithon*: the sessile community of organisms which grow on the surfaces of underwater stones.

To what extent did the 'simulated epilithon' resemble real epilithon? That is, to what extent was the conclusion justified? An earlier study by different authors [Lock *et al.* (1984) Oikos *42* 10–22] had shown clearly that real epilithon consists of cells embedded in a fibrous polysaccharide matrix – electronmicrographs revealing cells and microcolonies trapped within the matrix *and separated from one another by the material of the matrix.* So, in real epilithon, would not the recipients' receptor sites for pili be hidden by the matrix? Could the pilus even reach a recipient through a matrix which can exclude macromolecules? Moreover, given the presence of the matrix, would donor and recipient cells be likely to achieve direct wall-to-wall contact (of the type shown in Plate 8.1, top)? The simulated epilithon could answer none of these awkward (but crucial) questions because it lacked the essential feature of real epilithon: the matrix!

A genuine simulation would require (in addition to a matrix) a population of cells similar to that in real epilithon – with only a realistic fraction consisting of donors and recipients. However, in this experiment, each square centimetre of the filter was packed with 10^7 donors intimately mixed with 10^8 recipients (of known compatibility). From the authors' own figures, this was over 100 times greater than the total number of viable bacteria in real epilithon, and it consisted *solely* of donors and recipients; inexplicably, this arrangement was claimed to 'mimic nature'. Nature is not like this, and it is pointless to pretend that it is.

In this experiment it seems possible that donor–recipient contacts had already been initiated during mixing and filtration – i.e. even *before* the filter had been placed in its '*in situ*' test location in the river or canal.

Although seemingly concerned with 'natural habitats', this experiment in no way related to the real environment: it was little more than a laboratory-type mating experiment conducted outside the laboratory. Such experiments may be fun but they contribute nothing to science.

10.6   THE 'GREENHOUSE EFFECT'

In the natural carbon cycle (section 10.3.1; Fig. 10.1) large amounts of CO_2 are produced by animals, plants and microorganisms during respiration or fermentation; however, large amounts of CO_2 are used by the photosynthetic autotrophs (e.g. green plants and cyanobacteria).

Globally, the important balance between biological production and uptake of CO_2 is being upset e.g. by the burning of vast amounts of fossil fuel (petroleum, gas etc.) – leading to a rising level of CO_2 and consequent warming of the planet (the 'greenhouse effect').

Is the rising level of CO_2 likely to be offset by increased photosynthesis in plants (i.e. increased sequestration of CO_2)? In those plants studied, typical responses to increased CO_2 include e.g. enhanced root growth and a decrease in tissue nitrogen (i.e. a higher carbon:nitrogen ratio). Agricultural crop-type plants, in particular, show improved growth and better yields. However, a study by Körner and Arnone [Science (1992) 257 1672–1675] showed a less optimistic picture. These authors looked at the effect of increased CO_2 on experimental tropical ecosystems; they found no significant increase in above-ground biomass – and reported that the increased root growth was associated with an increased efflux of CO_2 from the soil to the atmosphere. This efflux of CO_2 appeared to be due mainly to the stimulated metabolism of microorganisms (including bacteria) in the *rhizosphere* (the root–soil environment). These results were interpreted to mean that raised levels of CO_2 may not necessarily result in increased sequestration of carbon but may bring about increased carbon cycling (cyclical exchange) between the atmosphere and terrestrial ecosystems. [Plants and increasing levels of CO_2: *Carbon Dioxide and Plant Responses*, D R Murray (1997) ISBN 0863 80213 3.]

Of course, the world's forests are a major natural 'sink' for CO_2. Unfortunately, ongoing de-forestation by man continually reduces the effectiveness of this sink. More optimistically, certain deep-rooted pasture grasses, imported into the South American savannas, appear to sequester considerable amounts of carbon in their particularly massive root systems; such grasses may therefore help to offset the effect of some of the non-biological emissions of CO_2 [Fisher *et al.* (1994) Nature 371 236–238].

Another greenhouse gas, methane (section 5.1.2.2), is oxidized by methylotrophs (section 6.4), but some escapes and contributes to the greenhouse effect. Oxides of nitrogen (which are also greenhouse gases) are produced during denitrification (section 5.1.1.2).

Recently, Harte [BC (1996) 5 1069–1083], referring to data from the Antarctic and Greenland ice cores, has pointed out that global warming may itself result in increased levels of greenhouse gases, i.e. that there is a potential positive feedback effect which is not reflected in our current models of future climatic change. Moreover, this author points to the many *synergies* which exist between different forms of environmental degradation; he also indicates that the assumption of a simple, *linear* (i.e. proportional) relationship between (human) population growth and environmental problems is overly optimistic. Harte's article is well worth reading, particularly by those who believe that environmental insults can be tackled piecemeal, i.e. in isolation from the environment as a whole.

10.7   THE PROBLEM OF RECOMBINANT BACTERIA IN THE ENVIRONMENT

The use of genetically engineered bacteria – e.g. for biological control (section 13.1.2) – causes unease owing to insufficient knowledge of the way in which such organisms may behave, or transfer their genes, in the natural environment. One approach to the problem is *biological containment*: arranging for a genetic 'self-destruct' mechanism to operate automatically when the organism's function has been completed. Such mechanisms are not 100% effective: some of the recombinant cells survive owing e.g. to mutation in the killing gene; however, elimination of the majority of recombinant cells seems a worthwhile objective. ['Suicide microbes' (review): Ramos *et al.* (1995) Biotechnology 13 35–37.]

10.8   UNCULTIVABLE/UNCULTURED BACTERIA

Of the existing species of bacteria, probably only a small proportion have been grown and isolated in the laboratory. It's likely that many species remain unknown simply because we have not yet offered them the right conditions for *in vitro* growth.

In the past, if we failed to culture and isolate a new organism (for example, one seen under the microscope) then only limited characterization would have been possible. Today, the methods of molecular biology enable us to both detect and characterize organisms which have not been cultured – or even seen! One approach depends on the fact that the 16S rRNA gene contains certain 'highly conserved' sequences of nucleotides – which occur universally in members of the domain Bacteria – as well as more variable regions which differ between e.g. genera and species. Primers complementary to the conserved sequences (*broad-range primers*) can be used in PCR (section 8.5.4) to amplify *any* accessible bacterial 16S rRNA genes (e.g. in clinical or environmental samples). The PCR products can then be sequenced (section 8.5.6) and the *variable* regions of the sequence compared with the 16S rRNA database of thousands of known species from all the major bacterial groups; the amplified gene may thus indicate the presence of an organism phylogenetically related to a known species or group. Even samples known to contain a *mixed* bacterial population can be examined in this way; in this case the (mixed) PCR products (various 16S rRNA genes) can be amplified by cloning (sections 8.5.1 and 8.5.4.4) prior to sequencing the DNA from individual clones.

11   Bacteria in medicine

Note 1.   Relevant information in other chapters:

    Chapter 7.   Details of protein synthesis etc. necessary for an understanding of the mode of action of various antibiotics.

    Chapter 8.   Principles of the methodology (e.g. PCR, NASBA) on which are based certain diagnostic/typing procedures; recombinant streptokinase.

    Chapter 12. Food poisoning; food hygiene.

    Chapter 14. Aseptic technique and basic practical bacteriology.

    Chapter 15. Sterilization, disinfection; antibiotics: mode of action, activity in macrophages, resistance, sensitivity tests.

    Chapter 16. Identification and epidemiological typing of bacteria.

Note 2.   Of the two categories of prokaryotic cell (section 1.1.1), only certain species of the domain Bacteria have been associated with human disease. Species of the domain Archaea are (apart from the methanogens) adapted to extreme environmental conditions (e.g. high temperature, low pH, high levels of electrolyte); for these organisms the human body would be too cold, too alkaline or too dilute.

11.1   BACTERIA AS PATHOGENS

Some diseases are due to 'errors' in the body's chemistry, but in many diseases symptoms result from the activities of certain microorganisms, or their product(s), on or within the body; any microorganism which (given suitable circumstances) can cause disease is called a *pathogen*. Among the many diseases of microbial origin, some are due to fungi, some to viruses, some to protozoa, and some to bacteria; a number of the latter diseases are described briefly at the end of the chapter.

In some diseases the link between disease and pathogen is highly specific: such a disease can be caused only by the appropriate species, or by particular strains of that species. Anthrax, for example, is caused only by certain strains of *Bacillus anthracis*; these strains contain plasmids (section 7.1) which encode (i) the anthrax toxin and (ii) a capsule which protects the pathogen. In other cases a disease may be due to any of several different causal agents; an example is gas gangrene: a disease which can be caused by one (or more) of several different species of *Clostridium*.

Sometimes a disease is due to an organism which does not usually behave as a pathogen and which may actually be a member of the body's own microflora (Table 11.1); for example, species of *Bacteroides*, which are common

Table 11.1   Human microflora: some of the bacteria commonly associated with the human body

Location	Species of
Colon	*Bacteroides* *Clostridium* *Escherichia* *Proteus*
Ear	*Corynebacterium* *Mycobacterium* *Staphylococcus*
Eye (conjunctiva)	*Staphylococcus* (coagulase-negative) *Corynebacterium* *Propionibacterium*
Mouth	*Actinomyces* *Bacteroides* *Streptococcus*
Nasal passages	*Corynebacterium* *Staphylococcus*
Nasopharynx	*Streptococcus* *Haemophilus* (e.g. *H. influenzae*)
Skin	*Propionibacterium* *Staphylococcus* Others (according e.g. to personal hygiene and environment)
Urethra	*Acinetobacter* *Escherichia* *Staphylococcus*
Vagina (adult, pre-menopausal)	*Acinetobacter* *Corynebacterium* *Lactobacillus* *Staphylococcus*

in the intestine, can sometimes give rise to peritonitis following accidental or surgical trauma in the lower intestinal tract. Organisms such as these are called *opportunist pathogens*.

Disease does not *necessarily* follow exposure to a given 'causal agent'. In fact, the occurence (or otherwise) of disease typically depends on various factors – including the degree of resistance of the host and the *virulence* (capacity to cause disease) of the pathogen.

Hospitals – containing concentrations of sick people – may themselves act as sources of infection; a disease acquired in hospital is called a *nosocomial* disease [nosocomial diseases (various aspects): Emmerson & Ayliffe (eds) BCID (1996) 3 159–306].

11.2   THE ROUTES OF INFECTION

The skin is normally an effective barrier to pathogens, but skin may be broken – e.g. by wounding, surgery or the 'bites' of insects etc. Wounds may admit any of a variety of potential pathogens capable of causing systemic disease (disease affecting the entire body) or localized disease. Bacterial pathogens which can enter via 'bites' include the causal agent of bubonic plague.

Mucous membranes – such as those of the intestinal, respiratory and genitourinary tracts – tend to be more vulnerable than skin, and infections commonly begin at these sites. In pneumonia and whooping cough, for example, infection begins at the respiratory surfaces, while in cholera and typhoid it begins at the intestinal mucosa.

Exposure to infection during surgery is generally minimized by adherence to an aseptic technique. Further preventive measures which may be taken e.g. during implantation/orthopaedic work include the incorporation of antibiotics into medical device polymers [Tunney, Gorman & Patrick (1996) RMM 7 195–205] and the use of antibiotic-impregnated bone cement [Wininger & Fass (1996) AAC 40 2675–2679].

11.2.1   Adhesion as a factor in infection

In many diseases there is an early phase in which the pathogen adheres to particular sites in the host. The need for attachment becomes clear when we consider, for example, that the common sites of infection, the mucous membranes, are continually flushed by their own secretions and may be subject to movements such as peristalsis – factors which tend to discourage the establishment of a pathogen. Adhesion may help a pathogen to compete more effectively with the host's own microflora.

Adhesion is essential e.g. for the virulence of *enterotoxigenic* strains of *E. coli* (so-called ETEC – Table 11.2); these strains adhere to the duodenal mucosa and produce enterotoxin(s) responsible e.g. for many of the cases of travellers' diarrhoea. Adhesion in this case is usually due to specific types of fimbria (section 2.2.14.2) which are encoded by plasmids (section 7.1).

Fimbria-mediated adhesion is also important e.g. for the adherence and invasion of *Haemophilus influenzae* type b; once within the host (e.g. in the bloodstream), the pathogen can become non-fimbriate – apparently by a random and reversible genetic switching mechanism [van Alphen & van Ham (1994) RMM 5 245–255].

Tooth decay (*dental caries*) is promoted by bacteria which adhere to tooth surfaces and gum margins and which contribute to *dental plaque*: a film composed mainly of bacteria, bacterial products, and salivary substances. *Streptococcus mutans*, a common component of plaque, forms extracellular water-insoluble glucans which assist bacterial adhesion; waste products of bacterial metabolism (e.g. lactic acid) cause localized demineralization in the teeth, permitting bacterial penetration.

Adhesion appears to be important also in the further decay of *filled* teeth, bacterial colonization occuring in the small gap between the filling material and the wall of the cavity [Buchmann *et al*. (1990) MEHD 3, 51–57] (Plate 2.1, *centre*); from this site, bacteria may penetrate the dentinal tubules and bring about destruction of the dentine.

11.2.2 Bacterial invasion of mammalian cells

Certain pathogenic bacteria can enter mammalian cells, and some can spread to adjacent cells in a tissue. Details are now emerging of the mechanisms involved. The uptake (internalization) of bacteria seems usually or always to require a positive contribution by the mammalian cell involving e.g. re-arrangement of actin molecules within the cytoplasm beneath the site of bacterial attachment; thus, e.g. actin re-arrangement is elicited by *Salmonella typhimurium* [Francis *et al*. (1993) Nature 364 639–642] and by *Neisseria gonorrhoeae* [van Putten & Duensing (1997) RMM 8 51–59].

Invasion by *Shigella* has been studied in tissue cultures. The results suggest that, *in vivo*, uptake of *Shigella* occurs via the *basolateral* surface of intestinal epithelial cells (rather than via the lumen-facing 'apical' surface). Uptake, which involves activity in the proteins actin and myosin, occurs by invagination of the mammalian cell membrane and the subsequent encapsulation of the bacterium in a membrane-bounded sac (the *phagosome*) within the cytoplasm of the mammalian cell. In typical phagocytosis the phagosome would then fuse with a *lysosome* (a vesicle containing degradative enzymes) to form a digestive sac (the *phagolysosome*) within which a bacterium would be killed. However, *Shigella* can escape from the phagosome, undergo cell division, and spread to adjacent host cells by the mechanism described in section 11.3.3.2. Phagocytosis of shigellae can be fatal for the mammalian cell (see e.g. section 11.5.2). [Pathogenicity of *Shigella*: Sasakawa (1995) RMM 6 257–266.]

Listeria monocytogenes can enter macrophages and non-phagocytic cells. Invasion requires e.g. interaction between a specific cell-surface protein of the pathogen, *internalin A*, and *E-cadherin* – a mammalian cell-surface protein normally involved in mammalian cell-to-cell binding. After uptake, *L. monocytogenes* can escape from the phagosome by secreting a protein, *listeriolysin O*, which degrades the phagosomal membrane; listeriolysin O is one of a family of *thiol-activated cytolysins* – toxins which can lyse the cholesterol-containing membranes of eukaryotic cells [thiol-activated cytolysins: Morgan, Andrew & Mitchell (1996) RMM 7 221–229]. The bacterium, liberated into the host cell's cytoplasm, can grow and spread to adjacent cells by the actin-based motility described for *Shigella* (section 11.3.3.2). On passing into an adjacent cell, the bacterium becomes encapsulated in a sac bounded by a *double* membrane (one membrane from each cell); escape from this sac involves e.g. secretion of an enzyme (a *lecithinase*) which hydrolyses the lecithin (phosphatidylcholine) component of the double membrane. [Pathogenicity of *L. monocytogenes*: McLauchlin (1997) RMM *8* 1–14.]

11.3   PATHOGENESIS: THE MECHANISM OF DISEASE DEVELOPMENT

How does a pathogen cause disease? Different pathogens act in different ways. Some produce toxins (or other substances) which disrupt specific physiological processes, while others invade particular cells or tissues (and may also form toxins).

Even when localized in the body, infections often have systemic effects. Such a generalized physiological response to infection has commonly been referred to by terms such as *septicaemia* or *sepsis*. An attempt has been made to give a precise meaning to the term 'sepsis' so that it can be usefully employed as a descriptor in clinical trials of various treatments for sepsis; for this purpose, sepsis is defined in terms of the *systemic inflammatory response syndrome* (SIRS) [American College of Chest Physicians/Society of Critical Care Medicine (1992) CCM *20* 864–874]. SIRS involves at least two of the following:

Temperature $> 38°C$ or $< 36°C$
Heart rate > 90 beats per minute
Respiration > 20 breaths per minute, or $PaCO_2 < 4.3$ kPa
WBCs $> 12\,000/mm^3$, $< 4000/mm^3$, or $> 10\%$ immature forms

SIRS can arise from various causes, including e.g. ischaemia, tissue necrosis or trauma, as well as from infection; when SIRS is a response to *infection* it is an indication of *sepsis*.

In some diseases, symptoms result from an 'over-reaction' of the body's own defence mechanisms.

A few examples of pathogenesis are given in the following sections and in Table 11.2.

11.3.1   Toxin-mediated pathogenesis

In some diseases the symptoms are caused by an *exotoxin*: a specific protein toxin which is released by the pathogen and which (in those cases where known) affects specific site(s) in the body; some of these proteins are phage-encoded (e.g. cholera toxin, diphtheria toxin), and some are plasmid-encoded (e.g. anthrax toxin).

11.3.1.1   Cholera

Cholera involves e.g. vomiting and a profuse diarrhoea which eventually becomes virtually water. The pathogen (certain strains of *Vibrio cholerae*) multiplies in the gut and forms a type of exotoxin: an *enterotoxin* (cholera toxin, CT). CT acts on mucosal cells in the small intestine, stimulating the enzyme adenylate cyclase; the resulting increased intracellular levels of cAMP (Fig. 7.12) appear to stimulate secretion of chloride ions into the gut lumen. Probably as a related effect, there is also a net loss of HCO_3^-. Outflow of water to the gut gives rise to the characteristic rice-water stools of cholera.

Infections by strains of *V. cholerae* which do not synthesize CT may give rise to (less severe) symptoms as a result of other toxins (the Zot and Ace toxins) encoded by the pathogen.

[Mechanisms of action of enterotoxins: Sears & Kaper (1996) MR 60 167–215.]

Following ingestion of *V. cholerae*, colonization of the small intestine and the development of disease depend on two main *virulence factors* (section 11.5): (i) appendages which mediate adhesion to the mucosal surface – the so-called 'toxin co-regulated pili' (TCP), and (ii) CT. The genes for CT and TCP are co-ordinately (jointly) regulated by transcriptional regulator proteins (ToxR, ToxS and ToxT) which apparently promote CT and TCP expression when receiving environmental signals within the intestinal tract.

The genes for CT (and also those for the Zot and Ace toxins) occur in the genome of a recently discovered filamentous phage, CTXΦ [Waldor & Mekalanos (1996) Science 272 1910–1914]. Genes for TCP occur in 'pathogenicity islands' (section 11.5.7). The apparent ability of signals in the host's intestinal environment to induce synthesis of TCP has recently lead to insight into the transmissibility of virulence factors under *in vivo* conditions (section 11.5.8).

See also 'cholera' in section 11.10.

11.3.1.2 Botulism

Botulism involves muscle paralysis, and death may result e.g. from mechanical (muscular) failure of the respiratory system. The pathogen (strains of *Clostridium botulinum*) forms a type of exotoxin – a *neurotoxin* – which acts at nerve–muscle junctions, inhibiting the release of acetylcholine and (hence) inhibiting nervous stimulation of the muscles. Disease can result from the ingestion of pre-formed toxin – usually in toxin-contaminated foods such as cooked meats, sausage, and improperly canned vegetables; that is, the pathogen itself need not be ingested. [Incidence, treatment etc. of botulism (review): Roblot *et al*. (1995) RMM 6 58–62.]

11.3.1.3 Tetanus

Tetanus ('lockjaw') involves uncontrollable contractions of the skeletal muscles, often leading to death by asphyxia or exhaustion. The disease develops e.g. when deep, anaerobic wounds are contaminated with the pathogen, *Clostridium tetani*. *C. tetani* produces a neurotoxin (*tetanospasmin*)

Table 11.2 Pathogenesis of some diseases caused by *Escherichia coli*[1,2]

Group	Disease
EIEC[3]	Usually food-borne. The pathogen adheres to, invades and destroys epithelial cells in the ileum/colon, causing dysentery; at least some virulence factors are plasmid-encoded (section 11.3.3.2). The pathogen can also cause a *watery* diarrhoea; at least one plasmid-encoded enterotoxin is formed.
EHEC/VTEC[4]	Usually food-borne. Minimum infective dose apparently < 100 cells. The pathogen forms one or both of two phage-encoded toxins whose activity closely resembles that of the toxin of *Shigella dysenteriae* type 1 (section 11.3.1.6); they are called shiga-like toxins I and II (=SLTI, SLTII) or VT1 and VT2, respectively. Symptoms range from mild diarrhoea to severe, bloody diarrhoea (*haemorrhagic colitis*), possibly reflecting differing responses to the toxins from different parts of the gut. Particularly in children, infection can result in *haemolytic uraemic syndrome* (HUS): typically, bloody diarrhoea followed by acute renal failure. [HUS: Taylor (1995) JINF *30* 189–192.] The toxins are apparently responsible for the haemorrhagic colitis and HUS.
ETEC[5]	Usually food- and/or water-borne. ETEC is a common cause of travellers' diarrhoea and of diarrhoea in children in the developing countries. The pathogen has plasmid-encoded fimbriae (section 2.2.14.2) – e.g. the so-called *colonization factors* CFI (CFAI) and CFII (CFAII) – with which it adheres specifically to the epithelium of the small intestine. ETEC forms heat-stable toxins (STI, STII) and heat-labile toxins (LTI, LTII). STI binds to the brush border and activates host intracellular mechanisms, resulting in stimulation of Cl^- secretion and/or inhibition of NaCl absorption – leading to watery diarrhoea. The mechanism of STII activity is not known; HCO_3^- (rather than Cl^-) seems to be secreted. LTI resembles cholera toxin (section 11.3.1.1) in its activity. LTII rarely occurs in a human host.
EPEC[6]	?Food and/or water-borne. The pathogen adheres to epithelium in the small intestine and destroys brush-border microvilli; interaction between EPEC and epithelial cells is a multi-stage process [Donnenberg, Kaper & Finlay (1997) TIM *5* 109–114]. Diarrhoea, mainly in infants, may result partly from reduction in absorption due to loss of microvilli.
EAggEC/EAEC[7]	Strains in this group cause persistent diarrhoea in children, particularly in developing countries. The pathogen forms toxins and adhesins, but the mechanism of pathogenesis is not known. Strains of EAggEC may be characterized *in vitro* by the 'aggregative' pattern of adherence to (mammalian) HEp-2 cells [see Elliott & Nataro (1995) RMM *6* 196–206]. O111:H12 strains in Brazil have been reported to have properties of EAggEC [Monteiro-Neto *et al.* (1997) FEMSML *146* 123–128].

[1] Strains of E. coli which cause diarrhoea/gastroenteritis have been classified into groups; five of the main groups are listed here. A strain within a given group may be referred to by a serological designation based on its O (and sometimes H) antigens (section 16.1.5.1).

Genes responsible for pathogenicity occur in the chromosome and/or in plasmids, phage genomes or transposons; often such genes occur in clusters (called *pathogenicity islands* or *pais*) in large plasmids or in chromosomes. [Genetics of E. coli virulence: Mühldorfer & Hacker (1994) MP 16 171–181.]

[2] Extensive details of the toxins formed by diarrhoea-causing E. coli are given by Sears & Kaper [(1996) MR 60 167–215].

[3] Enteroinvasive E. coli. Common strains are O124, O143 and O152.

[4] Enterohaemorrhagic E. coli, also called verocytotoxic E. coli (VTEC) from the toxicity of these bacteria for Vero cells (kidney cells of the African green monkey). Strains include O26 and O157:H7. [Detection and control of E. coli O157:H7 in foods: Meng et al. (1994) TIFS 5 179–185.] (See also section 14.6.1.) [Duration of faecal shedding of O157:H7 in children: Swerdlow & Griffin (1997) Lancet 349 745–746.]

[5] Enterotoxigenic E. coli. Strains include O6, O8, O63, O115 and O148.

[6] Enteropathogenic E. coli. Strains include O55, O111 (but see EAggEC in table), O114, O128.

[7] Enteroaggregative E. coli.

---

which acts on certain cells (interneurones) in the central nervous system; by inhibiting the release of a neurotransmitter (glycine) from these cells, tetanospasmin permits the simultaneous contraction of both muscles in a protagonist–antagonist pair – producing spastic (rigid) paralysis.

11.3.1.4   Botulinum toxin and tetanospasmin: similarities

Interestingly, although tetanospasmin and botulinum toxin give rise to very different clinical symptoms, these toxins have important similarities; thus, both toxins bind to nerve cells, both are taken up by the cells (internalized), and both produce their effects by acting as enzymes (zinc-endopeptidases) which cleave specific proteins in the nerve cells, thereby inhibiting the normal release of neurotransmitter substances [mechanism of action of tetanus and botulinum neurotoxins: Montecucco & Schiavo (1994) MM 13 1–8]. In that these toxins are now known to be zinc-endopeptidases, it may be possible to find specific chemical inhibitors for therapeutic use in botulism and tetanus; in an analogous case, the mammalian zinc-endopeptidase angiotensin-converting enzyme is inhibited by captopril, a drug sometimes used therapeutically for hypertension (high blood pressure).

11.3.1.5   Staphylococcal food poisoning

Staphylococcal food poisoning (Table 12.2) is due to the effects of enterotoxins produced by several species of *Staphylococcus* (mainly *S. aureus*). There are five distinct types of toxin (types A–E) which characteristically cause vomiting, and often diarrhoea, shortly after contaminated food is eaten. The mode of action of the toxins is not known. One suggestion is that vomiting may result from an effect on intestinal nerve receptors and/or from gut-wall neuropeptides causing release of histamine and leucotrienes from mast cells [detection and effects of staphylococcal enterotoxins: Tranter & Brehm (1994) RMM 5 56–64]. As these enterotoxins are now known to be superantigens (section 11.5.4.1) it seems probable that pathogenesis will be found to involve the effects of cytokine(s).

11.3.1.6 Shiga toxin and shiga-like toxins (verotoxins)

These enterotoxins, formed by some strains of *Shigella* and by EHEC (Table 11.2), respectively, have similar mechanisms. They bind to sites in the gut and are taken up by receptor-mediated endocytosis. Apparently, the toxins affect protein synthesis and e.g. inhibit the absorption of NaCl; however, the mechanism responsible for the fluid loss in dysentery remains unclear but may involve tissue damage/inflammation, the toxin serving to exacerbate the symptoms e.g. by causing vascular damage and (thus) promoting the formation of bloody stools etc. [Sears & Kaper (1996) MR *60* 167–215].

11.3.2 Pathogenesis involving other bacterial products

Aggressins are products which can promote the invasiveness of a pathogen. For example, certain bacteria, including *Streptococcus pyogenes* and most coagulase-positive staphylococci, produce *hyaluronate lyase* ('spreading factor'): an enzyme which cleaves hyaluronic acid, a component of the intercellular cement in animal tissues; in at least some cases this enzyme may assist bacterial penetration of an infected site. Another example is *streptokinase* (see section 8.5.10).

Bacterial products can also contribute to pathogenesis in a more mechanical way – as e.g. in cystic fibrosis (section 11.3.2.1).

11.3.2.1 Cystic fibrosis

Cystic fibrosis (CF) is an inheritable disease involving defective transmembrane transport of chloride. The lungs are typically congested with a thick, dehydrated mucus and may become infected with organisms such as mycobacteria, *Burkholderia cepacia* [review: Wilkinson & Pitt (1995) RMM 6 1–16] and *Pseudomonas aeruginosa*; *P. aeruginosa* can produce a viscous slime (alginate) that inhibits phagocytosis and promotes congestion – thus making for a poor prognosis. High levels of salt (NaCl) on CF airway epithelia may permit bacterial colonization by inhibiting normal antibacterial activity [Smith *et al.* (1996) Cell *85* 229–236].

Most strains of *P. aeruginosa* have genes for alginate, but strains isolated from the general environment typically do not express these genes; in CF patients, however, conditions in the lung seem to *select* for mucoid (i.e. alginate-producing) strains.

In *P. aeruginosa*, the conversion of non-mucoid strains to mucoidy apparently occurs if a specific sigma factor (section 7.5) – termed AlgU – becomes available for transcription of the alginate genes. AlgU can become constitutively (i.e. permanently) active e.g. when a mutation inactivates the *mucA* gene (the activity of AlgU being inhibited by the binding of MucA). A further negative regulator of AlgU activity, MucB, occurs in the periplasmic region. [Mucoidy of *P. aeruginosa* in cystic fibrosis: Schurr *et al.* (1996) JB *178* 4997–5004.]

11.3.3   Pathogenesis involving destruction of host cells or tissues

11.3.3.1   Typhoid

Typhoid, caused by *Salmonella typhi*, involves e.g. intestinal symptoms and systemic illness (*sepsis* – section 11.3). After ingestion, the organisms penetrate the ileal mucosa, entering the bloodstream (via the lymph system) and multiplying e.g. in the liver, gall bladder and spleen. The intestine can be secondarily invaded from the gall bladder. Intestinal inflammation can be so intense (e.g. in Peyer's patches in the ileum) that it causes local necrosis (death) of tissue (with formation of typhoid ulcers) and may cause perforation and haemorrhage.

11.3.3.2   Dysentery

Dysentery is caused by *Shigella* spp, EHEC and EIEC (Table 11.2), but *Shigella* and EIEC invade and spread among the host's mucosal cells by the same mechanism. The bacteria invade and destroy mucosal cells in the ileum/colon. The spread of bacteria within and between these host cells – an essential part of pathogenesis – requires the presence of a plasmid-encoded protein (VirG) in the bacterial outer membrane. Within the cytoplasm of an invaded intestinal cell, VirG induces the host cell to deposit actin filaments on the bacterial surface; ongoing deposition seems to propel the bacterium through the host cell's cytoplasm – thus facilitating the spread of these bacteria through the tissues. Another outer membrane protein (OmpT) – a protease which cleaves (and inactivates) VirG – appears to occur in most strains of *E. coli* but is absent in EIEC and *Shigella* spp; the *absence* of OmpT seems to be necessary for the invasive properties of these pathogens [Nakata *et al.* (1993) MM *9* 459–468]. (Interestingly, similar actin-based intra-host cell motility is shown by an unrelated pathogen, *Listeria monocytogenes* [Cossart & Kocks (1994) MM *13* 395–402].)

11.3.3.3   Oroya fever

Oroya fever, which occurs in parts of South America, involves e.g. fever and anaemia; the causal agent, *Bartonella bacilliformis*, is transmitted via the 'bites' of sandflies. *B. bacilliformis* grows in and on erythrocytes (red blood cells) and in the endothelial cells of the host; bacterial growth leads to the destruction of erythrocytes etc. and to associated symptoms.

11.3.4   Endotoxic shock (septic shock)

The *endotoxins* of Gram-negative bacteria have been regarded as macromolecular complexes containing certain elements of the cell envelope: (i) lipopolysaccharides (LPS – section 2.2.9.2), (ii) proteins, and (iii) phospholipids. The toxic component of LPS is lipid A.

The often-fatal condition *endotoxic shock* appears to involve the activity of blood-borne endotoxins (following lysis of Gram-negative bacteria). Endotoxins act on macrophages and other cells of the immune system, stimulating the secretion of certain physiologically potent agents (*cytokines*) such as interleukin-1 (IL-1) and tumour necrosis factor (TNF); these cytokines can recruit others, resulting in the classical symptoms of shock (e.g. a fall in blood pressure) and the blocking of blood vessels by white cells. Death may occur as a consequence of the progressive failure of the body's functions.

A vaccine, consisting of anti-endotoxin monoclonal antibodies, has been tried unsuccessfully [Baumgartner (1994) RMM 5 183–190].

Endotoxins have been classified as *modulins* (section 11.5.4).

11.3.5   *Helicobacter pylori*-associated intestinal disease

Helicobacter pylori (see Appendix) has been causally connected with e.g. gastritis and peptic-ulcer disease, although the details of pathogenesis are not fully understood. However, it has been found that lipopolysaccharides (section 2.2.9.2) from *H. pylori* include antigens identical to the Lewis x and Lewis y antigens which occur e.g. in human gastric mucosa, and this has suggested the possibility that anti-lipopolysaccharide antibodies may promote autoimmune inflammation by binding to the mucosal Lewis antigens. [Molecular mimicry of *H. pylori* LPS: Appelmelk *et al.* (1997) TIM 5 70–73.]

11.3.6   The Jarisch–Herxheimer reaction

This potentially fatal reaction may follow the first effective dose of an antimicrobial agent given to combat diseases caused by certain bacteria (particularly spirochaetes) or protozoa. Symptoms, which include an initial rise in temperature, are associated with a cascade of cytokines (e.g. TNF, IL-6, IL-8) that are presumed to be responsible for at least some of the pathophysiological events; the mechanism of the sudden release of cytokines is not understood. [Review: Griffin *et al.* (1994) BCID 1 65–74.]

11.4   THE BODY'S DEFENCES

11.4.1   Constitutive defences

Constitutive defences are non-specific defence mechanisms which are typically operative all the time.

To any potential pathogen, the normal healthy body presents a variety of obstacles and barriers. The skin, for example, is more than a simple physical barrier to infection. To most bacteria it is a hostile environment: water is scarce, and sites are occupied by the well-adapted skin microflora, some of which produce antibacterial fatty acids from lipids secreted by the sebaceous glands.

Mucous membranes, too, have their own defences: the secretions which bathe these tissues actively discourage the establishment of a pathogen, both by their mechanical flushing action and by their antibacterial substances; tears, for example, contain *lysozyme* (Fig. 2.7). Then there is the resident microflora with which any would-be pathogen must compete if it is to become established.

If a pathogen penetrates the outer defences it is immediately faced with the inner defences. Within the tissues and circulatory systems, certain specialized cells (*phagocytes*) engulf and destroy particles of 'foreign' matter – including many types of microorganism; these scavenging cells (which include e.g. *macrophages*) can usually prevent the establishment of a pathogen in the (nutrient-rich) tissues of the body.

If the pathogen persists it may cause *inflammation*: reddening, swelling, warmth and pain at the affected site. Inflammation can be caused by various agents – including e.g. heat and chemical irritants as well as microorganisms – and although non-specific, some of its effects can inhibit a pathogen. For example, inflammation involves an increased outflow of plasma (and of certain types of cell) from small blood vessels into the affected tissues (which therefore swell); the pathogen is therefore exposed to increased amounts of certain antimicrobial factors which occur in normal plasma. Inflammation is thus an important part of the body's generalized reaction to pathogens.

11.4.1.1 Complement

Normal plasma contains *complement*: a number of different proteins which, in the presence of certain types of molecule, undergo a 'cascade' of reactions involving sequential activation; on activation, various components of the system carry out specific physiological functions (Table 11.3).

Activation of the complement system can be triggered in various ways, the type of trigger determining which of two pathways are followed (Fig. 11.1). For example, the lipopolysaccharide (LPS) of a Gram-negative bacterium (section 2.2.9.2) can trigger the alternative pathway. Activation by a Gram-negative bacterium can have several consequences. For example, the binding of component C3b makes the cell more susceptible to phagocytosis by macrophages. Moreover, if components C5b–9 (the *membrane attack complex*) bind to the cell surface they can form a hole in the outer membrane which, in some cases, can lead to cell lysis (*immune cytolysis*). Lysis results not from the breaching of the outer membrane but from the opportunity that this

Table 11.3 Complement: some important anti-bacterial roles

Component	Role
C3a, C5a	Elicits release of histamine from mast cells/basophils, increasing the permeability of certain blood vessels to plasma/cells (part of the inflammatory response)
C3b	Immune adherence: increases the binding of the cells etc. to phagocytes (enhancement of phagocytosis)
C5a	Chemotaxis: attracts cells of the immune system to an infected site (part of the inflammatory process)
C5b6789	Lysis of Gram-negative bacteria under certain conditions (see section 11.4.1.1)

provides for lysozyme to reach peptidoglycan (the mechanically strong component of the cell envelope); thus, e.g. a potential Gram-negative pathogen (such as *Haemophilus influenzae*) on the conjunctiva would be at risk of lysis since the fluid which bathes the eye contains both lysozyme and complement.

Complement is thus an early, rapid and potent form of non-specific defence.

11.4.1.2 Interferons

Interferons (IFNs) are proteins secreted by certain types of animal cell in response to viruses, some bacteria, or antigens (see later). IFN-α (formed by blood mononuclear cells) and IFN-β (formed by fibroblasts) inhibit virus replication in those cells which bind these IFNs; such inhibition commonly affects a range of viruses, regardless of the inducing agent, and seems to involve interruption of protein synthesis. IFN-γ (formed by T lymphocytes) promotes the activation of macrophages, enhancing their antimicrobial potential, and enhances the general efficiency of the immune system by promoting cell-surface changes in certain host cells.

11.4.2 The adaptive response

11.4.2.1 Antibodies, antibody formation and vaccination

In addition to constitutive defences, the body can also respond *specifically* to a given pathogen. An individual cell can be recognized by its 'chemical fingerprint' – i.e. the molecules (such as lipopolysaccharides) which characterize the cell. Such molecules may act as *antigens*, i.e. their presence in the body may cause certain white blood cells (particular strains of B lymphocyte) to secrete proteins called *antibodies*; an antibody can combine specifically with the antigen that induced it. (In an adult, there are millions of different strains of B lymphocyte, each strain being able to recognise a different type of antigen and to secrete the matching antibody.) All antibodies are *immunoglobulins* – a class of protein present in blood and other body fluids (Fig. 11.2). The five main classes of immunoglobulin are IgA (found mainly in saliva, tears, on mucosal surfaces); IgD; IgE (found mainly bound to mast cells and basophils); IgG (quantitatively, the major immunoglobulin in plasma); IgM (a large molecule (a pentamer) contained almost exclusively within the blood vascular system).

Antibodies induced by a given pathogen can combine with that pathogen, but of what use is this? Suppose that antibodies bind to cell-surface antigens (such as lipopolysaccharides). When this happens the cell becomes more susceptible to phagocytosis because phagocytic cells (such as macrophages) have binding sites for the so-called 'Fc portion' of the antibody (Fig. 11.2) in an antigen–antibody complex. Moreover, most antigen–antibody complexes will automatically activate the complement system (section 11.4.1.1) and bind components of that system – the sequence of binding in this case following the classical pathway (Fig. 11.1); the binding of C3b to the Fc portion of a cell-bound antibody and/or to the bacterial surface causes even stronger bacterium–phagocyte binding (*immune adherence*) because phagocytes (such as macrophages) also have receptor sites for C3b. The phenomenon in which cells or other antigens are made more susceptible to phagocytosis is called *opsonization*.

Another consequence of bacterium–antibody binding is *antibody-dependent cell-mediated cytotoxicity* (ADCC): a phenomenon in which antibody-coated bacteria are *lysed* (rather than phagocytosed) by certain types of white blood cell (e.g. 'natural killer' (NK) cells, polymorphs) – apparently without involvement of complement.

Antibodies to toxins (*antitoxins*) can neutralize the activity of their corresponding toxins by binding to them; the resulting toxin–antitoxin complexes are removed by certain white blood cells which have receptors for the Fc portion of the antibody.

The 'down side' of antibodies is exemplified by certain *hypersensitivity* reactions – including *allergies* to antibiotics such as penicillin. In penicillin allergy, the *first* contact with antibiotic elicits above-normal amounts of IgE antibodies, many of which bind to receptors on mast cells and basophils – leaving their antigen-binding sites free at the cell surface. In subsequent contact, penicillin (the antigen) binds to its IgE antibodies on the surfaces of mast cells and basophils; this causes *degranulation* of the cells, i.e. release of various physiologically active substances (such as histamine) with risk of potentially fatal *anaphylactic shock* (involving e.g. vasodilation of terminal arterioles and bronchospasm).

Antibody formation. The actual mechanism of antibody formation varies according to the type of antigen. Thus, some large molecules containing

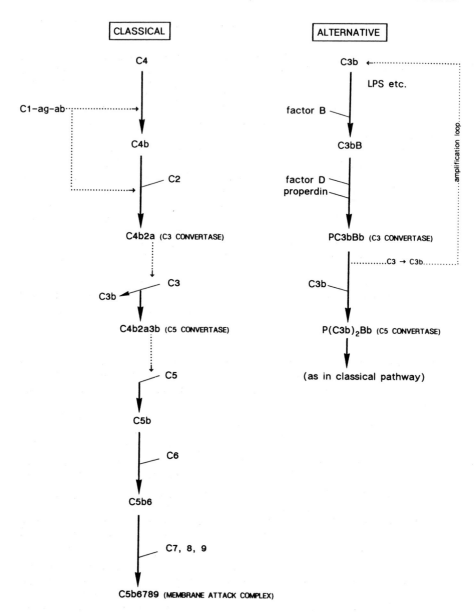

Fig. 11.1 The activation of complement (section 11.4.1.1) via the classical and alternative pathways (simplified scheme). Both pathways can lead to the same physiological effects, but they are triggered in different ways. C1, C2 etc. denote particular components of the complement system; 'a' and 'b' denote fragments of components produced by enzymic cleavage during the activation process. (For clarity, the diagram does not show *all* the cleavage products; for example, C4 is cleaved to C4a and C4b, but only the C4b fragment is considered here.) Dotted lines indicate those

cases in which a given complex acts enzymically to cleave certain components of the system.

Classical pathway. Activation of this pathway can be triggered e.g. by many types of antigen–antibody complex (section 11.4.2.1); such a complex may be formed, for example, by the binding of an antibody to a cell-surface bacterial antigen. Initially, C1 binds to the so-called Fc portion of the antibody (Fig. 11.2) in an antigen–antibody (ag–ab) complex. Thus activated, C1 cleaves C4. C4b binds C2, and C4b2 is cleaved, by activated C1, to C4b2a (a 'C3 convertase'). C3 convertase splits molecules of C3 to the fragments C3a to C3b.

If activation had been triggered by an antigen–antibody complex on a bacterial surface, C3b fragment(s) may bind to the 'Fc portion' of the antibody (Fig. 11.2) and/or to the bacterial surface. C3b promotes phagocytosis of those cells or complexes to which it binds because specific receptor sites for C3b occur at the surface of macrophages and other phagocytic cells; C3b binds strongly to these sites (*immune adherence*). C3b may also contribute to a C5 convertase (the next stage of activation).

C3b is a short-lived molecule, but low concentrations of C3b ('tickover' C3b) are normally present (even without complement activation) owing to its spontaneous formation through low-level hydrolysis of C3. This accounts for the availability of C3b for the initiation of the alternative pathway.

The cascade of reactions continues as shown. The complex C5b67 can bind to membranes and form a *membrane attack complex* (MAC) by adding C8 and C9. If MAC develops e.g. on the outer membrane of a Gram-negative bacterium (section 2.2.9.2), it forms a pore or channel through the membrane; the wall of the channel consists of six or more molecules of C9. In some cases (see section 11.4.1.1) this can cause lysis of the bacterium.

The complex C5b67 may bind at or near the site where activation of complement was initially triggered. Alternatively, this complex may bind to *another* cell in the vicinity – so that this cell may suffer lysis if the MAC is formed; such lysis is called *reactive lysis* or *bystander lysis*.

Alternative pathway. Activation of this pathway can be triggered e.g. by lipopolysaccharides (LPS; section 2.2.9.2); a low concentration of C3b is normally available for initiation (see above). Factors B and D, and properdin, are all proteins found in normal plasma. Note that the C3 and C5 convertases found in this pathway differ from those in the classical pathway. Note also the positive feedback (amplification) loop for C3b.

Fragments C3a and C5a. These fragments (not shown), called *anaphylatoxins*, can act on mast cells and basophils, causing release of mediators of inflammation such as *histamine*. Histamine affects the permeability of certain small blood vessels, allowing increased outflow of plasma and cells to e.g. infected tissues. Additionally, C5a acts as a chemotactic factor, attracting macrophages and other cells to the affected site.

Regulation of the complement system. The complement system can form some highly potent physiological agents, and since the activation process includes various stages of amplification it must be rigorously controlled. For this reason there are specific inhibitors of key stages in the activation process – e.g. molecules which inhibit C1 and which cleave/inactivate C3b.

Fig. 11.2   Antibodies: the basic structure of a (monomeric) immunoglobulin (Ig) molecule (diagrammatic). The glycoprotein Ig molecule consists of four polypeptide chains: two identical heavy chains (*thick lines*) and two identical light chains (*thin lines*) – the four chains being linked by disulphide bonds to form the Y-shaped molecule. The two heavy chains are adjacent for part of their length (the stem of the 'Y' – the so-called *Fc portion*); they diverge at the 'hinge region' (*circle*) to contribute to the two limbs of the Y-shaped molecule. A light chain completes each of the two limbs of the Y. Each limb is termed a *Fab fragment*. (Both Fab fragments can be cleaved from the molecule by the enzyme *papain*. Another enzyme, *pepsin*, cleaves (as one piece) both Fab fragments and the hinge region; this (single) piece is called the *F(ab')$_2$ portion*.)

An antigen-binding site occurs at the free end of each Fab fragment.

Among antibodies (section 11.4.2.1), IgG, IgD and IgE have the monomeric form shown in the diagram. IgA is a dimer (two monomers joined by their Fc portions). IgM is a pentamer (five monomers joined, radially, via their Fc portions).

IgG (molecular weight ~ 150 000) accounts for about 75% of the immunoglobulins in plasma, most being of the IgG1 subclass. IgG antibodies are important e.g. as opsonins and antitoxins in extravascular tissues as well as in the bloodstream; these antibodies can cross the placenta and protect the fetus and neonate. IgG antibodies generally predominate in the secondary response to antigen (section 11.4.2.1).

IgM (molecular weight ~ 970 000) forms about 5–10% of the plasma immunoglobulins. IgM antibodies are particularly good agglutinators of those antigens which display a pattern of repeated antigenic determinants (e.g. LPS); in man, they do not cross the placenta. IgM antibodies are typically the first to be formed in the primary response to antigens, and they are the main class of antibodies formed against thymus-independent antigens (such as pneumococcal polysaccharides).

Isotype refers to a class (e.g. IgG), subclass (e.g. IgG1) or smaller category of immunoglobulins. Note that a given isotype occurs in *all* normal individuals of the species.

Allotype refers to a variant form of Ig molecule; a given allotype does not occur in all normal individuals of the species – it occurs in only those individuals who have the relevant allele in their genotype (an *allele* being one of several variant forms of a given gene).

Idiotype refers to a specific Ig molecule defined in terms of the (highly individual) structure of its 'variable' region (i.e. its antigen-binding region).

repeated subunits – e.g. bacterial capsular polysaccharides – can elicit antibodies directly from specifically reactive B lymphocytes without help from T lymphocytes; this type of antigen is called a *thymus-independent antigen* (because T cells mature in the thymus gland).

Most antigens – e.g. smaller molecules, soluble proteins – elicit antibodies from B cells only with help from T cells; these are *thymus-dependent antigens*.

To elicit antibodies, a thymus-dependent antigen must first be taken up and *processed* by a so-called *antigen-presenting cell* (APC, e.g. a dendritic cell, a macrophage, or an antigen-specific B cell). At the end of the processing (which involves enzymic degradation), the antigen – linked to a *class II MHC molecule* (synthesized in the APC) – is *presented*, at the APC's surface, to a 'helper' (T_H) T cell which is specifically reactive to that particular antigen. The T_H cell recognises, and binds to, the combination of class II molecule and processed antigen at the surface of the APC, the processed antigen binding to the *T cell receptor* (TCR). The APC then secretes *interleukin 1* (IL-1) – one of a diverse family of soluble intercellular signalling molecules called *cytokines*. The T_H cell responds by secreting IL-4, IL-5 and IL-6; these interleukins cause cell division and differentiation in those specifically reactive B cells which have bound the given antigen. Some of these B cells differentiate to form *plasma cells* – which secrete antibodies matching the given antigen; initially, plasma cells typically secrete antibodies of the IgM class of immunoglobulins, but the antibodies secreted later are of another class, often IgG – though of the same *specificity*. The other specifically reactive B cells become *memory cells* – which secrete little or no antibody but which persist in the bloodstream for long periods (e.g. years); these cells are 'primed' to react to any *subsequent* exposure of the body to the given antigen. On subsequent exposure to that antigen, there is a much more vigorous production of antibodies (the *secondary response*); this occurs because there is now an *expanded clone* of B cells – specifically reactive to the given antigen – owing to cell division and memory cell formation during initial contact with the antigen.

Thymus-*in*dependent antigens such as the capsular polysaccharides of *Haemophilus influenza* type b also promote antibody production and cell division in B cells. However, memory cells are not formed in response to these antigens. Moreover, the antibody response to these antigens in adults differs greatly from that in infants and young children (up to about 2 years of age); the insufficiency of the response to *these* antigens in neonates and infants is apparently due to the immaturity of their B cells. This is important in the context of vaccination – see below.

Vaccination. The production of antibodies to particular antigens (of a pathogen) can be stimulated by introducing those antigens into the body (*vaccination*); the object of vaccination is to help the body defend itself in the event of a subsequent attack by a given pathogen. A *vaccine* – containing antigens of the given pathogen (e.g. killed cells, or cell components) – is introduced by injection or, sometimes, orally [oral vaccines (problems and approaches): Dougan (1994) Microbiology *140* 215–224]. Oral administration is *obligatory* in some cases. For example, good protection against bacterial *enteric* pathogens is obtained only when antigens act via the mucosal immune system of the *gut*; thus, the development of vaccines against ETEC (Table 11.2) is based on oral (rather than parenteral) administration [Gaastra & van der Zeijst (1996) RMM *7* 165–177]. The body responds to a vaccine by producing the corresponding antibodies. Subsequent exposure to the given pathogen stimulates vigorous production of specific antibodies owing to the *secondary response* (see above).

Vaccines for whooping cough (= pertussis; causal agents: *Bordetella pertussis*, *B. parapertussis*) are either whole-cell vaccines (WCVs) or acellular vaccines (ACVs); the latter include inactivated pertussis toxin and (usually) fimbrial or fimbria-like components. ACVs have been widely used in Japan since 1981. New ACVs are now being used in Europe. These ACVs show reduced side-effects (compared with WCVs), and are as effective as WCVs in preventing disease; however, they may give less protection against disease caused by *B. parapertussis* [Willems & Mooi (1996) RMM *7* 13–21].

Pathogens with *polysaccharide* capsules can cause various diseases in young children (section 11.5.1) because such children cannot respond adequately to polysaccharide antigens (see antibody formation, above). [Responsiveness of infants to capsular polysaccharides: Rijkers *et al.* (1996) RMM *7* 3–12.] For these young children, vaccines effective against such pathogens have been made by linking the polysaccharide to a *protein* – forming a *conjugate* vaccine which can elicit protective antibodies at e.g. 4 months of age; this is a timely schedule because infants are generally protected for the first 6 months or so by antibodies derived from the mother. A successful conjugate vaccine is available against e.g. diseases (such as meningitis, epiglottitis) caused by *Haemophilus influenzae* type b (Hib). [Ten years' experience with Hib conjugate vaccines in Finland: Eskola & Käyhty (1996) RMM *7* 231–241.]

Live vaccines against the new O139 strain of *Vibrio cholerae* (section 11.10) are being developed by deletion of the pathogen's genes for e.g. certain virulence factors [Waldor & Mekalanos (1994) JID *170* 278–283].

The vaccine against tuberculosis, BCG (bacille Calmette–Guérin), is a live, attenuated strain of *Mycobacterium bovis* given by intracutaneous injection. The protective efficacy of BCG varies greatly geographically, apparently reflecting the differing degrees of exposure to environmental mycobacteria.

Just as components of a pathogen can act as antigens, so too can many bacterial products, including toxins. The combination of a toxin with its antibody abolishes its harmful properties and assists in its elimination from the body. However, in certain toxin-mediated diseases (e.g. tetanus) death can occur before the body has had time for an effective antibody response. Vaccination against a toxin (such as tetanospasmin – section 11.3.1.3) involves the administration of a modified form of the toxin (a *toxoid*) which is no longer harmful but which has kept its specific antigenic characteristics; a toxoid thus induces the formation of anti-toxin antibodies.

Vaccination is typically given *prophylactically*, i.e. with the object of preventing disease. However, in e.g. botulism (section 11.3.1.2) an *antiserum* (in this case, a serum containing pre-formed anti-toxin antibodies) is often used in the *treatment* of disease. The use of pre-formed antibodies in this way is called *passive immunization*.

[Challenges in vaccinology: Poolman (1996) RMM 7 73–81.]

Monoclonal antibodies. If a B lymphocyte is 'fused' with a tumour cell, the resulting *hybridoma* can replicate as a tumour cell and secrete antibodies of the particular type specified by that B cell; we can thus obtain a large population of *identical* (monoclonal) antibodies. Monoclonal antibodies (mAbs) have many uses, including e.g. the detection of specific microorganisms by using 'labelled' monoclonal antibodies to detect their unique antigen(s).

11.4.2.2 Cell-mediated immunity

Cell-mediated immunity (CMI) is a form of defence involving cells rather than antibodies. It includes anti-cancer mechanisms, but there are also antimicrobial aspects; for example, *cytotoxic T cells* can recognize, bind to, and lyse virus-infected host cells (completely independently of antibodies).

The presence of specific antigen can lead to the *general* enhancement of phagocytosis by a mechanism which involves the release and uptake of effector molecules (*cytokines*) among cells of the immune system. Thus, when a T cell binds antigen on an APC (section 11.4.2.1), the T cell releases various cytokines; these include not only interleukins 4, 5 and 6 (which stimulate B cells) but also γ-interferon and interleukin 2 – which activate any responsive macrophage in the vicinity. On activation, a macrophage becomes more aggressive towards engulfed bacteria, exposing them to increased amounts of antibacterial substances (e.g. hydrogen peroxide); it can kill certain bacteria (e.g. some species of *Mycobacterium*) which are normally not killed in non-activated phagocytes. (*Legionella* spp can inhibit phagocyte activation [Dowling *et al.* (1992) MR 56 32–60].)

In some diseases however, CMI appears actually to contribute to the symptoms. In tuberculosis, for example, at least some of the tissue damage which occurs at the *tubercles* (localized sites of established infection) appears to be due to the leucocytes (white blood cells) which concentrate at these sites.

11.5 THE PATHOGEN: VIRULENCE FACTORS

Bacterial pathogenicity is attracting fresh interest. The resurgence of old problems (e.g. tuberculosis) – and increasing resistance to antibiotics – may be reason enough, but the revived interest may also reflect new approaches (e.g. genetic engineering) which are now available for seeking solutions to some of these problems. A detailed discussion of pathogenicity has been given by Smith [(1995) BR 70 277–316].

Overtly aggressive products such as toxins and aggressins are clearly *virulence factors*. However, so too are those products and strategies which help a pathogen to become established in the host and to evade the host's defences. Some of these factors are considered below. Certain virulence factors can be induced in the pathogen via signal transduction pathways (section 7.8.6). (See also section 11.5.8.)

11.5.1   Capsular camouflage

Some pathogens have anti-phagocytic capsules. For example, in certain strains of *Streptococcus pyogenes* the capsule is made of hyaluronic acid, a component of animal tissues; this 'camouflage' seems to give some protection against phagocytosis. The streptococcal M protein (section 2.2.13) is also an inhibitor of phagocytosis. The poly-D-glutamic acid capsule of *Bacillus anthracis* (causal agent of anthrax) appears to have a similar role.

Polysaccharide capsules occur in e.g. *Haemophilus influenzae* type b (Hib), *Neisseria meningitidis* and *Streptococcus pneumoniae*. Such capsules generally do not prevent phagocytosis in those adults, and older children, who have a normal immune system; defence against these pathogens requires (i) adequate amounts of antibodies to the capsular polysaccharide, (ii) the involvement of the complement system (section 11.4.1.1) for enhanced opsonization, and (iii) the activity of phagocytic cells. However, infants, and children up to about 2 years of age, give a poor response to polysaccharide antigens – and in this age group these pathogens may cause e.g. meningitis (Hib, *N. meningitidis*) and respiratory tract infections (*S. pneumoniae*); it is for this reason that the *conjugate* vaccines (section 11.4.2.1 (vaccination)) were developed.

11.5.2   Agents which kill cells of the immune system

Substances which kill phagocytes (*leucocidins*) are secreted e.g. by certain pathogenic staphylococci and streptococci; the staphylococcal *Panton–Valentine leucocidin*, for example, specifically lyses macrophages and polymorphonuclear leucocytes (PMNs).

Some Gram-negative pathogens secrete so-called *RTX toxins* which act as virulence factors primarily by affecting cells of the immune system. Thus, e.g. at high concentrations, the HlyA haemolysin of *E. coli* can lyse various cells, particularly PMNs, by creating lesions in the target cell's membrane; an RTX toxin of *Pasteurella haemolytica* (a causal agent of bovine respiratory disease) can lyse bovine PMNs. At lower concentrations, RTX toxins can e.g. inhibit antibody formation and contribute to pathogenicity by provoking adverse inflammatory effects. [RTX toxins (review): Coote (1996) RMM 7 53–62.]

Following phagocytosis, the intestinal pathogen *Shigella flexneri* is apparently able to kill macrophages by a mechanism involving the plasmid-encoded bacterial IpaB protein. IpaB binds to, and activates, the macrophage's IL-1β-converting enzyme (ICE), the activated form of ICE (a cysteine protease) then initiating *apoptosis* (programmed cell death) in the macrophage [Chen *et al.* (1996) EMBO Journal 15 3853–3860].

11.5.3 Antigenic variation

Certain pathogens can change their cell-surface chemistry – and, hence, their antigens; this may help, even temporarily, to avoid the effects of specific antibodies. For example, in relapsing fever (caused by a species of *Borrelia*) there are several cycles of fever and remission, and bacteria isolated from the blood during each period of fever are found to have different surface antigens. In *B. hermsii*, antigens involved in antigenic variation are encoded by a linear (i.e. not ccc) multicopy plasmid; antigen switching appears to involve site-specific recombination (section 8.2.2) between different individual plasmids in a given cell. SSR is also involved in the mechanism for phase variation in *Salmonella* (Fig. 8.3c).

11.5.4 Modulins

Modulins are virulence factors which can damage the host by inducing (inappropriate) synthesis of various *cytokines* [Henderson, Poole & Wilson (1996) MR 60 316–341]. Cytokines are protein signalling molecules which normally co-ordinate the activities of various components of the immune system (e.g. in the inflammatory response); they are thus necessary for effective defence against pathogenic microorganisms. Cytokines have a range of functions (Table 11.4) and are potent physiological agents; hence, inappropriate synthesis of cytokines may produce pathological effects – as e.g. in endotoxic shock (section 11.3.4; see also section 11.3.6).

Modulins include endotoxins (section 11.3.4); fragments of peptidoglycan (which induce e.g. IL-1β and IL-6 in monocytes); superantigens (section 11.5.4.1); and shiga-like toxins (which induce IL-1α, IL-6 and TNF-α in macrophages). Cytokines IL-1, IL-4, IL-6, TNF-α and IFN-α are induced in mononuclear cells by *protein A*, a cell-surface protein found in most strains of *Staphylococcus aureus*; protein A also acts as an antiphagocytic virulence factor by binding the Fc portion of IgG antibodies.

11.5.4.1 Superantigens

A superantigen is a protein, secreted by certain pathogens, which can bind *non-specifically* to the receptor on many strains of T lymphocyte (T cell) and, simultaneously, to an MHC class II molecule on an APC (see section 11.4.2.1).

Table 11.4 Examples of functions of some cytokines[1]

Cytokine	Secreted by (e.g.)	Function(s)
IL-1 (interleukin 1)	Macrophages, B cells dendritic cells, fibroblasts	Activates T cells; induces leucocyte-binding molecules in endothelial cells (i.e. cells lining blood vessels); induces IL-6; induces a rise in temperature (fever); induces (further) IL-1
IL-4	T cells (after activation by antigen-presenting cells)	Promotes differentiation of B cells to plasma cells; inhibits synthesis of IL-1 and TNF by macrophages
IL-6	T cells and other cells	Promotes differentiation of B cells; induces fever
IL-8	Macrophages, fibroblasts, endothelial cells, monocytes	Chemotactic factor: attracts immune cells to the site
TNF (tumour necrosis factor)	Macrophages, lymphocytes	Activates lymphocytes; induces IL-6; induces fever; induces leucocyte-binding molecules in endothelial cells; induces (further) TNF
INF-γ (γ-interferon)	T cells (after activation by APCs)	Activates macrophages

[1] Note that these functions – listed in isolation, and out of context – are, in reality, integrated with the activities of a highly complex and dynamic network consisting of a large number of agents. Some functions can be carried out by more than one type of cytokine, and some cytokines have a range of different functions – which can vary according to the type of cell to which they bind. Moreover, a given cytokine may be synthesized by a number of different types of cell, under differing conditions. A further complication is that cytokines may work together (synergism) or oppose one another (antagonism).

Such binding mimics the effect of a specific antigen, leading to activation of the T cell and release of cytokines. The secretion of many molecules of superantigen by the pathogen causes *polyclonal* activation of T cells, i.e. the activation of many different strains of T cell (as though a number of different antigens had been present). Such a massive release of cytokines can produce e.g. symptoms of shock; moreover, polyclonal T cell activation can cause suppression of the immune system.

Superantigens include the enterotoxins (A–E) of *Staphylococcus aureus*, the toxic-shock-syndrome toxin-1 (TSST-1) of *S. aureus*, and streptococcal superantigen A (SSA) of *Streptococcus pyogenes*. When injected into mice, *S. aureus* enterotoxin B elicited the cytokines TNF, IL-1, IL-6 and IFN-γ.

[Superantigens (review): Fleischer (1995) RMM **6** 49–57.]

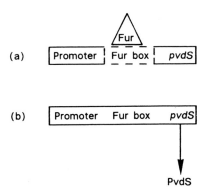

Fig. 11.3 Simplified scheme for the iron-regulated synthesis of pyoverdin in *Pseudomonas aeruginosa* (section 11.5.5). (a) In the presence of adequate concentrations of iron, the transcription repressor, the Fur (ferric uptake regulation) protein (product of the *fur* gene), binds to a highly conserved region of DNA (the 'Fur box'), inhibiting transcription of gene *pvdS*. (b) When the iron concentration is insufficient, Fur no longer binds to the Fur box – thus permitting transcription of *pvdS*. The product of *pvdS* (PvdS) is a sigma factor (see sections 7.5 and 7.8.9) which is needed for the transcription of other genes involved in the biosynthesis of pyoverdin.

11.5.5 Aerobactin and pyoverdin – iron chelators

Pathogens need iron, e.g. for the synthesis of certain enzymes and/or cytochromes – but iron can be scarce in the host organism; for example, in the mammalian bloodstream, the host's iron-chelating glycoprotein, *transferrin*, binds ferric iron and transports it into cells. To cope with this problem, many pathogens encode their own iron-uptake system: a *siderophore* (a low-molecular-weight ferric iron chelator, which the pathogen secretes) and associated cell-surface receptors at which the iron–siderophore complex is bound. Thus, e.g. extra-intestinal strains of pathogenic *E. coli* (such as those causing neonatal meningitis) often form *aerobactin*: a hydroxamate siderophore which binds iron and then binds to a receptor site on the outer membrane; aerobactin seems able to compete successfully with transferrin. [Acquisition of transferrin-bound iron by pathogens (review): Cornelissen & Sparling (1994) MM *14* 843–850.]

In *Pseudomonas aeruginosa*, iron shortage triggers the synthesis of a yellow-green fluorescent siderophore, *pyoverdin*, which stimulates the pathogen's growth even in the presence of transferrin. A mechanism for the iron-regulated induction of pyoverdin (Fig. 11.3) has been proposed by Leoni *et al.* [(1996) JB *178* 2299–2313].

Salmonella typhimurium encodes a catecholamide siderophore – *enterochelin* – whose expression, like that of pyoverdin in *P. aeruginosa*, involves regulation by the Fur protein. In *S. typhimurium* the Fur protein is also a regulator of the acid tolerance response (section 7.8.2.6).

11.5.6   Adhesins

The ability of a pathogen to adhere to the host's tissues is an important virulence factor which was mentioned in section 11.2.1. Bacterial cell-surface structures which promote adhesion are called *adhesins*.

In many cases adhesins are fimbriae (section 2.2.14.2) and are often encoded by plasmids. Among strains of ETEC (Table 11.2) there are more than 10 different types of fimbrial adhesin – but there are also several types of non-fimbrial adhesin [Gaastra & van der Zeijst (1996) RMM *7* 165–177].

11.5.7   Pathogenicity islands

A pathogenicity island is a cluster or 'cassette' of several to many genes which encode a particular set of virulence characteristics; such islands occur in the chromosome and/or in extrachromosomal elements such as plasmids.

One example is the 35 kb region in EPEC (Table 11.2) encoding the means by which this pathogen adheres to intestinal cells and destroys their microvilli; this island, which is called LEE (locus of enterocyte effacement), is inserted in the *E. coli* chromosome downstream of the *selC* gene. In a different (uropathogenic) strain of *E. coli* (so-called UPEC), the 70 kb PAI-1 island encodes e.g. a haemolysin; interestingly, the site of insertion of PAI-1 in the chromosome of UPEC is exactly the same as that of LEE in EPEC. Thus, a non-pathogenic strain of *E. coli* receiving the LEE or PAI-1 cassette may develop as either EPEC or UPEC, respectively. However, in general, the acquisition of a pathogenicity island does not necessarily guarantee that the recipient cell will become a pathogen because e.g. the imported gene(s) must function in the context of the chromosome as a whole.

In at least some cases, a pathogenicity island has a GC% (section 16.2.1) which differs from that of the rest of the chromosome; this has suggested that pathogenicity islands have been acquired by 'horizontal gene transfer' (i.e. by processes such as transduction and conjugation).

[Review: Groisman & Ochman (1996) Cell *87* 791–794.]

11.5.8   Influence of the host environment on the transfer of virulence factors between infecting bacteria

Recent work indicates that signals within the mammalian host environment can promote the transfer of toxigenicity between strains of infecting bacteria. Thus, under *in vitro* (laboratory) conditions, phage CTXΦ, which encodes cholera toxin (section 11.3.1.1), infects certain CT-negative strains of *V. cholerae* at only very low rates. However, within the mouse intestine, lysogenic conversion of these strains by CTXΦ (which could be derived e.g. from a co-infecting, CT-positive strain) occurs with a nearly 1 million-fold greater efficiency; this increased rate of lysogenic conversion appears to be due to the induction of TCP (section 11.3.1.1) in response to signals in the intestinal environment – TCP being probable receptor sites for phage CTXΦ. This is relevant to our ideas of the general transmissibility of virulence factors: a given virulence factor which appears to be non-transmissible in the laboratory may well be readily transmissible in the appropriate *in vivo* environment [see Mel & Mekalanos (1996) Cell *87* 795–798].

11.6   THE TRANSMISSION OF DISEASE

In some diseases the pathogen does not normally spread from one individual to another. Examples include e.g. tetanus and gas gangrene; in these diseases the pathogen is usually a wound contaminant which typically originates in the soil rather than in another, infected individual.

Other diseases spread from one person (or animal) to another, either directly or indirectly. In relatively few (bacterial) diseases does transmission require direct contact between an infected individual and a healthy one, and these diseases generally involve pathogens which cannot survive for long periods of time outside the body; an example is the venereal disease syphilis, caused by *Treponema pallidum*.

Most diseases spread indirectly from person to person, usually in a way related to the normal route of infection (section 11.2). For example, pathogens which infect via the intestine are commonly transmitted in contaminated food or water; gastroenteritis, dysentery, typhoid and cholera are usually spread in this way. Such transmission generally involves some connection between the food or water and the faeces of a patient who is suffering from the disease; contamination may occur, for example, when the hands of a food-handler carry traces of faecal matter, when a housefly lands alternately on faeces and food, or when sewage has leaked into a source of drinking water. Often the pathogen can be traced back – via the food or water – to individual(s) suffering from the disease. Sometimes, however, the source of a pathogen is a person who is *not* suffering from the disease but who is nevertheless playing host to the pathogen and acting as a *reservoir* of infection; such an apparently healthy individual who is a source of pathogenic organisms is called a *carrier*. One notorious carrier, a cook by the name of Mary Mallon, transmitted typhoid fever to nearly thirty people before she was traced, and the name 'typhoid Mary' is sometimes used to refer to an actual or suspected carrier in outbreaks of typhoid and other diseases.

Pathogens which infect via the respiratory tract are often transmitted by so-called *droplet infection*. When a person coughs or sneezes, or even speaks loudly, minute droplets of the mucosal secretions are expelled from the mouth; these droplets can contain pathogens if such pathogens are present on the respiratory surfaces. Since the smallest droplets can remain suspended in air for some time, they can be inhaled by other individuals and can therefore act as vehicles for the transmission of pathogens. Diphtheria, whooping cough and pulmonary tuberculosis are examples of diseases which can be transmitted in this way.

A few bacterial pathogens are transmitted from one person to another by a third organism called a *vector*. For example, bubonic plague (caused by *Yersinia pestis*) is transmitted by fleas, while epidemic typhus (caused by *Rickettsia prowazekii*) is typically transmitted by lice; tularaemia (caused by *Francisella tularensis*) can be transmitted e.g. by biting flies or by ticks. Oroya fever (section 11.3.3.3) is another example.

11.7   LABORATORY DETECTION AND CHARACTERIZATION OF PATHOGENS

Rapid detection/identification/typing of a pathogen is particularly important in certain diseases, and in some cases this is possible by nucleic-acid-based methods: section 16.1.6.

Uncultivable/uncultured bacteria suspected of involvement in disease may be investigated by a PCR-amplification procedure: see section 10.8.

Traditional methods of detection/characterization are still widely used; some of these are outlined below.

11.7.1   Culture, and the detection of toxin

Attempts are commonly made (where possible) to culture the pathogen from a sample of e.g. sputum, pus, urine or faeces; before culture, the urine may be centrifuged (and the sediment used as an inoculum), while a faecal sample may be dispersed in Ringer's solution (and an aliquot used as an inoculum). The choice of medium, and the conditions of incubation, will depend on the inoculum and on the nature of the suspected pathogen; where appropriate, an enrichment medium (section 14.2.1) is used.

If an exotoxin is suspected, attempts may be made (by various methods) to detect the toxin in appropriate samples (e.g. contaminated food). [Detection of staphylococcal enterotoxins: Tranter & Brehm (1994) RMM 5 56–64.]

11.7.1.1   Blood culture

Blood culture is used e.g. to detect blood-borne microorganisms, mainly bacteria, in diseases such as typhoid. Essentially, 5–10 ml of the patient's blood (taken aseptically – section 14.3) is added to 50–100 ml of a medium such as trypticase soy broth; the medium usually contains a blood anticoagulant, and may also contain e.g. particular enzymes to inactivate antibiotic(s) carried over in the blood. The inoculated medium is incubated, and is examined, daily, by subculture to an appropriate solid medium.

Nucleic-acid-based techniques (such as PCR – section 8.5.4) can be used to identify organisms isolated by blood culture, or to detect/identify organisms in clinical samples of blood or CSF. These methods involve the use of e.g. species-specific PCR primers and/or labelled probes; as well as giving a rapid result, such methods are also able to detect non-cultivable/antibiotic-damaged bacteria. However, there are several problems – e.g. (i) traditional culture is needed for antibiotic-susceptibility testing (section 15.4.11.1); (ii) heparin (a blood anticoagulant) and haemin inhibit PCR; (iii) unlike traditional methods, existing nucleic-acid-based techniques do not indicate the intensity of the bacteraemia (i.e. concentration of bacteria).

[Re-evaluation of blood culture: Jones (1995) RMM 6 109–118.]

11.7.2   Immunofluorescence microscopy

This technique permits detection of a specific type of organism in a smear (section 14.9) containing a mixture of organisms. Antibodies (section 11.4.2.1), specific to the given organism, are first linked chemically to a fluorescent dye (e.g. fluorescein); the *conjugate* (i.e. suspension of dye-linked antibodies) is added to the smear, unbound conjugate is rinsed off, and the slide is examined by *epifluorescence microscopy*. In epifluorescence microscopy, ultraviolet radiation is beamed onto the specimen from above; visible light from any bound, fluorescent antibodies is seen, in the usual way, via the objective lens of the microscope.

11.7.3   Complement-fixation tests (CFTs)

A CFT may be used to detect specific antibodies in a sample of serum; the presence of antibodies specific to a given pathogen may indicate past or present infection with that pathogen.

Complement (section 11.4.1.1) is bound ('fixed') by most types of anti-gen–antibody complex (section 11.4.2.1); hence, if fixation occurs when serum is added to a mixture of specific antigen and complement then this typically indicates that the serum contains antibody which matches the antigen. Fixation in a given test mixture naturally lowers the amount of free complement remaining in the mixture; such a decrease (and its extent) – which indicates the presence (and amount) of specific antibody in the serum sample – is monitored in a CFT.

The sample of serum is first heated (56°C/30 minutes) to destroy the patient's own complement, and is then serially diluted. A known, standard amount of complement is added to each dilution. To each dilution is then added a standard amount of specific antigen, and the whole is incubated for 18 hours at 4°C. The presence/absence/quantity of specific antibody in a

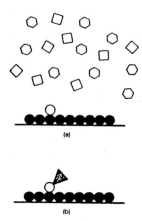

Fig. 11.4 The *principle* of enzyme-linked immunosorbent assay (ELISA). Here, an attempt is being made to detect a low concentration of a particular antibody (open circle) among high concentrations of other antibodies (open squares, hexagons) in a sample of serum.

(a) Specific antigen (solid circles) is immobilized e.g. on the inner surface of a plastic test-tube, and is exposed to the serum. The single molecule of specific antibody binds to its corresponding antigen. Other antibodies, which do not bind, are subsequently removed by washing.

(b) The single, bound antibody is then detected by means of anti-immunoglobulin antibodies – i.e. antibodies whose corresponding antigens are themselves antibodies, and which thus bind specifically to other antibodies; before use, each anti-immunoglobulin antibody is chemically linked to a particular enzyme. An enzyme–anti-immunoglobulin antibody complex (stippled triangle) has bound to the solitary bound antibody, so that the latter is now labelled with an enzyme; by detecting the enzyme one can now detect the antibody. The enzyme is detected by adding a suitable substrate which, in the presence of the enzyme, yields a measurable product which is *amplified* by continued enzymic action.

given dilution of serum is then determined from the amount of free (unfixed) complement remaining in that dilution. The residual complement is assayed by adding (to each dilution) a *haemolytic system*: a suspension of 'sensitized' erythrocytes (red blood cells) which lyse in the presence of complement. If, in a given dilution of the serum, *no* erythrocytes lyse (the cells sedimenting to form a discrete 'button') then that dilution contains no residual complement – maximum antigen–antibody binding having occured; that is, in that dilution of serum, the concentration of antibodies was sufficient, when complexed with antigen, to fix all the complement. If, for example, all the erythrocytes are lysed in the highest concentration of serum then the test is *negative*, i.e. the serum contained maximum residual complement and (therefore) no detectable specific antibodies.

Adequate controls must be included in any CFT.

The classical Wassermann test for syphilis is one example of a CFT.

11.7.4   Enzyme-linked immunosorbent assay (ELISA)

ELISA is a higly sensitive test for specific antigens or antibodies. The *principle* is shown in Fig. 11.4.

11.7.5   *In situ* hybridization (ISH)

By using labelled *probes* (section 8.5.3) we can often detect pathogens *in situ*, i.e. actually within the infected tissue. The probes themselves contain species- or strain-specific sequences of nucleotides which hybridize with the complementary sequences in DNA or RNA targets in the pathogen.

The *direct* detection of pathogens by ISH has two main advantages. First, ISH can detect a pathogen before the pathogen could be cultured. This is particularly important when culture time is long (e.g. weeks for *Mycobacterium tuberculosis*) and is also important for pathogens which cannot be cultured (e.g. *M. leprae*). Species of *Mycobacterium* have been detected by ISH in both fresh-frozen and paraffin-wax sections of infected tissue.

Second, ISH can be useful for diagnosis in immunodeficient (e.g. AIDS) patients whose poor immune response may preclude reliable serological tests for antibodies to specific pathogens. Moreover, in general, antibodies may not be detectable at all stages of an infection.

ISH can distinguish virulent from non-virulent strains of a pathogen by using probes which detect genes for specific virulence factors.

Various aspects of the methodology of ISH are discussed in a special issue of *Histochemical Journal* [(1995) *27* 1–99].

11.7.6   Characterization of the pathogen

If obtained in pure culture, the pathogen may be examined by the procedures and tests outlined in Chapter 16; this typically permits identification to the level of species (or serotype, section 16.1.5.1). Commonly, a pathogen's pattern of resistance to a range of antibiotics (its *antibiogram*) is determined – often by a disc diffusion test (section 15.4.11.1).

11.8   PREVENTION AND CONTROL OF TRANSMISSIBLE DISEASES

Once we know how a disease spreads, it's often possible to devise methods for preventing or limiting its spread. Clearly, any disease which spreads only by direct physical contact can be prevented simply by avoiding such contact with infected persons. For other transmissible diseases, prevention or control may involve blocking the route of infection from one person to another. Thus, the spread of diseases such as typhoid and cholera can be halted by measures such as: (i) improvement in personal hygiene – e.g. washing hands after visiting the lavatory; (ii) protection of food etc. from flies and other insects likely to carry pathogens, and reduction in the numbers of such insects by the use of insecticides; (iii) protection of drinking water from contamination by sewage, and the effective treatment of communal water supplies with antimicrobial agents such as chlorine; (iv) disinfection of small quantities of untreated water before consumption – e.g. by boiling or by treatment with a disinfecting agent such as halazone; (v) detection, isolation and treatment of carrier(s) (section 11.6), if involved.

Diseases which spread by droplet infection (section 11.6) are generally more difficult to deal with. The physical exclusion of droplets (e.g. with face masks) is usually impracticable, and one of the main control measures in diseases such as diphtheria involves the protection of susceptible individuals by vaccination (section 11.4.2.1). In such diseases there may be some advantage in effectively isolating (*quarantining*) sick individuals until they are no longer able to act as sources of the pathogen.

Diseases spread by vectors can be controlled by eliminating the vector or by reducing its numbers. Such control is applicable e.g. in typhus and bubonic plague.

11.9   A NOTE ON THE TREATMENT OF BACTERIAL DISEASES

The availability of a wide range of antibiotics (section 15.4) means that most bacterial diseases can be treated by *chemotherapy*, i.e., therapy involving the use of *chem*ical agents (such as antibiotics). ('Keemotherapy' presumably involves the use of *keem*ical agents!) Considerations involved in the use of a given antibiotic include: (i) effectiveness against the pathogen; (ii) occurrence of resistant strains of the pathogen; (iii) antagonism with other antibiotics (section 15.4.10); (iv) possible side-effects (including allergic reactions); (v) (for pregnant women): possible teratogenicity, i.e. the risk of causing problems in a fetus.

In some diseases (e.g. botulism, diphtheria) treatment commonly includes administration of *antitoxin*, i.e. a preparation containing antibodies to the relevant toxin.

In gas gangrene, treatment includes surgical removal of dead/infected tissue, and may include the use of hyperbaric oxygen.

11.10   SOME BACTERIAL DISEASES

The following brief, note-form descriptions give some idea of the range of types of infection which can involve bacteria. A few of the diseases (e.g. conjunctivitis) can be caused by agents other than bacteria, but most of the diseases listed are caused only by bacteria – and often only by specific pathogens.

The name of the disease is followed by (i) the name(s) of the causal agent(s), (ii) the characteristic symptoms, and (iii) the route(s) of infection, where known.

Anthrax *Bacillus anthracis* (virulent strains). Localized pustule on skin (anthrax boil) or, rarely, lung infection (woolsorters' disease) or intestinal anthrax; sepsis can occur in untreated cases. Infection via skin wounds, inhalation, ingestion.

Bacterial vaginosis Disease associated with increased numbers of e.g. species of *Bacteroides* and *Gardnerella*. Malodorous discharge. Vagina typically less acidic and more 'reducing' than is normal. 'Clue cells' (vaginal epithelial cells coated with small Gram-variable rods) typical in smears.

Botulism *Clostridium botulinum* (virulent strains). Generalized weakness, defective vision, respiratory paralysis (section 11.3.1.2). In adults: commonly, ingestion of pre-formed toxin; less commonly(?): intestinal infection with toxigenic *C. botulinum* [Sonnabend *et al.* (1987) Lancet *i* 357–360]. Rarely, wound contamination with *C. botulinum*. In infants: a *toxicoinfection* in which *C. botulinum* produces toxin in the intestinal tract.

Brucellosis *Brucella* spp. Headache, malaise, intermittent fever which may persist for years. Infection via mouth (e.g. unpasteurized milk), conjunctivae or wounds. In animals, the reproductive system is generally affected, often leading to abortion.

Cellulitis Usually, strains of *Staphylococcus* or *Streptococcus*. Diffuse, spreading inflammation typically affecting subcutaneous tissues. Infection via wounds etc.

Cholera *Vibrio cholerae* (virulent strains). Intestinal infection. Copious watery stools (section 11.3.1.1) and dehydration. Infection via oral–faecal route, usually via contaminated water. Until recently, cholera was caused only by the *V. cholerae* O1 serogroup (distinguished on the basis of O antigens) – which includes the cholerae (= 'classical') and El Tor biotypes; none of the other 137 O-serogroups of *V. cholerae* (the 'non-O1' serogroups) cause cholera. Recently, a new, *cholera-causing* non-O1 serogroup (designated O139 Bengal) has emerged in South India; immunity to O1 *V. cholerae* does not protect against O139, but anti-O139 vaccines are being developed. [*V. cholerae* O139 Bengal: Nair *et al.* (1996) RMM **7** 43–51.]

Conjunctivitis Various, e.g. *Staphylococcus aureus*, other staphylococci, *Haemophilus influenzae*, *Streptococcus pneumoniae*. Inflammation of the conjunctivae (mucous membranes of the eye). [Ocular bacteriology, including conjunctivitis (review): Willcox & Stapleton (1996) RMM **7** 123–131.]

Cystitis Various, e.g. *Escherichia coli* (strains which have P fimbriae adhere to uroepithelium), *Proteus* spp. Inflammation of the urinary bladder. Infection: downwards from the kidneys or (more commonly) upwards from the urethra, with *E. coli* and *Proteus* spp being common causal agents in the latter type.

Diphtheria *Corynebacterium diphtheriae* (virulent strains). Fever. Membrane forms usually in the tonsillar and/or adjacent regions, inhibiting breathing and/or swallowing. Toxin may cause e.g. myocarditis. A carrier state (section 11.6) is recognized. Droplet infection (section 11.6) or ingestion of contaminated food, milk etc. [Microbiology and epidemiology of diphtheria: Efstratiou & George (1996) RMM *7* 31–42.]

Dysentery (bacillary)   *Shigella* spp, EHEC, EIEC. Intestinal pain; watery stools, usually with fever and malaise, commonly followed (in *Shigella* and EHEC disease) by bloody/mucoid stools. Dehydration. Infection via oral–faecal route, e.g. contaminated food, water.

Erysipelas   *Streptococcus pyogenes*. A form of cellulitis, often affecting the face; fever, prostration, sepsis may occur. Infection via wounds etc.

Food poisoning   Various, e.g. *Bacillus cereus, Campylobacter jejuni, Clostridium perfringens, Escherichia coli, Salmonella typhimurium, Staphylococcus aureus, Vibrio parahaemolyticus, Yersinia enterocolitica*. Acute gastroenteritis. Abdominal discomfort/pain. Usually diarrhoea (but may be little/none in staphylococcal food poisoning and in one type caused by *B. cereus*). Nausea and vomiting are common, but not invariably present. Oral–faecal route; ingestion of pathogen and/or toxin in contaminated food. (See section 12.3.)

Gas gangrene   Typically, *Clostridium perfringens* (type A), *C. septicum* and/or *C. novyi*. Spreading necrosis (death) which may start in any of various tissues. Tissues swell and contain pockets of gas (formed by bacterial metabolism). Infection via wounds, boils etc.

Gonorrhoea   *Neisseria gonorrhoeae* ('gonococcus'). Discharge from the genitourinary tract; other symptoms may occur. Sexual contact.

Legionnaires' disease   *Legionella pneumophila*. Malaise, myalgia, fever, pneumonia. Infection presumed to occur via contaminated aerosols.

Leprosy   *Mycobacterium leprae*. According to the type of leprosy, lesions in any of a variety of tissues (commonly including the skin) and/or destruction of nerves in the peripheral nervous system. [Various aspects of leprosy (multi-author papers): TRSTMH (1993) *87* 499–517. Epidemiology (recent insight): van Beers, de Wit & Klatser (1996) FEMSML *136* 221–230.]

Leptospirosis   *Leptospira interrogans* (a spirochaete). Mild form: fever, headache, myalgia; typically impaired kidney function. Severe form (*Weil's disease, infectious jaundice*): above symptoms with e.g. vomiting, diarrhoea, liver enlargement, jaundice, haemorrhages, meningitis. Infection via wounds or mucous membranes; the pathogen occurs e.g. in water contaminated with urine from infected animals.

Listeriosis   *Listeria monocytogenes* (virulent strains). Meningitis (particularly in the immunocompromised); abortion/stillbirth. Infection via contaminated foods. In animals, abortion/stillbirth (infection e.g. via incorrectly made silage – section 13.1.1.1). [Listeriosis: McLauchlin (1997) RMM *8* 1–14.]

Lyme disease   *Borrelia burgdorferi*. Typically, recurrent arthritis with/without neurological and/or cardiac symptoms. Transmitted via the bites of infected ixodid ticks.

Meningitis   Various, e.g. *Neisseria meningitidis, Haemophilus influenzae* type b (common cause in infants and children), *Streptococcus pneumoniae*. Inflammation of the meninges (membranes which cover the brain and spinal cord). Droplet infection, head wounds etc.

Plague (bubonic)   *Yersinia pestis*. Fever, haemorrhages, *buboes* (inflamed, swollen, necrotic lymph nodes); sepsis and e.g. meningitis. Infection via the 'bite' of a flea. *Pneumonic* plague, involving e.g. severe pneumonia, is spread by droplet infection.

Pneumonia   Various, e.g. *Streptococcus pneumoniae*, *Haemophilus influenzae* type b, *Staphylococcus aureus*, *Klebsiella pneumoniae*. Fever, difficulty in breathing, chest pain, cough. Droplet infection etc.

Pseudomembranous colitis   A potentially fatal condition involving e.g. watery diarrhoea, fever and pseudomembranous patches of inflammatory exudate, mainly on the colonic mucosa; it is commonly preceded by antibiotic therapy (particularly with e.g. clindamycin), and the causal agent is usually or always *Clostridium difficile*. *C. difficile* forms toxins A (enterotoxin) and B (cytotoxin); possibly, toxin A stimulates secretion after gaining access to toxin-B-damaged cells [Sears & Kaper (1996) MR *60* 167–215]. [Laboratory diagnosis of *C. difficile*-associated disease: Brazier (1995) RMM *6* 236–245.]

Q fever   *Coxiella burnetii*. Acute form: e.g. fever, muscular pain, pericarditis, respiratory symptoms; chronic form: endocarditis, osteomyelitis, miscarriage. Inhalation, ingestion of contaminated milk, mother–fetus transmission (rare), transfused blood etc. [Minireview: Raoult (1996) JMM *44* 77–78.]

Scarlet fever   *Streptococcus pyogenes*. Commonly in children: sore throat, fever, swelling of cervical lymph nodes, rash. Droplet infection, ingestion of contaminated milk etc.

Syphilis   *Treponema pallidum*. A lesion (*chancre*) at the site of infection (typically genital mucosa) followed by a skin rash and, after months/years in untreated cases, lesions in e.g. heart, central nervous system etc. Infection by direct, particularly sexual, contact.

Tetanus   *Clostridium tetani* (virulent strains). Sustained, involuntary muscular contraction (section 11.3.1.3). Typically, wound contamination.

Toxic shock syndrome   Commonly, *Staphylococcus aureus* (strains producing e.g. toxic-shock-syndrome toxin-1; TSST-1). Vomiting, diarrhoea, fever, symptoms of shock; death in severe cases. Infection via e.g. tampons in menstruating women, wounds.

Trachoma   *Chlamydia trachomatis*. Conjunctivitis, scarring of conjunctival tissue, inturned eyelids, abrasion of the cornea by the eyelashes, causing ulceration. Infection occurs contaminatively.

Tuberculosis (pulmonary)   Usually *Mycobacterium tuberculosis*. Lesions develop in the lungs, and the pathogen may spread to various parts of the body via the lymphatic system and/or bloodstream. Infection e.g by inhalation. Currently, tuberculosis is a major cause of death in the world. In the USA, the rising incidence of this disease has been attributed to: (i) social factors, (ii) inefficient treatment programmes, and (iii) the rising numbers of (immunologically compromised) people with AIDS [epidemiological aspects of tuberculosis: Bloom & Murray (1992) Science *257* 1055–1064]. It has been reported that *M. tuberculosis* can promote HIV replication *in vivo* and in an *in vitro* model [Goletti *et al.* (1996) JIM *157* 1271–1278]. [Mycobacterial disease (various aspects): BCID (1997) 4 (issues 1 and 2, July). The role of the laboratory in the prevention/control of multidrug-resistant tuberculosis: Desmond (1997) RMM *8* 81–90.]

Tularaemia   *Francisella tularensis*. Chills, fever, nausea, headache, vomiting, prostration; sometimes severe gastrointestinal symptoms/sepsis. Contaminative infection via wounds or mucous membranes, or by the 'bite' of a tick or biting fly etc.

Typhoid   *Salmonella typhi*. Fever, transient rash, intestinal inflammation, sepsis, sometimes with tissue necrosis and intestinal haemorrhage. A carrier state (section 11.6) is recognized, *S. typhi* typically localizing in the gall bladder. Oral–faecal route, e.g. contaminated food or water.

Typhus (classical, epidemic)   *Rickettsia prowazekii*. Headaches, sustained fever and a rash which may haemorrhage, muscular pain. Contamination of a wound or 'bite' with louse faeces containing *R. prowazekii*; possibly, inhalation of dried, infected louse faeces.

Urethritis   Various, e.g. *Neisseria gonorrhoeae* (in gonorrhoea), or, in non-gonococcal (= non-specific) urethritis, *Chlamydia trachomatis*, *Mycoplasma hominis*. Inflammation of the urethra. Commonly, infection involves sexual contact.

Whipple's disease   *Tropheryma whippelii* (a Gram-positive rod which has not been cultured). Symptoms associated with intestinal and other regions – including malabsorption of nutrients, diarrhoea, involvement of central nervous system. Diagnosis depends mainly on histological examination of tissues. [Histology in Whipple's disease: von Herbay *et al.* (1996) Virchows Archiv *429* 335–343.]

Whooping cough   *Bordetella pertussis*. Paroxysms of coughing, each followed by an inspiratory 'whoop'. Droplet infection.

12 Applied bacteriology I: food

12.1 BACTERIA IN THE FOOD INDUSTRY

Bacteria are used for: (i) fermentations in the dairy industry; (ii) the processing of raw materials in the manufacture of coffee and cocoa; (iii) the manufacture of food additives; (iv) other processes, such as vinegar production. (Bacterial involvement in the production of food for farm animals is considered in Chapter 13.)

12.1.1 Dairy products

The manufacture of butter, cheese and yoghurt involves a *homolactic fermentation* (section 5.1.1.1) in which the lactose in milk is metabolized to lactic acid.

Butter ('cultured creamery butter') is usually made from pasteurized cream to which a *lactic acid starter* culture has been added; the starter contains e.g. *Lactococcus* (*Streptococcus*) *cremoris* and/or *L.* (*S.*) *lactis* as the main lactic acid producers, together with *L.* (*S.*) *lactis* subsp *diacetylactis* and/or *Leuconostoc cremoris* as the main contributors of diacetyl. Lactic acid and diacetyl contribute to the flavour, diacetyl giving the characteristic 'buttery' odour and taste. Following fermentation (in which the pH falls to about 4.6) the cream is cooled and churned to de-stabilize the fat globules. The aqueous phase (*buttermilk*) is strained off, and the butter (often salted) may be stored for long periods at e.g. −25°C. 'Sweet cream butters' are made without a starter.

Cheese is made by the coagulation and fermentation of milk, the different types of cheese reflecting differences e.g. in the source of milk (cows', goats' milk etc.), in the type of milk (whole milk, skimmed milk etc.) and in the types of microorganism used – commonly species of *Lactococcus* and *Lactobacillus*. Fermentation lowers the pH, thus helping in the initial coagulation of milk protein; additionally, minor products of fermentation (e.g. acetic and propionic acids) give characteristic flavours. (In Swiss cheeses such as Emmentaler and Gruyère, the typical flavour of propionic acid is due to the use of *Propionibacterium* spp.) The cheese (coagulated protein, fat) is separated from the aqueous phase (*whey*), and may then undergo various processes such as salting and ripening.

Yoghurt is usually made from pasteurized low-fat milk which is high in milk solids. The milk is inoculated with *Lactobacillus bulgaricus* and *Streptococcus thermophilus* and incubated at 35–45°C for several hours; the pH falls to about 4.3, coagulating the milk proteins. The bacteria act co-operatively: *L. bulgaricus* breaks down proteins to amino acids and peptides – which stimulate the growth of *S. thermophilus*; formic acid produced by *S. thermophilus* stimulates the growth of *L. bulgaricus*, which forms most of the lactic acid.

12.1.2   Coffee and cocoa

The manufacture of coffee from ripe coffee fruits requires the initial removal of a sticky mucilaginous mesocarp from around the two beans in each fruit. The outer skin of the fruit is disrupted, and the whole is left to ferment. The mucilage is degraded by the fruit's own enzymes and by microbial extracellular enzymes. As well as e.g. yeasts (single-celled fungi), the important organisms in this process include pectinolytic species of e.g. *Bacillus* and *Erwinia*. After fermentation, the beans are washed, dried, blended and roasted.

Cocoa is made from the seeds (beans) of the cacao plant, the fruit of which is a pod containing up to 50 beans covered in a white mucilage. The mucilage is fermented by yeasts, which produces ethanol, and the ethanol is oxidized to acetic acid by certain aerobic bacteria; other organisms are also present. Once free of mucilage, the beans – which darken during the week-long fermentation – are dried and roasted.

12.1.3   Food additives

Monosodium glutamate, the ubiquitous 'flavour enhancer', is manufactured from L-glutamic acid; this latter product is obtained commercially from certain bacteria – e.g. strains of *Corynebacterium glutamicum* grown aerobically on substrates such as molasses or hydrolysed starch. These organisms form glutamic acid from 2-oxoglutaric acid; because they lack the appropriate enzyme, commercial strains of *C. glutamicum* cannot metabolize 2-oxoglutaric acid in the TCA cycle (Fig. 5.10), so that maximum conversion to glutamic acid can occur. To allow secretion of the glutamic acid, the cell envelope is made more permeable by certain procedures; this is necessary to prevent feedback-inhibition of glutamic acid synthesis.

Xanthan gum is an extracellular polysaccharide slime synthesized by strains of *Xanthomonas campestris*. It has many commercial uses – e.g. as a gelling agent, a gel stabilizer, a thickener and a crystallization inhibitor in various foods.

12.1.4   Vinegar

Vinegar is made by the *acetification* of various ethanol-containing products – e.g. wine, cider, beer. Manufacture involves a carefully controlled process in which the ethanol is oxidized, aerobically, to acetic acid by species of *Acetobacter*. The vinegar may be aged, filtered, bottled and pasteurized.

'Vinegar' made chemically (e.g. by the carbonylation of methanol) is called *non-brewed condiment*; it is cheaper than bacterially produced vinegar.

12.2   FOOD PRESERVATION

The various methods of food preservation aim to prevent or delay microbial and other forms of spoilage, and to guard against food poisoning; such methods therefore help the product to retain its nutritive value, extend its shelf-life and keep it safe for consumption. Physical methods of preservation (e.g. refrigeration) are generally preferred to chemical methods.

12.2.1   Physical methods of food preservation

12.2.1.1   *Pasteurization and UHT processing*

Pasteurization is a form of heat treatment used e.g. for milk, vinegar and certain foods; its object is to kill certain pathogens and spoilage organisms. Milk is held at a minimum temperature of 72°C for at least 15 seconds (high-temperature, short-time [HTST] pasteurization). Pasteurization kills the causal agents of many milk-borne diseases (such as salmonellosis and tuberculosis) as well as much of the natural milk microflora; it also inactivates certain bacterial enzymes (e.g. lipases) which would otherwise cause spoilage. The causal agent of e.g. Q fever (*Coxiella burnetii*) is not necessarily killed by HTST pasteurization.

In the UHT process, milk is typically pumped between heated plates and achieves a temperature of approx. 141°C for a holding time of 3–4 seconds. The UHT ('ultra heat-treated') milk is collected and packaged aseptically, and it normally has a shelf life (without refrigeration) of about 6 months. In an alternative method, heat is supplied by injecting steam under pressure; this procedure is particularly useful for viscous products such as custard.

Samples are taken at frequent intervals to test for contamination; rapid results are obtained e.g. by the Lumac system described in section 12.3.3.1.

12.2.1.2   *Canning*

Typically, suitably prepared food is put into metal containers ('cans' or 'tins') which are then exhausted of air, sealed, and heated – usually to well over 100°C. After heating, the cans are generally cooled by water; the cooling water must be microbiologically clean because, on rare occasions, the normally efficient seal (Fig. 12.1) has allowed contamination of food during the cooling process.

What determines the temperatures and times used in canning? To answer this we first look briefly at the way in which a lethal temperature inactivates a

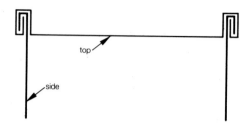

Fig. 12.1 Double-seaming, as used in the canning process (section 12.2.1.2). The edge of the can-end is wrapped around the body flange to form a seam of five thicknesses of metal. Not shown (but often present) are corrugations in the body and ends of the can; these help to relieve strain during the heating process.

population of cells or endospores (see Fig. 12.2); this will also help to make clear the meanings of some important parameters used in the food industry: the D, z and F values.

After processing, canned food must not contain *Clostridium botulinum* (section 11.3.1.2) capable of growth and toxin production under the conditions of storage. To achieve this, the level of heating will depend on the type and pH

of the food and on the way in which the canned food is to be stored. 'Neutral' foods (such as potatoes, carrots, mushrooms) require, as a minimum, a *12D cook* (= *botulinum cook*), i.e. heating, at a given temperature, for 12 times the D period of *C. botulinum* endospores at that temperature. After such heating, the probability of contamination by *C. botulinum* is regarded as negligible.

Foods whose pH is below 4.6 may be given less heat treatment because it is generally supposed that *C. botulinum* does not grow or form toxin below this pH. (However, one early report [Raatjes & Smelt (1979) Nature *281* 398–399] described growth and toxin production in a laboratory medium (not in food) at values of pH down to 4.0.)

Large (catering-size) cans of *cured* meat (e.g. ham) may be processed at relatively low temperatures (e.g. 70°C in the centre of the can for several minutes) but such cans *must* be stored under refrigeration; this is because safety (i.e. lack of growth/toxin production by *C. botulinum*) depends on the combined effects of curing salts and low temperature.

Sweetened, condensed milk is not heated after can-sealing because the high content of sugar (low water activity: section 3.1.3) inhibits the growth of most bacteria.

The range of heat treatment of canned foods is summarized in Table 12.1.

Spoilage of canned foods. Heat treatment generally inactivates at least some of the enzymes which would otherwise cause spoilage in stored, canned foods. However, spoilage is sometimes caused by heat-stable enzymes and/or by certain bacteria, or endospores, which survive heat treatment. For example, the endospores of *Bacillus stearothermophilus* may survive a botulinum cook, and may cause spoilage in low-acid foods if cans are stored at temperatures

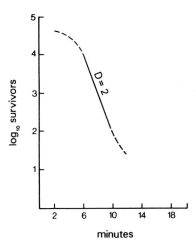

Fig. 12.2   The effect of a lethal temperature on a population of cells or endospores. In an idealized response, the number of survivors falls exponentially with time, at a given temperature, i.e. the graph is linear when survivors are plotted on a logarithmic scale. In practice, there are commonly variations from the linear response – for example, the non-linear 'shoulder' and 'tail' effects (*dashed line*).

Using this idea of 'logarithmic death', we can define three important parameters used in the food industry.

D value: the time (in minutes), at a given temperature, for a population of a particular type of bacterium or endospore to fall by 90% (i.e. by one unit on the log_{10} scale). In the linear part of the graph, the population has fallen by 1 log unit in 2 minutes, so that (in this case) $D = 2$. Notice that, in the linear part of the graph, the actual number of organisms killed in a given period of time depends on the initial number present; for example, 9000 are killed in 2 minutes when the population falls from 10 000 to 1000, and 900 are killed in 2 minutes when the population falls from 1000 to 100. That is, the *proportion* of cells/spores killed in a given time remains constant in the linear part of the graph.

The temperature at which a D value is obtained (by experiment) may be shown as a subscript: e.g. D_{70} = the value at 70°C. For cells of *Staphylococcus aureus*, D_{70} is typically less than 1 minute (much longer in some strains), and for *S. epidermidis* it is about 3 minutes; a D_{60} of 1 hour has been reported for *Salmonella seftenberg*, though for *Escherichia coli* the D_{60} is typically only a few minutes.

z value: the amount by which the temperature (°C) must be increased in order to decrease a given D value by 90% (i.e. to one-tenth of its value). For example, if – as in the graph – the temperature used to kill particular cells/spores gives a D value of 2, then the z value would be the increase in temperature needed to give a D value of 0.2 for the same type of cells/spores.

F value: the time (in minutes) required for adequate heat treatment at 121.1°C *calculated from* the time required for adequate heat treatment at another temperature (characterized by a particular D value), assuming a z value of e.g. 10°C. F values are used e.g. to compare heat treatments at different temperatures.

D, z and F values form a *basis* for designing a suitable heat treatment. Other factors also have to be considered – e.g. the time taken for heat to penetrate the contents of the can.

Table 12.1   Canned foods: the range of heat treatment

Type of food	Heat treatment	Comment
Potatoes, carrots, mushrooms and other 'neutral' foods	Botulinum cook as a minimum	Treatment necessary to kill the endospores of *Clostridium botulinum*; inactivation of spoilage organisms may involve further heating
Acidic produce	Minimum may be less than the botulinum cook	It is generally reported that *C. botulinum* does not grow or form toxin below pH 4.6 (but see text)
Catering-size cans of *cured* meat (e.g. ham)	Moderate heating (e.g. 70°C at the centre of the can for several minutes)	*Must* be stored under refrigeration
Sweetened, condensed milk	Nil	Low water activity (due to high concentration of sugar) inhibits the growth of most bacteria; pre-pasteurization of the milk helps to prevent spoilage by fungi

above about 30°C; in such spoilage, the food is soured but the cans are not swollen (such cans being called *flat sours*). Low-acid foods are often heated for more than the (minimum) botulinum cook in order to avoid this type of spoilage.

Canned foods usually remain safe and edible for some years if stored correctly.

12.2.1.3   Refrigeration

Temperatures such as 0–8°C (used for short-term storage) can delay spoilage by inhibiting the metabolism and growth of contaminating organisms and/or the activity of their extracellular enzymes; even so, psychrotrophic organisms (section 3.1.4) may cause spoilage. Note that some bacterial pathogens (e.g. *Listeria monocytogenes*, *Yersinia enterocolitica*) can continue to grow at 4°C.

Freezing (used for long-term storage) may kill some contaminants, and it also reduces the amount of available water (section 3.1.3). At sub-zero temperatures – such as –5 to –10°C – certain *fungi* may become important spoilage agents of e.g. meat.

12.2.1.4   Dehydration

Dehydration reduces the available water (section 3.1.3) to a point at which contaminants will not grow. The process may involve evaporation by heating – as e.g. in the manufacture of dried milk. Alternatively, the amount of available water may be reduced by adding sodium chloride (as in salted fish products) or syrups (as in preserved fruits).

12.2.1.5 Ionizing radiation

Ionizing radiation (e.g. high-energy electrons or *gamma*-radiation) is used in some countries for treating foods such as chicken and fish products, strawberries and spices.

12.2.2 Other methods of food preservation

12.2.2.1 Acidification

The traditional method of *pickling* preserves food by virtue of the lowered pH. This can be achieved either by adding acids (usually lactic acid, sometimes vinegar) or, in some cases, by fermenting the food; *sauerkraut* is cabbage which has been subjected to a natural lactic acid fermentation involving species of *Lactobacillus* and *Leuconostoc*.

12.2.2.2 Preservatives

Food preservatives are chemicals which can inhibit contaminants; some inhibit fungi as well as bacteria. They include benzoic acid (used e.g. in fruit juices, cordials), nitrites and sorbic acid. The use of preservatives is typically subject to governmental regulations.

The traditional method of *curing* (particularly pig meat) involves permeating the meat, at e.g. 4°C, with a solution that includes sodium chloride (which reduces water activity), and sodium nitrite (which, under appropriate conditions, inhibits bacterial growth and the germination/outgrowth of endospores).

In *smoking* (bacon, fish etc.) the food is exposed for hours/days to wood smoke; this reduces water activity in the food, and permeates the food with certain antimicrobial substances (e.g. phenolic compounds) present in the smoke.

12.3 FOOD POISONING AND FOOD HYGIENE

'Food poisoning' usually refers to acute *gastroenteritis* resulting from the ingestion of food contaminated with certain pathogens and/or toxins (Table 12.2); traditionally, the term has also included food-borne cases of botulism (sections 11.3.1.2 and 11.10).

Table 12.2 Food poisoning: some of the main types[1,6]

Causal agent	Foods commonly implicated[2]	Incubation period[3]; main symptoms; comments
Bacillus cereus[4]	Unrefrigerated meat dishes; contaminated spices Cooked rice, stored without refrigeration	8–16 hours; diarrhoea, abdominal pain, nausea 1–7 hours; vomiting, abdominal pain, nausea
Campylobacter jejuni, C. coli[7]	Undercooked meat, poultry; unpasteurized milk	1–7 days; abdominal pain, diarrhoea – watery or bloody; sometimes preceded by influenza-like symptoms; the pathogen cannot grow below 25°C
Clostridium botulinum	Meat, fish; home-canned mushrooms	See 'botulism' in section 11.10
Clostridium perfringens	Cooked, unrefrigerated meat dishes	8–24 hours; abdominal pain, diarrhoea; large numbers of bacteria must be ingested, the toxin being released when sporulation occurs in the gut
Escherichia coli (EIEC, EHEC/VTEC, ETEC, EPEC)	Undercooked meat; milk (and water)	See Table 11.2
Salmonella serotypes	Poultry (including cross-contamination from raw poultry): eggs	12–48 hours; diarrhoea, abdominal pain, vomiting, fever
Staphylococcus aureus	Various (ham, custards, desserts, sandwiches, poultry etc.)	<1–6 hours; nausea, vomiting, abdominal pain, often diarrhoea; the staphylococcal enterotoxins (section 11.3.1.5) can withstand 100°C for some time; the source of contamination is usually a food-handler
Vibrio parahaemolyticus	Shellfish	12–24 hours, but may be less than 2 hours; diarrhoea, abdominal pain[5]
Yersinia enterocolitica	Meat, milk	Fever, diarrhoea, abdominal pain, vomiting; the pathogen can grow at 4°C and below

[1] Other food-poisoning pathogens include *Shigella* spp; *Listeria monocytogenes* [McLauchlin (1997) RMM 8 1–14]; some viruses (e.g. rotaviruses).
[2] Only a few examples are given; pathogens/toxins can occur in unexpected places – for example, botulism has been caused by hazelnut yoghurt in which the hazelnut component had been insufficiently heat-treated.
[3] The approximate period between ingestion of contaminated food and the start of the symptoms.
[4] Produces two forms of disease, each caused by a distinct type of toxin.
[5] Disease, epidemiology and laboratory identification of *V. parahaemolyticus* (review) [Eko & Rotimi (1995) RMM 6 137–145].
[6] Evidence suggests that food-borne diarrhoeal disease can also be caused by *Aeromonas* spp [Thornley *et al.* (1997) RMM 8 61–72].
[7] *Campylobacter* infections: Leach [(1997) RMM 8 (3) (in press)], and Ketley [(1997) Microbiology 143 5–21].

Listeriosis (section 11.10) does not fit the description 'food poisoning' (above); however, the disease can be fatal, and food seems to be the main source of infection. Although *Listeria monocytogenes* can be isolated from various foods (e.g. some cheeses), an accurate assessment of the risk of listeriosis is hindered e.g. by the occurrence of some weakly pathogenic and even non-pathogenic strains of this species [Hof, Nichterlein & Kretschmar (1994) TIFS 5 185–190].

12.3.1   Bacterial food poisoning

Where do the pathogens come from? In some foods (e.g. poultry, eggs, shellfish) pathogens may be present at source but, in general, contamination may occur anywhere between the source and point of consumption; in the kitchen, pathogens are frequently transferred from one food to another (= *cross-contamination*), particularly from raw meat or poultry to cooked foods.

Sometimes the pathogen must *grow* on or in the food to form enough cells or toxin to cause disease. However, growth may not be necessary when, for example, disease can be caused by relatively few cells of the pathogen – e.g. 100 cells of *Shigella*, or less than 1000 cells of *Campylobacter*.

Food poisoning may involve an *infection* of the gut (= *food-borne infection*) and/or the action of toxins (= *food-borne intoxication*).

The risk of food poisoning involves the following factors.

1. The initial level of contamination on/in the food. In the production/ manufacture of food, the greater the initial contamination the less likely it is to be reduced to safe levels by routine processing. In prepared foods, contamination may be above or below the level needed to cause disease.
2. The type of processing, storage, distribution and preparation involved in the production of food. The risk of food poisoning depends on the ease with which a pathogen can gain access to, and grow in or on, food at each stage of the food chain, and on the success or failure of those processes used to eliminate pathogens. With meat and poultry, major food-poisoning pathogens should ideally be eliminated at farm level, i.e. in livestock, since control of contamination is much more difficult after slaughter and distribution of the products.
3. The dose/response relationship, particularly the smallest dose (of cells or toxin) able to cause disease. In fact, susceptibility to pathogens varies widely with age, acquired immunity and state of health; 'special-risk' groups, in which the risk is greater and the disease may be far more serious, include infants, the very old, pregnant women, and the immunocompromised (including AIDS patients). Because 'dose/response' can vary, acceptable (i.e. safe) levels of contamination are often based on epidemiological studies, limited experimental data, and a consensus of opinion [foods and microbiological risks: Baird-Parker (1994) Microbiology *140* 687–695].

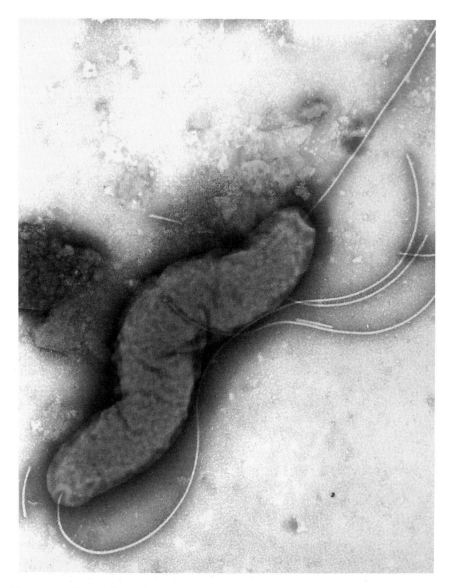

Plate 12.1 *Campylobacter jejuni* (×60 000): a common causal agent of acute infectious diarrhoea in all age groups (particularly young adults) in the developed countries, and (primarily) in infants and very young children in the developing countries. The pathogen is associated with animals and animal products, especially poultry; cross-contamination of other foods (section 12.3.1) appears to be an important mode of transmission.

Photograph courtesy of Dr Alan Curry, PHLS, Withington Hospital, Manchester, UK.

12.3.2   Food hygiene – domestic

In the kitchen, safe food becomes unsafe in various ways without necessarily showing warning signs (taste, smell, appearance); for example:

1. In general, handling food with unwashed hands; this may promote infection e.g. by the faecal–oral route. Handling ready-to-eat foods with unwashed hands after handling raw meat or poultry may allow cross-contamination (e.g. with *Salmonella* from poultry) and – particularly if foods are then left for some time at room temperature – growth of the pathogen to dangerous levels.
2. Using knives and/or work surfaces for raw meat and then for cold, pre-cooked foods without a thorough wash between uses.
3. Contamination by pathogens derived from the food handler – often the case with staphylococcal food poisoning.
4. Contamination by flies etc.
5. Undercooking contaminated meat or poultry – e.g. through inadequate thawing of frozen products prior to cooking.
6. Inadequate cooking of eggs stored (correctly) in the refrigerator; as they are initially colder than eggs kept (incorrectly) at room temperature, such eggs require cooking for a longer time.
7. Storage of food at too high a temperature (due e.g. to faulty refrigeration) – allowing bacterial growth and the conversion of low levels of contamination to dangerous levels. To work efficiently, a refrigerator should not be overloaded, i.e. air should circulate around the shelves; the coldest shelves should be 0–5°C. The freezer compartment should not be warmer than –18°C.
   Note that *Yersinia enterocolitica* (and e.g. *Listeria monocytogenes*) can grow at and below 5°C and may therefore increase in number under refrigeration.
8. Consuming food after the 'use by' date – or even before that date if storage conditions have not been appropriate.

12.3.3   Food hygiene – industrial

For food producers and manufacturers the guiding principle in hygiene is the exclusion or elimination of pathogens or the reduction of contamination to safe ('acceptable') levels. This involves e.g. rearing/selecting disease-free animals (for meat, poultry and some fish products), and the use of appropriate and reliable methods of preservation (section 12.2).

Because unsafe food on an industrial scale may affect large numbers of people, the manufacture and sale of food is subject to government regulations. One approach to food safety – used world-wide – is the HACCP system.

12.3.3.1   The hazard analysis critical control point (HACCP) system

The HACCP system is essentially the detailed surveillance of a production process, with control of known hazards at specific critical stages. It includes the following phases: (i) objective assessment of hazards, using a flow-chart which shows all aspects of the production process (and which may include the source of raw materials and/or the distribution/sale of final products); (ii) identification of *critical control points* (CCPs), i.e. stages in the process at which preventative or corrective action can be taken effectively against the known hazards; (iii) monitoring at CCPs to ensure that the process continues within pre-determined tolerance limits.

Bacteriological monitoring at CCPs may detect the presence/level of contamination or may detect the presence of a specific organism; either way, rapid assessment is needed to permit rapid corrective action when required. Traditional methods – particularly those based on colony counting – are typically too slow for timely intervention in a modern production process; some rapid methods of assessment are described briefly below.

Lumac® *Autobiocounter M 4000* (Perstorp Analytical LUMAC, Landgraaf, The Netherlands). This system is used to detect microbial contamination in samples of UHT milk etc.; the principle depends on the detection of ATP in the sample – the presence of ATP indicating contamination with metabolizing (living) organisms. Samples are pre-incubated (e.g. 30°C/48 hours) before testing to allow growth of contaminating organisms. After incubation, an aliquot (e.g. 50 μl) of the sample is transferred to an ATP-free cuvette (container) and treated first with an ATPase to hydrolyse the sample's background ATP. The sample is then treated with a reagent which degrades the cytoplasmic membrane of any contaminating cells – allowing leakage of ATP from such cells. ATP is detected by adding reagents (luciferin–luciferase) which produce light (*chemiluminescence*) in the presence of even minute amounts of ATP; emitted light is measured by a photomultiplier (light meter) and the result indicated digitally. The addition of reagents etc. is fully automated, and the test is complete in about 15–20 minutes.

In such an arrangement, each batch of product is held back until the test results become available – being released for distribution when tested samples are found to be satisfactory. Despite the delay (for sample pre-incubation), the manufacturing process itself can be continuous – requiring only that batches be retained pending test results.

Impedance monitoring. The food sample is dispersed in a suitable growth medium, and the number of bacteria in the sample is assumed to be proportional to the level of any ongoing metabolism detected in the medium. The level of metabolism is assessed by measuring the decrease in electrical impedance (increase in conductivity) in the medium with time; any fall in impedance is assumed to result from the metabolism of complex molecules and the formation of increased numbers of ions. There are some problems: particular contaminating organism(s) may fail to grow in the medium, and unknown factor(s) may inhibit normal metabolism in other species.

Carbon dioxide monitoring. In some cases, the level of contaminating organisms can be assessed by measuring their output of carbon dioxide; this method is useful e.g. for detecting those organisms which cannot be monitored efficiently by impedance changes.

Direct epifluorescent filter technique (DEFT). This method (see section 14.8.1) is particularly useful for milk.

Limulus *amoebocyte lysate (LAL) test*. This test, which detects or quantifies *lipopolysaccharide* (LPS) (section 2.2.9.2), is used to monitor contamination by Gram-negative bacteria. The test depends on the ability of LPS to coagulate a lysate of the blood cells (amoebocytes) of the horseshoe crab, *Limulus polyphemus*; LPS less than 10^{-9} g/ml may be detected. The LAL test registers *any* LPS-containing bacterium, and even isolated LPS; hence, both living and dead cells can give a positive test.

Monitoring with DNA probes. A labelled DNA *probe* (section 8.5.3) may be able to detect a particular type of bacterium in a food sample provided that the probe is sufficiently specific to the target organism. An extension of this idea is the use of PCR (section 8.5.4) to detect a given organism by using organism-specific primers – as is done in a medical context (section 8.5.4.1).

13 Applied bacteriology II: miscellaneous aspects

13.1 FEEDING ANIMALS, PROTECTING PLANTS

As well as influencing soil fertility (Chapter 10), bacteria make specific contributions to agriculture by helping us to feed farm animals and assisting in the protection of certain crops.

13.1.1 Silage and single-cell protein

13.1.1.1 Silage

Silage-making, the traditional way of preserving grass (and certain other crops), enables animals to be fed during the winter months when vegetation is relatively scarce. Essentially, finely chopped grass is stored anaerobically and allowed to ferment. Bacteria – e.g. *Lactobacillus* spp – are present on the vegetation and/or in the storage vessel (*silo*); they metabolize plant sugars (e.g. fructose, glucose, sucrose) mainly by a lactic acid fermentation (section 5.1.1.1). The production of lactic acid rapidly lowers the pH (to about 4.0), thereby inhibiting those organisms (particularly *Clostridium* spp) which would otherwise cause putrefaction; this helps to preserve the nutritive value of the crop. The acidity also helps to inhibit growth of *Listeria monocytogenes*, an organism which can grow in incompletely fermented silage (pH > 5.5); infection with *L. monocytogenes* may lead e.g. to abortion/stillbirth.

In *big-bale* silage-making, the vegetation is fermented in large black plastic bags rather than in a silo.

Various silage additives are used e.g. to increase the efficiency of the process and/or to enhance the performance of animals fed on the silage. One common additive includes strains of lactic acid bacteria that carry out a homolactic fermentation (section 5.1.1.1).

Ideas for improved silage additives include (i) use of specific strains of lactic acid bacteria particularly suited to given target crops; (ii) use of lactic acid bacteria that carry out a *heterolactic* fermentation – volatile fatty acids inhibiting the growth of fungi when silage is exposed to air at feeding time; (iii) use of genetically engineered lactic acid bacteria which can utilize polysaccharides in those silage crops which have low levels of soluble carbohydrates [new trends in silage inoculants: Weinberg & Muck (1996) FEMSMR 19 53–68].

13.1.1.2 Single-cell protein (SCP)

SCP refers to the cells of certain microorganisms (including bacteria, yeasts and microalgae) grown in large-scale cultures for use as a source of protein in the animal (and human) diet. In the early 1980s, when the price of protein was high, thousands of tons of bacterial protein were being manufactured each year for animal feed; the organism, *Methylophilus methylotrophus* (a methylotroph – section 6.4), was grown on a methanol substrate, and its yield was improved by incorporating an *E. coli* gene (encoding glutamate dehydrogenase) which improved the assimilation of ammonia.

13.1.2 Biological control

'Biological control' generally refers to the use, by man, of one species of organism to control the numbers or activities of another. Such exploitation is used on a commercial scale e.g. in agriculture and forestry. It commonly involves the use of certain microorganisms and/or their toxins to kill or disable the insect pests of certain crop plants; such microorganisms/toxins are called *microbial insecticides* (*bioinsecticides, biopesticides*).

Strains of *Bacillus thuringiensis* produce various types of insecticidal crystal protein (ICP) – designated Cry types I–IV (and subtypes); these products are important bioinsecticides which are used world-wide against insect pests on a range of crops. The genes encoding these ICPs are typically plasmid-borne.

Most ICPs are sporulation-specific, i.e. they are formed only in sporulating cells – as a *parasporal crystal* located near the endospore; synthesis is tied to sporulation because transcription of the relevant ICP genes requires sporulation-specific sigma factors such as σ^E – analogous to σ^E in *Bacillus subtilis* (see Fig. 7.14). The genes of CryIII ICPs differ in that they are transcribed during vegetative growth; these ICPs are overexpressed in mutant strains which are blocked in the phosphorelay (Fig. 7.14).

[Regulation of ICP production: Baum & Malvar (1995) MM *18* 1–12.]

Crystals applied to crops are soon degraded, necessitating repeat application; to avoid this problem, *B. thuringiensis* has been modified so that toxin accumulates within the toxin-producing cells – this (partly protected) intracellular toxin still being highly active [Lereclus *et al.* (1995) Biotechnology *13* 67–71].

Certain strains of *Bacillus thuringiensis, B. sphaericus* and *Clostridium bifermentans* form toxins which kill mosquitoes, and much effort (involving e.g. recombinant DNA technology) is being made to develop a product suitable for the control of mosquito-borne diseases such as malaria, filariasis and yellow fever [Porter, Davidson & Liu (1993) MR *57* 838–861].

13.2 BIOMINING (BIOLEACHING)

For many years, chemolithotrophic bacteria (particularly *Thiobacillus fer-rooxidans*) have been used commercially to extract certain metals from low-grade ores. Today, this process (*biomining, bioleaching*) is of greater interest because many of the sources of richer ore have been exhausted. An estimate of the world production of bioleached copper, for example, is in excess of 1 million tons per year [Ehrlich & Brierley (1990) *Microbial Mineral Recovery*. New York: McGraw-Hill].

Copper can be recovered e.g. from low-grade ores containing the mineral chalcopyrite ($CuFeS_2$). Essentially, a liquor containing sulphuric acid and chemolithotrophic bacteria is allowed to percolate through a mound of crushed ore, and is repeatedly re-cycled; ions (Fe^{2+}, S^{2-}) are leached from the ore and are oxidized by the bacteria (to Fe^{3+} and sulphate, respectively), resulting in further solubilization of Fe^{2+} and copper (as $CuSO_4$). The process thus becomes self-sustaining. Various nutrients are provided by the liquor and/or by the ore itself. Copper in the leachate (up to 5 grams per litre) is periodically removed (e.g. by electrolysis). Bacterial metabolism maintains a temperature of about 50°C, and the convective upflow of air preserves the necessary aerobic conditions.

Leaching of copper sulphide at very low pH (advantageous because it inhibits Fe III precipitation) has been achieved with e.g. strains closely related to *T. thiooxidans* – identification of the leaching organisms involving characterization of the spacer region between 16S and 23S rRNA genes (see Fig. 16.6) in DNA extracted from leachate [Vásquez & Espejo (1997) AEM 63 332–334].

An examination of the sulphur chemistry of bacterial leaching has suggested that the leaching of pyrite involves a cyclical degradative process in which the intermediate products include thiosulphate and polythionates; ferric ions (Fe^{3+}), produced by bacterial oxidation of Fe^{2+}, have an active role in oxidizing both pyrite and thiosulphate – the lithotrophic metabolism of the leaching bacteria maintaining a constant supply of these ions [Schippers, Jozsa & Sands (1996) AEM 62 3424–3431].

In recent years bioleaching has been used (e.g. in South Africa, Brazil, Australia) as a preliminary process in the extraction of gold from recalcitrant arsenopyrite ores; gold cannot be solubilized from these ores in the usual way (by the use of cyanide) because it is embedded in a matrix of pyrite and arsenopyrite (which protects the gold from cyanide). The crushed ore is therefore subjected to bacterial action (Plate 13.1); arsenopyrite (FeAsS) is solubilized by oxidation (to $FeAsO_4$ and H_2SO_4), thus exposing the gold – which can then be extracted by cyanidation. Gold is separated from the gold–cyanide complex by adsorption to carbon (e.g. charcoal). The spent, cyanide-containing liquor can be detoxified by bacteria, cyanide being oxidized to CO_2 and urea, and urea being oxidized to nitrate.

Plate 13.1 Bioleaching for gold (section 13.2): cells of *Thiobacillus* among particles of crushed arsenopyrite ore at the Fairview mine in Barberton, South Africa. Gold-bearing arsenopyrite concentrate is ground to a fine powder, mixed with water, acid and nutrients, and passed through a series of highly aerated bio-oxidation tanks containing chemolithotrophic bacteria (e.g. *Thiobacillus ferrooxidans*). These bacteria solubilize the arsenopyrite matrix – thus exposing the gold which can then be extracted by cyanide. As a result of bioleaching, the amount of gold recovered may be as much as twice that recoverable without bioleaching. Bioleaching is a low-energy process compared with the traditional methods of treating recalcitrant ores: roasting (high-temperature oxidation) and pressure leaching (aerated acid-digestion under pressure). Photograph reproduced by courtesy of Alex Hartley.

Studies on the genetics/physiology of organisms such as *T. ferrooxidans* should help us to develop strains with increased efficiency in bioleaching.

[Microbiology of metal leaching (symposium report): FEMSMR (1993) *11* 1–267. Mining with microbes (review): Rawlings & Silver (1995) Biotechnology *13* 773–778. *Bioleaching: Fundamentals and Applications*, Rawlings D E (ed.), Austin, Texas: R G Landes (in press, 1997).]

13.3 BIOLOGICAL WASHING POWDERS

'Biological' washing powders usually contain enzymes called *subtilisins*, produced by species of *Bacillus*. One of these, 'subtilisin Carlsberg' (obtained from *B. licheniformis*), can hydrolyse most types of peptide bond (in proteins) and even some ester bonds (in lipids). It is stable over a wide range of pH, and its stability does not depend on Ca^{2+}; this latter feature is important because washing powders often include agents which 'soften' water by chelating ions such as Ca^{2+}.

13.4 SEWAGE TREATMENT

Sewage includes domestic wastes (e.g. from drains and water-closets) and often varying amounts of agricultural and/or industrial effluent; it contains substances in suspension, in solution, and in colloidal form.

If discharged to rivers or lakes etc. sewage can be harmful in various ways. It can, for example, be a source of infection – promoting the spread of water-borne diseases such as cholera. Another problem is its content of dissolved organic matter; in metabolizing such nutrients, the large numbers of sewage bacteria can quickly use up the available oxygen in a locally polluted region – particularly in slow-moving or static waters. This can mean death for fish and other oxygen-dependent animals living in these waters. Additionally, such anaerobiosis permits the growth of sulphate-reducing bacteria (section 5.1.1.2; Fig. 10.3) and other organisms whose metabolic products include sulphide and other malodorous substances. Two major aims of sewage treatment are therefore (i) to eliminate (or reduce the numbers of) pathogens, and (ii) to diminish the oxygen-depleting ability of the sewage, i.e. to diminish its *biological oxygen demand* (BOD).

13.4.1 Aerobic sewage treatment

Raw (untreated) sewage first undergoes *primary* treatment which normally includes time in a settlement tank for removal of some of the particulate matter (separated as sludge). The so-called 'settled sewage' is then subjected to *secondary* treatment.

One form of secondary treatment is the familiar *trickle filter* (=biological filter) (Plate 13.2, *top*) in which sewage is sprayed, via holes in a horizontal

---

Plate 13.2 Sewage treatment (aerobic): old and new technology (see section 13.4.1). *Top*. Trickle filters at Bodmin, Cornwall, UK. *Bottom*. BAF units at St Austell, Cornwall. *Bottom, inset*: bacteria (approx. ×11 000) colonizing a particle of 'Biocarbone' filter medium used in some types of BAF unit.

Photographs of treatment plants in Cornwall courtesy of Mr Brian Lessware ABIPP, South-West Water, Exeter, UK.
Electronmicrograph of bacteria on a Biocarbone particle courtesy of Anjou Recherche – OTV, Compagnie Générale des Eaux, Maisons Laffitte, France.

rotating arm, onto a thick layer of crushed rock enclosed by a circular wall; percolating through the crushed rock, the sewage makes close contact with surfaces that bear a biofilm containing large numbers of ciliates (protozoa) and other organisms – including bacteria such as *Zoogloea ramigera*. The sprayed sewage carries with it dissolved oxygen, so that some organic matter in the sewage can be oxidized by the sewage organisms and by those in the biofilm on the rock surfaces; the process is a controlled form of mineralization (section 10.3.4). After treatment, the effluent has a much lower BOD, i.e. when discharged to a river etc., it will take less oxygen from the water. As the sewage percolates, large numbers of bacteria are consumed by the protozoa on the rock surfaces.

Another form of aerobic secondary treatment is the *activated sludge* process. Settled sewage enters a vessel containing activated sludge, i.e. a mass of organisms consisting mainly of bacteria (e.g. *Acinetobacter* spp, *Alcaligenes* spp, *Sphaerotilus natans*, *Zoogloea ramigera*) and protozoa (ciliates, flagellates and amoebae). Effluent and sludge are vigorously agitated and aerated for e.g. 6–12 hours so that much of the soluble organic matter in the effluent is oxidized or assimilated by the biomass; the BOD is thus greatly reduced. The final effluent is obtained after a further stage in a sludge-settling tank. During the process, microbial growth increases the mass of sludge, and some is kept for treating the next batch of sewage.

A more efficient secondary treatment process has appeared in the last decade or so: the *biological aerated filter* (BAF) (Plate 13.2, *bottom*). This consists essentially of a submerged bed of finer granular material (coated with biofilm – see Plate 13.2, *bottom, inset*) through which the sewage passes, downwards, while air is pumped in at or near the base of the bed; the use of small granules allows the system to function as a mechanical filter (for fine particulate matter) as well as a process for mineralizing the dissolved organic matter in the sewage. A BAF contains up to five times more biomass in the biofilm of the filter bed than that in a trickle filter of equivalent size so that, for a given treatment capacity, the BAF can be much smaller.

Under appropriate flow conditions (allowing adequate aeration), nitrifying bacteria (section 5.1.2; Fig. 10.2) can become established and functional in the BAF's biofilm; the system can also be operated anaerobically to encourage denitrification (sections 5.1.1.2 and 10.3.2; Fig. 10.2), so that a high proportion of the nitrogen in sewage can be eliminated by operating aerobic and anaerobic reactors in series. The type of bacterium reported to denitrify very efficiently under *aerobic* conditions [Patureau *et al.* (1994) FEMSME 14 71–78] may offer the possibility of carrying out nitrification and denitrification jointly in the same aerobic reactor, thus reducing the cost of treatment. Combined nitrification–denitrification has been discussed by Kuenen & Robertson [(1994) FEMSMR 15 109–117].

Periodic cleaning of the BAF is achieved by backwashing.

The three forms of aerobic sewage treatment are compared in Table 13.1.

Table 13.1 Sewage treatment: comparison of three aerobic processes[1]

Process	Mechanical sieving action	Enhanced aeration for efficient reduction of BOD	One-stage high performance[2]	In-built nitrification potential[3]
Trickle filter	+/−	−	−	−
Activated sludge	−	+	−	+/−
BAF	+	+	+	+

[1] See section 13.4.1.
[2] In terms of lowering BOD *and* clarification of effluent.
[3] In trickle filters, the available oxygen is used primarily for the oxidation of carbon. Those activated sludge plants which are specially designed to include nitrification tend to be larger than those which are not so designed; nitrification requires longer exposure to the sludge. Nitrification in BAF units depends on the maintenance of good aeration and appropriate loading of the system; under such conditions nitrifying bacteria can colonize the filter bed.

13.4.2 Anaerobic sewage treatment

Sewage containing a relatively high content of solids – e.g. farm wastes, sludge from some aerobic treatment processes – can be treated by *anaerobic digestion*. In this process, complex organic matter is broken down to simple substances which include a high proportion of gaseous products; it involves a wide range of bacteria which, collectively, carry out a spectrum of metabolic activities. Essentially, sewage is digested in a tank at about 35°C. Polymers, such as polysaccharides, are degraded by extracellular enzymes, and the resulting subunits (sugars etc.) are fermented (e.g. by species of *Bacteroides* and *Clostridium*) to products which include acetate, lactate, propionate, ethanol, CO_2 and hydrogen. Methanogens (section 5.1.2.2) produce methane from acetate and from the CO_2 and hydrogen; much of the bulk of sewage carbon is eliminated via CO_2 and methane.

Anaerobic digestion can yield a final product which is relatively odorless and rich in microbial biomass – a useful agricultural fertilizer. The gaseous product (*biogas, sewer gas*) may contain more than 50% of methane, and it can contribute most or all of the energy needs of the treatment process.

13.4.3 Sewage treatment and the environment

The technology for proper sewage treatment has been available (and used) for many years. Despite this, outfalls (long pipes) are still being built to convey raw (untreated) sewage into coastal waters; for those responsible, this may seem a 'cheap option', but for the *environment* it is certainly a most expensive option.

13.5   WATER SUPPLIES

Supplies of fresh (i.e. non-saline) water are obtained mainly from rivers (surface waters) and *aquifers* (underground layers of water-bearing rock); sub-surface water is called *groundwater*. Surface waters are typically much more polluted. The extent of reliance on surface waters varies geographically; thus, e.g. British Columbia (Canada) and Scotland rely heavily on surface waters, while Austria, Denmark and Portugal obtain almost all their supplies from groundwater.

Water intended for public supply must be treated to (i) eliminate pathogens (those causing water-borne diseases such as cholera), and (ii) eliminate, or decrease to safe levels, any harmful substances which may be present. Treatment also aims at a final product which is acceptable in terms of clarity, taste and odour.

13.5.1   Large-scale (urban) water supplies

Water undergoes various processes before entering the mains distribution system; the lower the quality of the 'raw' water (i.e. the water to be made potable) the more extensive the treatment. Thus, different combinations of processes are used for treating different types of water.

13.5.1.1   Treatment of groundwater

Groundwater (obtained from bore-holes and springs) is generally of good quality and may need little more than aeration, rapid sand filtration and disinfection (see later); however, appropriate treatment may be required if it contains significant levels of e.g. nitrates (see later).

13.5.1.2   Treatment of surface waters

Initially, raw water may be stored for a number of days; this e.g. allows sedimentation of some particulate matter. The water may then be microstrained through rotating stainless-steel mesh drums (pore size ~30 μm).

Certain types of raw water (e.g. water from polluted rivers – and also groundwater) may contain low levels of dissolved oxygen, and such water must be aerated e.g. by a cascade or fountain process which achieves good air–water contact; aeration avoids certain problems during subsequent treatment.

Much of the particulate matter can be removed by adding a coagulant (such as aluminium sulphate) which causes the particles to aggregate (forming *flocs*). To remove the flocs (*clarification*) the water is passed upwards through a *floc blanket clarifier*: a tank within which the flocs form a 'sludge blanket' below a layer of clarified water; the process is continuous, the clear water overflowing at the top of the tank and passing on to the next stage of treatment.

Water containing only fine particles (and dissolved substances) may be passed downwards through a *rapid sand filter*: a bed of sand (grain size ~1 mm) which acts essentially as a mechanical sieve; when the filtration rate falls, air is blown upwards through the sand (to dislodge the particulates) and this is followed by water (back-washing) to remove the solids.

A *slow sand filter* contains finer sand, the upper layer of which supports a biofilm of microorganisms (including bacteria and algae); thus, as well as acting as a mechanical sieve, the slow sand filter also effects biological purification e.g. by mineralizing some of the dissolved organic matter and by removing some nitrogen and phosphorus as biomass. Slow sand filters can eliminate certain taste- and odour-causing substances, and may also reduce the levels of any cyanobacterial toxins which may be present. Dangerous levels of these toxins are produced by 'blooms' of e.g. *Anabaena, Aphanizomenon, Microcystis, Nodularia* and *Oscillatoria* which can develop in bodies of water (including reservoirs) if conditions are right; such toxins may cause gastroenteritis, liver disease and other conditions in man and animals [Bell & Codd (1994) RMM 5 256–264].

The biofilm (which includes e.g. nitrifying bacteria – section 5.1.2) operates under aerobic conditions; hence, a slow sand filter will function correctly only if the water has been adequately aerated (see above).

Slow sand filters are not back-washed; instead, the uppermost layer is skimmed off and must be periodically replaced to maintain the depth of the filter.

Before disinfection (usually by chlorine) the pH of the water may need to be adjusted chemically. (pH affects the activity of chlorine.)

Disinfection. Water is disinfected, usually with chlorine, before entering the distribution system. Appropriate levels of chlorine are particularly effective against vegetative pathogenic bacteria of faecal origin – such as *E. coli* and species of *Campylobacter, Clostridium, Salmonella, Shigella* and *Vibrio*; hence, efficient chlorination is an excellent measure against cholera, typhoid and a range of other diseases.

Being a strong oxidizing agent, chlorine reacts with (and is 'used up' by) various impurities in the water; thus, some chlorine is initially lost in satisfying this *chlorine demand*. Once the demand has been satisfied, addition of further chlorine to the water will leave *free residuals* of chlorine; because chlorine reacts with water, the free residuals will include Cl_2, HClO and ClO^-. The recommended 'free residual' level of chlorine is usually about 0.5 parts per million (p.p.m.) (= 0.5 mg/litre). Chlorine is more effective in water at pH < 7 because dissociation of hypochlorous acid (HClO) – the more efficient disinfecting agent – tends to be suppressed under acidic conditions.

If the water contains ammonia, chlorine and ammonia will combine to form *chloramines* (= *combined residuals*). The chloramines are disinfectants which

Plate 13.3   Water supply from surface waters: damming a river can help to cope with fluctuations in river flow and demand for water. This small (44-metre high) dam on the West Okement River at Meldon, Dartmoor (Devon, UK), was completed in 1972; it created a reservoir of 23 hectares with a capacity of 3091 million litres. The catchment area is 1660 hectares of high moorland. The reservoir yields 24.5 million litres daily and serves a population of about 200 000 in nearby towns and rural areas. Water is

pumped to the treatment works using energy from a hydro-electric plant added in 1987.

Storage of water in a reservoir may improve water quality by allowing time for (i) bacterial mineralization of dissolved organic matter, and (ii) sedimentation of some particulate matter; this lowers the cost of treatment. However, under certain conditions, impounded water may support dense populations of algae/cyanobacteria, causing e.g. potential problems with toxins (see text). Moreover, deep, semi-static water can become thermally stratified; the lower, colder layer is then likely to become anaerobic and support the growth of sulphate-reducing (sulphide-producing) bacteria. Algal growth is sometimes controlled with algicides; stratification is often prevented by pumping water from base to surface.

---

decompose slowly, releasing chlorine; they are less effective but more persistent than chlorine. Ongoing addition of chlorine to the water will increase levels of chloramines to a maximum (which depends on the initial level of ammonia). Further addition of chlorine will oxidize the chloramines – so that levels of combined residuals will fall; after the *breakpoint* (Fig. 13.1), any chlorine added to the water will contribute to free residuals – i.e. to achieve free residuals the dose of chlorine must be beyond the breakpoint (= *breakpoint chlorination*) (Fig. 13.1).

A minority of water treatment plants add ammonia (as well as chlorine); in such cases, disinfection depends on combined residuals rather than on free residuals.

Superchlorination involves the use of relatively high concentrations of chlorine to eliminate undesirable odours and tastes; excess chlorine is then removed by adding sulphur dioxide (*sulphonation*) until the normal residual level of chlorine is reached.

Testing for free residuals can be carried out e.g. with tablets of DPD (*N,N*-diethyl-*p*-phenylenediamine). Chlorine levels are indicated by the degree of intensity of the red coloration produced with DPD; the intensity of coloration (and, hence, level of chlorine) is determined by comparison with a colour chart or 'comparator'.

Note that fully treated water may still contain pathogens. For example, certain viruses (e.g. Norwalk virus) and protozoa (e.g. thick-walled oocysts of *Cryptosporidium*) can survive routine levels of chlorination.

Ozone (compared with chlorine) is a stronger disinfectant, but it lacks residual action; its use can be followed by low-level chlorination to provide residual disinfection.

Tests for the efficiency of disinfection. Standards for the bacteriological quality of treated drinking water are provided by the World Health Organization (WHO), the US Environmental Protection Agency (USEPA), the European Commission (EC) and by appropriate bodies in various countries. Certain bacteria – particularly coliforms (see Appendix) and 'faecal streptococci' – are used as indicators of faecal pollution; their presence in treated drinking water

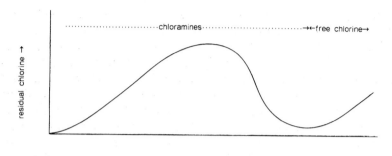

amount of chlorine added →

Fig. 13.1 Chlorination of water supplies (at about pH 7): chlorine–ammonia interaction and the breakpoint (section 13.5.1.2). The chlorine initially added may oxidize various substances/ions; this loss of chlorinating power (reflecting the *chlorine demand*) is shown by the shallowness of the slope at the start of the graph. Chlorine subsequently added reacts with ammonia to form monochloramine (NH_2Cl) and, as further chlorine is added, a proportion of dichloramine ($NHCl_2$). The rise in concentration of chloramines (= *combined residuals*) is indicated in the graph as a rise in 'residual chlorine'. As more chlorine is added, a point is reached at which chlorine starts to oxidize the chloramines (to N_2 and HCl) – so that their concentration (and that of 'residual chlorine') falls; after the combined residuals (chloramines) have reached their lowest level (the *breakpoint*), continued addition of chlorine brings about a rise in the level of *free residuals* (i.e. chlorine and the products of its reaction with water – section 13.5.1.2).

points to a failure of the treatment process or to contamination following treatment. Table 13.2 lists some recommended standards.

Coliforms and faecal streptococci are used as indicator organisms because they are common intestinal bacteria; if they are present in water samples then sewage-borne pathogens are *potentially* present in the water. It is not appropriate to test water samples routinely for each of the (many) different types of sewage-borne pathogen because e.g. a given pathogen may occur only intermittently. Moreover, normally, the indicator bacteria greatly outnumber pathogens, and they occur in large numbers in faeces (e.g. *E. coli* ~10^8 cells/gram); hence, these bacteria are easier to identify in samples, and it is possible to detect very low levels of faecal pollution.

Traditional tests for indicator bacteria include *membrane filtration*: a known volume of sample is filtered through a membrane filter (pore size e.g. 0.2 μm) and the membrane is then incubated, face upwards, on a pad saturated with an appropriate liquid selective medium; indicator bacteria (if present) form colonies on the membrane.

In the *multiple-tube test*, aliquots of sample are added to each of a number of tubes containing a lactose-based medium (e.g. MacConkey's broth – Table 14.1) and a Durham tube (Fig. 16.3); on incubation (37°C), any tube which has received at least one viable coliform (in the aliquot) will give a 'positive' test (lactose fermentation): acidification (shown by the pH indicator) and gas production (gas in the Durham tube). The number of positive and negative tubes is then used, in conjunction with statistical tables, to indicate the *most probable number* (MPN) of coliforms in the water sample. This is actually a *presumptive coliform count* because a 'positive' result in any given tube could be due to certain spore-forming bacteria (which can also ferment lactose and form gas); this is particularly important when testing *chlorinated* water supplies because spore-formers are more resistant than coliforms to chlorine. Confirmation that the result is due to *E. coli* – a thermotolerant (= 'faecal') coliform – requires two further tests which are carried out at 44°C/24 hours: the *indole test* (see section 16.1.2.5) and the *Eijkman test*. In the Eijkman test, each 'positive' tube is subcultured to a medium (e.g. lauryl tryptose lactose broth) that includes an agent which inhibits spore-forming bacteria; on incubation at 44°C, acid and gas from lactose, and indole from tryptophan, is taken as confirmation of *E. coli*.

Adequately chlorinated water should not contain indicator organisms. High counts of *E. coli* in water from the distribution system would indicate heavy or recent contamination, while low counts would suggest lighter or earlier contamination. Tracing sources of faecal contamination may be facilitated by information such as the recent report that faecal pollution from human and animal sources can be differentiated by differences in the antibiotic resistance patterns of faecal streptococci [Wiggins (1996) AEM *62* 3997–4002].

The problem of nitrates, pesticides etc. Nitrates occur in surface water and, increasingly, in groundwater; a major source is agricultural fertilizer. The upper limit recommended by the EC/UK (NO_3/litre) is 50 mg/litre; the USEPA value is 44.29 mg/litre (= 10 mg/litre as N). Water containing high levels of nitrate is sometimes blended with low-nitrate waters. Alternatively, the water may be stored for extended periods to permit denitrification (section 10.3.2). Removal of nitrate by ion-exchange processes is also practicable.

Pesticides and some other unwanted substances can be removed by adsorption to activated carbon.

Processes which deal with nitrates, pesticides etc. (and with hardness) are referred to as *tertiary treatment*.

13.5.1.3 Problems in the distribution system

Microbial growth/slimes in pipelines etc. can lower water quality and reduce pumping efficiency. Such growth has been linked with the presence of organic carbon in the water, and the increased availability of such carbon due to oxidation by chlorine and ozone. However, workers in Finland have recently shown that microbial growth can be low – despite high levels of carbon – if the concentration of phosphorus is low; this suggests that a more stringent removal of phosphorus from the water might permit more

Table 13.2 Bacteriological standards for treated drinking water: coliforms (see Appendix) and 'faecal streptococci'

	Maximum number			
	E. coli *or* thermotolerant coliforms[1]	*Total coliforms*	*Faecal coliforms*[1]	*'Faecal streptococci'*[2]
EC		0/100 ml	0/100 ml	0/100 ml
UK		0/100 ml	0/100 ml	0/100 ml
WHO	0/100 ml			

[1] 'Thermotolerant' and 'faecal' are synonymous; these coliforms produce acid and gas from lactose at 44°C in 24 hours.
[2] 'Faecal streptococci' refer to *Enterococcus faecalis* and related organisms.

cost-effective ways of controlling the problem [Miettinen, Vartiainen & Martikainen (1996) Nature *381* 654–655].

Pathogens may enter treated water via defective pipes, pumps, valves, service reservoirs/water towers etc., and may survive/grow if the level of residual disinfectant is inadequate. Such events are likely to involve a complex interaction of diverse parameters [coliform regrowth in drinking water: LeChevallier, Welch & Smith (1996) AEM *62* 2201–2211].

13.5.2 Small-scale water supplies

Small rural supplies (e.g. from streams or wells) can be disinfected by ultraviolet radiation after passage through an efficient fibre filter. Alternatively, disinfection of the filtered water can be effected by calcium hypochlorite tablets in a 'chlorinator' fitted in series with the incoming supply; the tablets gradually dissolve, releasing chlorine.

Small amounts of water (on expeditions etc.) can be disinfected by boiling or by tablets of *halazone* (*p*-carboxy-*N,N*-dichlorobenzenesulphonamide).

Novel methods have been used to improve the microbiological quality of drinking water in rural areas of some developing countries. For example, multi-layered local fabric, when used as a filter, can retain a high proportion of plankton-associated *Vibrio cholerae* and this procedure is likely to reduce the incidence and severity of cholera in countries such as Bangladesh [Huq *et al.* (1996) AEM *62* 2508–2512]. In Kenya, solar heating has been used to disinfect drinking water, and clinical trials of this method are currently in progress to test its efficacy in preventing water-borne diarrhoeal disease among children of the Maasai community [Joyce *et al.* (1996) AEM *62* 399–402].

13.6   PUTTING PATHOGENS TO WORK

Pathogens such as *Clostridium botulinum* and *C. tetani* produce highly potent neurotoxins which cause severe or fatal disease. Botulinum toxin, for example, inhibits release of the neurotransmitter *acetylcholine* – with consequent reduction in muscle activity, or muscle paralysis. However, this precise effect of the toxin has been put to good use in the treatment of e.g. hyperactive muscle disorders such as strabismus ('squint'). [Botulinum neurotoxin (medical uses): Schantz & Johnson (1992) MR *56* 80–99; Jankovic & Hallett (eds) *Therapy with Botulinum Toxin*, New York: Marcel Dekker, 1994.]

13.7   PLASTICS FROM BACTERIA: 'BIOPOL'

The natural bacterial storage polymer poly-β-hydroxybutyrate (PHB: section 2.2.4.1) is the basis of a range of biodegradable thermoplastics ('Biopol' – trade name of Zeneca, Great Britain). The homopolymer (PHB) is formed under appropriate conditions when *Alcaligenes eutrophus* uses glucose as the sole source of carbon. When the growth medium is appropriately supplemented, *A. eutrophus* forms co-polymers of hydroxybutyrate and hydroxyvalerate; the proportion of hydroxyvalerate (controlled by adjusting the composition of the growth medium) determines the properties of these co-polymers. The intracellular granules of polymer (either PHB or co-polymer) are harvested and purified to a fine, white powder.

Biopol can be used for containers, mouldings, fibres, films and coatings, and can be worked by blow-moulding and injection-moulding processes. While stable in normal use, Biopol is fully degradable after suitable disposal.

Recently, genes encoding enzymes of the PHB biosynthetic pathway in *A. eutrophus* have been modified and expressed in plastids of the green plant *Arabidopsis thaliana* – in which PHB was formed; this may allow the eventual use of such *transgenic* plants for the commercial production of bioplastics [review: Poirier, Nawrath & Somerville (1995) Biotechnology *13* 142–150]. Currently, efforts are being made to develop a transgenic form of the plant oilseed rape (*Brassica napus*) for the commercial production of bioplastics.

[Useful review on bioplastics: *Biotechnology and Bioengineering* (1996) *49* 1–14.]

14 Some practical bacteriology

14.1 SAFETY IN THE LABORATORY

A student new to bacteriology should be constantly aware that he or she is dealing with living organisms – which may include actual or potential pathogens. Good bacteriology is safe bacteriology, and it is wise to get to know the safety rules of the laboratory before carrying out any practical work; the following rules deserve special attention.

1. While working in the laboratory wear a clean laboratory coat to protect your clothing. Do not wear the coat outside the laboratory.
2. Put *nothing* into your mouth. It is potentially dangerous to eat, drink or smoke in the laboratory. For pipetting, use a rubber bulb (teat) or a mechanical device such as a 'pi-pump'; do not use your mouth for suction. If necessary, use self-adhesive labels.
3. Keep the bench – and the rest of the laboratory – clean and tidy.
4. Dispose of all contaminated wastes by placing them (not throwing them) into the proper container.
5. Leave contaminated pipettes, slides etc. in a suitable, active disinfectant for an appropriate time before washing/sterilizing them.
6. Avoid contaminating the environment with *aerosols* containing live bacteria/spores. An aerosol consists of minute (invisible) particles of liquid or solid dispersed in air; aerosols can form e.g. when bubbles burst, when one liquid is added to another, or when a drop of liquid falls onto a solid surface – things which can happen during many bacteriological procedures (see e.g. Fig. 16.2). Particles of less than a few micrometres in size can remain suspended in air for some time and may be inhaled by anyone in the vicinity; clearly, aerosols can be a potential source of infection. Bacteriological work is sometimes carried out in special cabinets (described later) – partly in order to avoid the risk of infection from aerosols.
7. Report all accidents and spillages, promptly, to the instructor or demonstrator.
8. Wash your hands thoroughly before leaving the laboratory.

14.2 BACTERIOLOGICAL MEDIA

A *medium* (plural: *media*) is any solid or liquid preparation made specifically for the growth, storage or transport of bacteria; when used for growth, the medium generally supplies all necessary nutrients. Before use, a medium must be *sterile*, i.e. it must contain no living organisms. (Methods for sterilizing media are given in Chapter 15.)

Before discussing the different media, it will be helpful to give again the meanings of a few words which are used very commonly in bacteriology; this is best done by giving the following outline of a simple laboratory procedure. To grow an organism such as *E. coli*, the bacteriologist takes an appropriate sterile medium and adds to it a small amount of material which consists of, or contains, living cells of that species; the 'small amount of material' is called an *inoculum*, and the process of adding the inoculum to the medium is *inoculation*. (The tools and procedures used for inoculation are described in sections 14.3 to 14.5.) The inoculated medium is then *incubated*, i.e. kept under appropriate conditions of temperature, humidity etc. for a suitable period of time. Incubation is usually carried out in a thermostatically controlled cabinet called an *incubator*. During incubation the bacteria grow and divide – giving rise to a *culture*; thus, a culture is a medium containing organisms which have grown (or are still growing) on or within that medium.

A liquid medium may be used in a test tube (which is stoppered by a plug of sterile cotton wool, or which has a simple metal cap – see e.g. Fig. 16.3) or in a glass, screw-cap bottle; a *universal bottle* (Fig. 14.1) is a cylindrical bottle of about 25 ml capacity, while a *bijou* is smaller (about 5–7 ml).

Most solid media are jelly-like materials which consist of a solution of nutrients etc. 'solidified' by *agar* (a complex polysaccharide gelling agent obtained from certain seaweeds). A solid medium is commonly used in a plastic *Petri dish* (illustrated in Fig. 16.2) – usually the size which has a lid diameter of about 9 cm. The medium, in a molten (liquid) state, is poured into the Petri dish and allowed to set; a Petri dish containing the solid medium is called a *plate*. (A *vented* Petri dish has three very small projections equally spaced around the edge of the inside of the lid; this keeps the lid of the (closed) Petri dish slightly raised, thus facilitating equilibrium between the air/gases inside and outside the Petri dish.)

14.2.1 Types of medium

For many chemolithoautotrophic bacteria the medium can be a simple solution of inorganic salts (CO_2 being used for carbon).

Nutritionally undemanding heterotrophs (such as *E. coli*) need only the common organic substances found in *basal media* (Table 14.1). Many bacteria will not grow in basal media, but may do so after the addition of substances such as egg, serum or blood; media which have been supplemented in this way are called *enriched media*.

A *selective medium* is one which supports the growth of certain bacteria in preference to others. An example is MacConkey's broth (Table 14.1) – in which the bile salts inhibit *non*-enteric bacteria but do not inhibit enteric

Table 14.1 Some common types of bacteriological medium[1]

Medium	Composition of medium: typical formulation (% w/v in water)
Basal medium	
Peptone water	Peptone (soluble products of protein hydrolysis) 1%; sodium chloride 0.5%
Nutrient broth[2]	Peptone 1%; sodium chloride 0.5%; beef extract 0.5–1%
Nutrient agar	Nutrient broth gelled with 1.5–2% agar
Differential medium	
MacConkey's agar	MacConkey's broth (see below) gelled with 1.5–2% agar
Enriched medium	
Blood agar	Nutrient agar (or similar medium) containing 5–10% defibrinated or citrated blood
Chocolate agar	Blood agar heated to 70–80°C until the colour changes to chocolate brown
Serum agar	Nutrient agar (or similar medium) containing 5% (v/v) serum
Enrichment medium	
Selenite broth	Peptone 0.5%; mannitol 0.4%; disodium hydrogen phosphate 1%; sodium hydrogen selenite ($NaHSeO_3$) 0.4%
Selective medium	
Deoxycholate–citrate agar (DCA)	Meat extract and peptone 1%; lactose 1%; sodium citrate 1%; ferric ammonium citrate 0.1%; sodium deoxycholate 0.5%; neutral red (pH 8.0 yellow to pH 6.8 red) 0.002%; agar 1.5%
MacConkey's broth	Peptone 2%; lactose 1%; sodium chloride 0.5%; bile salts (e.g. sodium taurocholate) 0.5%; neutral red 0.003%
Transport medium	
Stuart's transport medium	Salts; agar 0.2–1.0% (semi-solid or 'sloppy' agar); sodium thioglycollate; methylene blue (as redox indicator)

[1] A major source of information on several thousand microbiological media is *Handbook of Microbiological Media* (Atlas, RM & Parks, LC) CRC Press, 2nd edn, 1996; ISBN 0-8493-2638-9.
[2] In bacteriology, 'broth' may refer to any of various liquid media, but, when used without qualification, it commonly refers to nutrient broth.

species; this medium can be used e.g. to isolate enteric bacteria from a mixture of enteric and non-enteric bacteria when both types are present in an inoculum. (To some extent, *all* media are selective in that no medium can give equal support to the growth of every type of bacterium.)

An *enrichment medium* allows certain species to outgrow others by encouraging the growth of wanted organism(s) and/or by inhibiting the growth of unwanted species. Hence, if an inoculum contains only a few cells of the required species (among a large population of unwanted organisms), growth in a suitable enrichment medium can increase ('enrich') the proportion of required organisms. For example, selenite broth (Table 14.1) inhibits many types of enteric bacteria (including e.g. *E. coli*) but does not inhibit *Salmonella typhi*, the causal agent of typhoid. Suppose, for example, that we need to detect *S. typhi* in a specimen of faeces from a suspected case of typhoid. The specimen may contain only a few cells of *S. typhi*, so that it may be difficult or impossible to detect them among the vast numbers of non-pathogenic enteric bacteria. However, if an inoculum from the specimen is incubated in selenite broth, the proportion of cells of *S. typhi* increases to the point at which they can be detected more readily.

A *solid* medium is used, for example, to obtain the *colonies* (section 3.3.1) of a particular species. Many solid media are simply liquid media which have been solidified by a gelling agent such as gelatin or agar; agar is the most commonly used gelling agent because (i) it is not attacked by the vast majority of bacteria, and (ii) an agar gel does not melt at 37°C – a temperature used for the incubation of many types of bacteria. (By contrast, gelatin can be liquefied by some bacteria, and it is molten at 37°C.) One widely used agar-based medium is *nutrient agar* (Table 14.1), a general-purpose medium used for culturing (i.e. growing) many types of bacteria; it can also be enriched and/or made selective by the inclusion of appropriate substances.

Blood agar is an agar-based medium enriched with 5–10% blood; it is used e.g. for the culture (growth) of nutritionally 'fastidious' bacteria such as *Bordetella pertussis* (causal agent of whooping cough), and also to detect *haemolysis* (section 16.1.4.1). *Chocolate agar* is made by heating blood agar to 70–80°C until it becomes chocolate brown in colour; it is more suitable than blood agar for growing certain pathogens (e.g. *Neisseria gonorrhoeae*).

MacConkey's agar is an example of a *differential medium*, i.e. one on which different species of bacteria may be distinguished from one another by differences in the characteristics of their colonies etc. On MacConkey's agar, lactose-utilizing enteric bacteria (such as *E. coli*) form *red* colonies because they produce acidic products (from the lactose) which affect the pH indicator in the medium; enteric species which do not use lactose (e.g. most strains of *Salmonella*) give rise to colourless colonies.

Sorbitol MacConkey agar (SMAC) (medium CM813, Oxoid, Basingstoke, UK) resembles MacConkey's agar but contains sorbitol instead of lactose; it is useful for detecting the pathogenic (EHEC/VTEC) O157 strain of *E. coli* – which does not ferment sorbitol and which therefore forms colourless colonies on this medium. (Most strains of *E. coli* ferment sorbitol and form pink colonies.) The addition of a supplement consisting of cefixime (a 3rd generation cephalosporin antibiotic) and potassium tellurite (product SV48, Mast Diagnostics, Merseyside, UK) has been found to increase the isolation rate of O157 by suppressing the growth of other non-sorbitol-fermenters (such as *Proteus*).

Some solid media contain neither agar nor gelatin. For example, *Dorset's egg* is made by heating (and, hence, coagulating) a mixture of homogenized hens' eggs and saline; it is used e.g. as a *maintenance medium* – i.e. a medium used first for the growth and then the 'storage' of a given organism. Dorset's egg has been widely used for the storage of *Mycobacterium tuberculosis*.

Many media contain substances (e.g. peptone, tap water) whose *exact* composition is usually unknown. Sometimes it is necessary to use a medium in which all the constituents, including those in trace amounts, are quantitatively known; such a *defined medium* is prepared from known amounts of pure substances – e.g. inorganic salts, glucose, amino acids etc. in distilled or de-ionized water. A defined medium would be used e.g. when determining the nutritional requirements of a given species of bacterium.

A *transport medium* is used for the transportation (or temporary storage) of material (e.g. a swab) which is to be subsequently examined for the presence of particular organism(s); the main function of the medium is to maintain the viability of those organism(s), if present. A transport medium need not support growth; in fact, growth may be disadvantageous since waste products formed may adversely affect the survival of the organisms. One such medium, *Stuart's transport medium* (Table 14.1), is suitable e.g. for a range of anaerobic bacteria and for 'delicate' organisms such as *Neisseria gonorrhoeae*.

14.2.2   The preparation of media

Most media can be obtained commercially in a dehydrated, powdered form. Such media are commonly dissolved in the appropriate volume of water, sterilized, and dispensed to suitable sterile containers; as an alternative, some media are dispensed to containers before sterilization.

For most agar-based media (e.g. nutrient agar) the powdered medium is mixed with water and steamed to dissolve the agar; the whole is then sterilized in an *autoclave* (section 15.1.1.3) and subsequently allowed to cool to about 45°C, a temperature at which the agar remains molten. To prepare a *plate*, some 15–20 ml of the molten agar medium is poured into a sterile Petri dish which is left undisturbed until the agar sets. Blood agar plates are made by mixing molten nutrient agar (at about 45–50°C) with 5–10%, by volume, of (e.g. citrated) blood before pouring the plates.

For some uses (e.g. streaking: section 14.5.2), the surface of a newly made plate must be 'dried' – i.e. *excess* surface moisture must be allowed to evaporate; this is often achieved by leaving the plate, with the lid partly off, in a 37°C incubator for about 20 minutes. Spread plates (section 14.5.2) may also be dried.

To prepare a nutrient agar *slope* or *slant* (Fig. 14.1) the molten agar medium is allowed to set in a sterile bottle or test tube which has been placed at an angle to the horizontal.

Some types of medium cannot be sterilized by autoclaving because one or more of their constituents are destroyed at the temperatures reached in an autoclave. Such media include e.g. DCA (Table 14.1), which is steamed but not autoclaved, and those media which contain glucose or other heat-labile sugars; in preparing the latter type of medium the sugar solution is sterilized

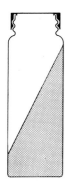

Fig. 14.1   A slope (also called a *slant*). The medium (stippled) has a large surface area available for inoculation; the thickest part of the medium is known as the *butt*. Slopes are commonly made of agar-based or gelatin-based media, and a slope is usually prepared in a universal bottle (as shown), in a bijou or in a test tube. Slopes are used e.g. for storing a purified strain of bacteria. A sterile slope is inoculated with an inoculum from a pure culture of the bacterial strain, and the slope is then incubated at a suitable temperature to allow growth; the slope can then be stored in a refrigerator at 4–6°C until needed e.g. as a source of inoculum.

separately by filtration (section 15.1.3) before being added to the rest of the (autoclaved) medium.

14.3   ASEPTIC TECHNIQUE

Instruments and media etc. must be *sterile* before use (section 15.1); if we are not *sure* of their sterility we will simply not know what is happening in our practical work. Additionally, *during* bacteriological procedures, instruments and materials must be protected from contamination by organisms that are constantly present in the environment. *Aseptic technique* involves the pre-use sterilization of all instruments, vessels, media etc., and avoidance of their subsequent contact with non-sterile objects – such as fingers, or the bench top etc.

Vessels containing sterile contents are kept closed except for the minimum time needed for access. Before opening a vessel (e.g. a sterile bottle, or one containing a pure culture), the rim of the screw-cap (or equivalent) is passed briefly through the bunsen flame to prevent any live contaminating organisms from falling into the vessel when the cap is removed; this procedure is called *flaming*, and it is used e.g. whenever an inoculum is withdrawn from a culture, or when a sterile medium is being inoculated. Flaming of the bottle's rim is carried out immediately before the vessel is closed. Flaming is generally not used when working with Petri dishes, and is never used when the contents of a vessel are likely to catch fire.

The risk of contamination in the laboratory may be further reduced by treating bench tops etc. with a suitable disinfectant, and by filtering the air to remove cells and spores of bacteria and fungi etc. Sometimes bacteriological work is done in a *safety cabinet* (= *sterile cabinet*). In a *class II* cabinet, sterile (filtered) air constantly flows down onto the work surface, and air passes to the exterior after further filtration (Fig. 14.2). Work is conducted via an open panel in the front of the cabinet. A *class III* cabinet is a gas-tight cabinet in

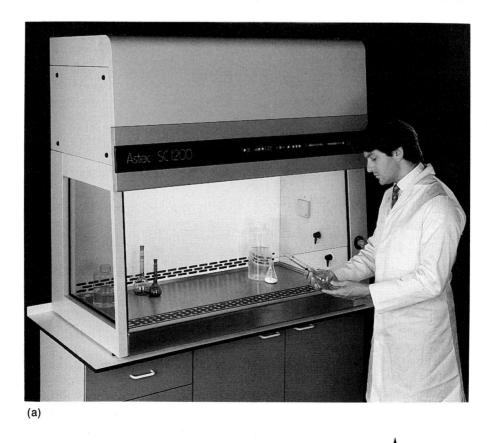

(a)

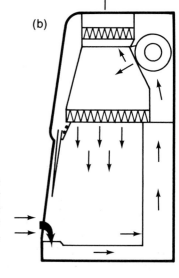

(b)

Fig. 14.2 (a) A class II safety cabinet, and (b) the pattern of air-flow (arrows) during use; filtered air passes downwards onto the working surface, and air passes outwards through a filter at the top of the cabinet. (Courtesy of Astec Environmental Systems, Weston-super-Mare, Avon, UK.)

which air is filtered before entry and before discharge to the environment; work is conducted via arm-length rubber gloves fitted into the front panel, and access to the interior of the cabinet is via a separate two-door sterilization/disinfection chamber. Class II cabinets are common in college laboratories; class III cabinets are used e.g. when working with highly pathogenic microorganisms.

14.4   THE TOOLS OF THE BACTERIOLOGIST

In most cases bacteria can be handled with one of the simple instruments shown in Fig. 14.3. A loop or straight wire is sterilized immediately before use by flaming: the wire portion of the instrument is heated to red heat in a bunsen flame and is then allowed to cool.

If a sterile loop is dipped into a suspension of bacteria and withdrawn, the loop of wire retains a small circular film of liquid containing a number of bacterial cells – and this can be used as an inoculum; the size of this inoculum will depend on (i) the concentration of cells in the suspension, and (ii) the size of the wire loop (which often carries 0.01–0.005 ml of liquid) – clearly, two factors which can be controlled. Even smaller amounts of liquid can be manipulated with a straight wire since this picks up only the minute volume of suspension which adheres to the wire's surface.

The loop and straight wire can also be used for picking up small quantities of solid material – e.g. small amounts of growth from a bacterial colony – simply by bringing the wire loop, or the tip of the straight wire, into contact with the material; the *amount* of material which adheres to the wire will be unknown, but usually this is not important. Liquid or solid inocula carried by a loop or straight wire can be used to inoculate either a liquid or a solid medium (section 14.5).

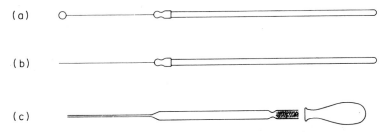

Fig. 14.3   Basic tools of the bacteriologist. (a) A loop: a piece of platinum, nickel-steel or nichrome wire, bent into a closed loop at the end and held in a metal handle of about 10–12 cm in length. (b) A straight wire: the metal handle carries a straight piece of wire of about 5–8 cm in length. (c) A Pasteur pipette: an open-ended glass tube, the narrow end of which has an internal diameter of about 1 mm; the wider end is plugged with cotton wool, before sterilization, and a rubber bulb (teat) is fitted immediately before use.

Both the loop and straight wire must always be flamed immediately after use so that they do not contaminate the bench or environment. Spattering, with aerosol formation, may occur when flaming a loop or straight wire containing the residue of an inoculum; for this reason, flaming is often carried out with a special bunsen burner fitted with a tubular hood. Alternatively, flaming may be carried out in a safety cabinet.

Larger volumes of liquid may be handled by means of Pasteur pipettes or graduated pipettes; suction is obtained either from a rubber bulb (teat) or from a mechanical device – the mouth is never used. Pipettes used in bacteriology are usually plugged with cotton wool (Fig. 14.3), before being sterilized, in order to avoid contamination from the bulb or from the mechanical pipetting device during use. Pipettes are usually sterilized (in batches) inside metal canisters or in thick paper envelopes; when a pipette is removed from the container, only the plugged end should be held so as to avoid contaminating the rest of the pipette. Pasteur pipettes are commonly used once only and then discarded into a jar of suitable disinfectant. Graduated pipettes which have been contaminated with bacteria are immersed in a disinfectant until they are safe to handle, when they can be washed, sterilized and re-used.

14.5   METHODS OF INOCULATION

14.5.1   Inoculating a liquid medium

To inoculate a liquid medium with a *liquid* inoculum, the loop (or straight wire) carrying the inoculum is simply dipped into the liquid medium, moved slightly, and then withdrawn. Inoculation can also be carried out with a Pasteur pipette. With a *solid* inoculum, the loop or straight wire may be rubbed lightly against the inside of the vessel containing the medium – to ensure that at least some of the inoculum is left behind when the instrument is withdrawn.

14.5.2   Inoculating a solid medium

Solid media may be inoculated in a variety of ways, particular methods being used for particular purposes.

Streaking (Fig. 14.4) is used when individual, well-separated colonies are required, and the (liquid or solid) inoculum is known to contain a large number of cells. In this method the inoculum is progressively 'thinned out' in such a way that individual, well-separated cells are left on at least some areas of the plate – usually in the third, fourth or fifth streakings (Fig. 14.4); on incubation, each well-separated cell gives rise to an individual colony.

In *stab inoculation*, a solid medium – e.g. the butt of a slope (Fig. 14.1) – is inoculated with a straight wire by plunging the wire vertically into the

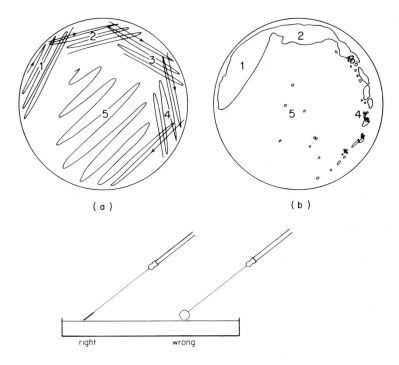

Fig. 14.4 Streaking: inoculating a plate to obtain individual colonies. (a) A loop carrying the inoculum is moved from side to side (i.e. streaked) across a peripheral region of the plate, following the path shown at 1. The loop is then flamed and allowed to cool; the *sterile* loop is now streaked across the medium as shown at 2. Streakings 3, 4 and 5 are similarly made, the loop being flamed and cooled between each streaking and after the last. (b) After incubation, those areas of the plate on which large numbers of cells had been deposited show areas of confluent growth, as at 1, 2 and 3; well-separated cells give rise to individual colonies, as at 5.

When streaking with a loop, the plane of the wire loop should *not* be vertical. Starting from the wrong position (see lower diagram), the correct position is achieved by rotating the loop's handle through 90°; in the 'right' position (lower diagram), the loop – *lightly* in contact with the surface of the medium – is moved towards and away from the viewer during streaking.

medium; the inoculum (at the tip of the wire) is thus distributed along the length of the stab. This procedure is used e.g. for inoculating deep, microaerobic or anaerobic parts of a medium.

A *spread plate* is made by spreading a small volume of liquid inoculum (e.g. 0.05–0.1 ml) over the surface of a solid medium by means of a sterile L-shaped glass rod (a 'spreader').

A *flood plate* is made by flooding the surface of a solid medium with a liquid inoculum and withdrawing excess inoculum with a sterile Pasteur pipette.

If the inoculum in a flood plate, or a spread plate, contains enough cells, incubation will give rise to a *lawn plate*: a plate in which the surface of the medium is covered with a layer of confluent growth (section 3.3.1).

A plate is sometimes inoculated with a *swab*: typically, a compact piece of cotton wool attached securely to one end of a thin wooden or plastic stick or a piece of wire. A sterile swab is used e.g. for sampling organisms at a given site (such as the throat). After exposure, the swab is drawn lightly across the surface of a plate of suitable medium – taking care that all areas of the cotton wool make contact with the medium. The swab may be used to inoculate the entire surface of a plate; alternatively, it may be used to inoculate a small, peripheral area, the inoculum then being further distributed by streaking.

14.6 PREPARING A PURE CULTURE FROM A MIXTURE OF ORGANISMS

Some of the basic techniques of bacteriology can be illustrated by following through a common procedure such as the *isolation* of a particular strain or species from a mixture of organisms. The following describes the isolation of *E. coli* from a sample of sewage (which usually contains a range of enteric and non-enteric organisms).

A loopful of sewage is streaked onto a plate of MacConkey's agar (section 14.2.1), and the plate is then incubated for 18–24 hours at 37°C. (Plates are commonly incubated upside-down, i.e. with lid below; if incubated the right way up, water vapour from the medium may condense on the inside of the lid and drop onto the surface of the medium – with possible disruption of the streaked inoculum.) During incubation, well-separated cells of *any* species capable of growing on the medium will each give rise to an individual colony. After 18–24 hours on MacConkey's agar, *E. coli* forms round, red colonies of about 2–3 mm in diameter – but not all colonies with this appearance will necessarily be those of *E. coli*. The next step is to choose several such colonies for further examination; since *E. coli* is very common in sewage, at least one of the selected colonies is likely to be that of *E. coli*.

Before identification can be attempted it is necessary to ensure that each of the selected colonies contains cells of only one species. There is always the possibility that a given colony – even a well-separated one – may contain the cells of two different species; this may occur if, during streaking, two different cells had been deposited (by chance) very close together on the surface of the medium. To resolve this doubt, each colony is subcultured; *subculturing* is a process in which cells from an existing culture or colony are transferred to a fresh, sterile medium. To subculture a given colony, the surface of the colony is touched lightly with a sterile loop so that a minute quantity of growth adheres to the loop; the growth is then streaked (in this case) onto a plate of sterile MacConkey's agar. (Some bacteria form very small colonies, and in such cases it is often easier to subculture by touching the surface of the colony with the tip of a straight wire; the inoculum is then carried (on the straight wire) to a fresh medium where it is streaked with a sterile loop.) Each plate, inoculated from a single colony, is then incubated. If each red colony had been that of a single species, we should now have several pure cultures – at least one of which is likely to be that of *E. coli*. (To increase the chances that a culture is pure it may be subcultured again.) Each pure culture can now be examined by the identification process outlined in Chapter 16.

14.6.1 Immunomagnetic separation (IMS)

IMS is a comparatively recent technique which can help to separate one type of organism from a mixture of organisms by exploiting the specificity of cell-surface antigens. In this method, antibodies to the required organism are used to coat small magnetic beads (Dynabeads, trade name of Dynal, Skøyen, Oslo, Norway). The coated beads and the sample are mixed thoroughly, and the beads are then drawn to one side of the containing vessel by a magnetic field; the beads will thus have captured any cells whose antigens have bound to the antibodies. After washing, the beads can be used to inoculate an appropriate growth medium.

IMS has been used e.g. in an epidemiological study for detecting the *E. coli* O157 serotype (Table 11.2, EHEC) in bovine faeces [Chapman *et al.* (1994) JMM *40* 424–427].

14.7 ANAEROBIC INCUBATION

Anaerobic bacteria are incubated under anaerobic conditions. This can be achieved by using an *anaerobic jar* – one form of which is the McIntosh and Fildes' jar: a strong, metal cylindrical chamber with a flat, circular, gas-tight lid. The jar is loaded with a stack of inoculated plates, and the lid is secured with a screw-clamp. The jar is then connected to a suction pump via one of two valves in the lid; after a few minutes the valve is closed. Hydrogen (in a rubber bladder) is then passed into the jar via the other valve; this valve is then closed. Evacuation and re-filling may be repeated several times. On the inside of the lid is a gauze envelope containing a catalyst (e.g. palladium-coated alumina pellets) which promotes chemical combination between hydrogen and the last traces of oxygen. The jar is then placed in an incubator for an appropriate period of time. (Within the jar, plates are stacked the right way up; if stacked upside-down, the agar may be sucked from the base of the Petri dish by the vacuum.)

Another (more modern) form of anaerobic jar is a stout cylindrical vessel of strong, transparent plastic with a flat, gas-tight lid. The jar is loaded with plates; water is then added to a small packet of chemicals which is dropped into the jar immediately before the lid is secured with a screw-clamp. The chemicals liberate hydrogen which, in the presence of a catalyst, combines with all the oxygen in the jar. Since, in this case, there is no vacuum, the plates can be inserted upside-down (i.e. lid-side down) so as to avoid the problem of condensation.

Most anaerobic jars contain a redox indicator which indicates the state of anaerobiosis in the jar. In metal jars the indicator is placed in a small glass side-arm, while in plastic jars an indicator-soaked pad is usually visible through the wall of the jar.

Some anaerobes can be grown (without an anaerobic jar) in media such as *Robertson's cooked meat medium* (minced beef heart, beef extract (1%), peptone (1%), sodium chloride (0.5%) and a reducing agent, e.g. L-cysteine or thioglycollate); the medium, sterilized by autoclaving, is stored in screw-cap universal bottles which are sometimes equilibrated under oxygen-free conditions after inoculation and before closure.

Anaerobic cabinets allow samples/cultures to be handled/incubated under oxygen-free conditions with control of e.g. temperature, humidity and CO_2 concentration; items inside the cabinet can be manipulated by means of gas-tight gloves fixed into the front panel.

14.8   COUNTING BACTERIA

The total number of (living and dead) cells in a sample is called the *total cell count*; the number of living cells is the *viable cell count*. Counts in liquid samples are usually given as the number of cells per millilitre (or per 100 ml).

14.8.1   Total cell count

The total cell count in a liquid sample (e.g. a broth culture) can be estimated by direct counting in a counting chamber (Fig. 14.5).

Another method – the *direct epifluorescent filter technique* (DEFT) – is used e.g. for counting organisms in milk. Essentially, the milk is passed through a membrane filter, and the cells retained on the filter are stained with a fluorescent dye; ultraviolet radiation is then beamed onto the filter, and the (fluorescent) cells are seen (through a microscope) as bright particles against a dark background. (For DEFT, the milk is pre-treated to disrupt fat globules etc. which would otherwise block the filter.)

The total cell count can also be estimated by comparing the *turbidity* of the sample with that of each of a set of tubes (*Brown's tubes*) containing suspensions of barium sulphate in increasing concentration; the tubes range from transparent (tube 1), through translucent, to turbid and opaque (tube 10). For a given species of bacterium the turbidity of a particular tube corresponds to the turbidity of a suspension of cells of known concentration. The sample is examined in a tube of size and thickness equivalent to those containing the standard suspensions; the turbidity of the sample is matched, visually, with that of a particular tube, and the concentration of the sample is then read from a table supplied with the tubes.

14.8.2   Viable cell count

Most methods of estimating the viable cell count involve the inoculation of a solid medium with the sample (or diluted sample). After incubation, the number of cells in the inoculum can be estimated from the number of colonies which develop on or within the medium. It is always assumed that each colony has arisen from a single cell; the number of cells which actually give rise to colonies depends at least partly on the type of medium used and on the conditions of incubation.

In the *spread plate* (or *surface plate*) method, an inoculum of about 0.05–0.1 ml is spread over the surface of a sterile plate, as described earlier; the plate is 'dried' (section 14.2.2), incubated, and the viable cell count is estimated from (i) the number of colonies, (ii) the volume of inoculum used, and (iii) the degree (if any) to which the sample had been diluted. If a sample is suspected of containing many cells – e.g. 10^6 cells/ml – it can be diluted in 10-fold steps, and an inoculum from each dilution spread onto a separate plate; at least one dilution will give a countable number of colonies.

In the *pour plate* method, the (liquid) inoculum is mixed with a molten agar-based medium (at about 45°C) which is then poured into a Petri dish and allowed to set; on incubation, colonies develop within (as well as on) the medium, and the viable count is calculated as in the spread plate method.

Yet another method for viable count is *Miles and Misra's method* (Fig. 14.6).

A sample likely to contain small numbers of bacteria (e.g. water from a *clean* river) may be passed through a sterile membrane filter of pore size about 0.2 μm – which retains cells on the upper surface; a volume of, say, 100 ml or more may be filtered. The membrane is then placed (cell-side uppermost) onto an absorbent pad saturated with a suitable medium; on incubation, nutrients diffuse through the membrane, and colonies develop from those cells capable of growth under such conditions. The viable cell count can then be estimated from (i) the number of colonies on the membrane, and (ii) the volume of sample filtered.

14.8.3   Counting cells in (or on) solids

Sometimes we need to count the bacteria in a sample of food or fabric etc. In order to do this, the particles of food etc. must be broken up and/or the bacteria must be separated from the sample; ideally, clumps of bacteria should also be broken up. The sample and a diluent (such as Ringer's solution) can be sealed into a sterile plastic bag (a *stomacher*) which is then subjected to mechanical agitation; in this way, at least some of the bacteria can be brought into suspension in the diluent.

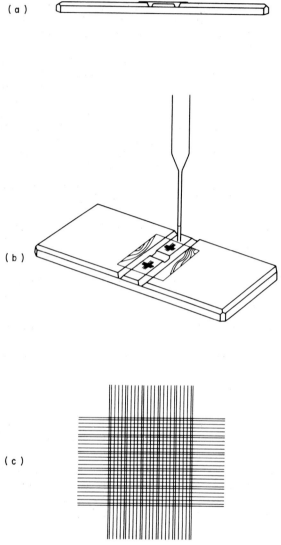

Fig. 14.5 A typical counting chamber (haemocytometer). The instrument, seen from one side at (a), consists of a rectangular glass block in which the central plateau lies precisely 0.1 mm below the level of the shoulders on either side. The central plateau is separated from each shoulder by a trough, and is itself divided into two parts by a shallow trough (seen at (b)). On the surface of each part of the central plateau is an etched grid (c) consisting of a square which is divided into 400 small squares, each $1/400$ mm^2. A glass cover-slip is positioned as shown at (b) and is pressed firmly onto the shoulders of the chamber; to achieve proper contact it is necessary, while pressing, to move the cover-slip (slightly) against the surface of the shoulders. Proper (close) contact is indicated by the appearance of a pattern of coloured lines (Newton's rings), shown in black and white at (b).

Using the chamber. A small volume of a bacterial suspension is picked up in a Pasteur pipette by capillary attraction; the thread of liquid in the pipette should not be more than 10 mm. The pipette is then placed as shown in (b), i.e. with the opening of the pipette in contact with the central plateau, and the side of the pipette against the cover-slip. With the pipette in this position, liquid is automatically drawn by capillary attraction into the space bounded by the cover-slip and part of the central plateau; *the liquid should not overflow into the trough.* (It is sometimes necessary to tap the end of the pipette, *lightly*, against the central plateau to encourage the liquid to enter the chamber.) A second sample can be examined, if required, in the other half of the counting chamber. The chamber is left for 30 minutes to allow the cells to settle, and counting is then carried out under a high power of the microscope – which is focused on the grid of the chamber. Since the volume between grid and cover-slip is accurately known, the count of cells per unit volume can be calculated.

A worked example. Each small square in the grid is $1/400$ mm^2. As the distance between grid and cover-slip is $1/10$ mm, the volume of liquid over each small square is $1/4000$ mm^3 – i.e. $1/4\,000\,000$ ml.

Suppose, for example, that on scanning all 400 small squares, 500 cells were counted; this would give an average of $500 \div 400 (= 1.25)$ cells per small square, i.e. 1.25 cells per $1/4\,000\,000$ ml. The sample therefore contains $1.25 \times 4\,000\,000$ cells/ml, i.e. 5×10^6 cells/ml. Several counts may be made and averaged.

If the sample had been diluted before examination (because it was too concentrated), the count obtained must be multiplied by the dilution factor; for example, if diluted 1-in-10, the count should be multiplied by 10.

N.B. The chamber described above is the *Thoma chamber*; in a *Helber chamber* the distance between central plateau and cover-slip is 0.02 mm.

14.9 STAINING

Staining is often used to detect, categorize or identify bacteria, or to observe specific bacterial components; in most cases the cells are killed and 'fixed' before being stained.

Dyes etc. are usually applied to a thin film of cells on a glass microscope slide. Cells from a pure culture may be examined as follows. A loopful of water is placed on a clean slide, and (using the loop) a speck of growth from a colony is mixed ('emulsified') with the water to form a suspension of cells. Using the loop, the suspension is spread over an area of one or two square centimetres and allowed to dry – forming a *smear*. The smear is then fixed by passing it quickly through a bunsen flame twice; it is then ready for staining and subsequent examination under the microscope.

A smear may also be made directly from the sediment of centrifuged urine, or pus from an abscess.

14.9.1 The Gram stain

The background to this stain is given in section 2.2.9. One of many versions of the procedure is as follows.

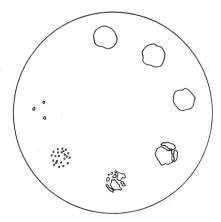

Fig. 14.6   Miles and Misra's method (the 'drop method') for estimating viable cell counts. The sample is first diluted, usually in 10-fold steps – i.e. 1/10, 1/100 . . . etc. One drop (of known volume) from each dilution is then placed at a separate recorded position on the surface of a 'dried' plate of suitable medium. The drops are allowed to dry, and the plate is incubated until visible growth develops. Drops which contained large numbers of viable cells give rise to circular areas of confluent growth. Any drop which contained less than about 15 viable cells will give rise to a small, countable number of colonies; by assuming that each colony arose from a single cell, the viable count can be estimated from (i) the number of colonies, (ii) the volume of the drop, and (iii) the dilution factor.

The same Pasteur pipette can be used throughout *if* the drops are taken first from the highest dilution and then from the next highest dilution, and so on.

A Pasteur pipette can be *calibrated*, i.e. the volume of drops delivered by the pipette can be measured. This is done by drawing into the pipette a measured volume (e.g. 1 ml) of water, and counting the number of drops formed when this volume is discharged; if 1 ml is discharged in, say, 30 drops, then each drop is about 1/30 ml.

A heat-fixed smear (see above) is stained for 1 minute with *crystal violet*; it is then rinsed briefly under running water, treated for 1 minute with *Lugol's iodine* (a solution of iodine and potassium iodide in water), and briefly rinsed again. Decolorization is then attempted by treating the stained smear with a solvent such as ethanol (95%), acetone or iodine–acetone. This is the critical stage: with the slide tilted, the solvent is allowed to run over the smear only for as long as dye runs *freely* from it (about 1–3 seconds); the smear is then *immediately* rinsed in running water. At this stage, any Gram-negative cells will be colourless; Gram-positive cells will be violet. The smear is now counterstained for 30 seconds with dilute *carbolfuchsin* to stain (red) any Gram-negative cells present. After a brief rinse, the smear is blotted dry and examined under the oil-immersion objective of the microscope (magnification about 1000×).

Certain species of bacteria do not give a clear or constant reaction to the Gram stain – sometimes reacting positively, sometimes negatively; these bacteria are said to be *Gram-variable*. To avoid the problems of Gram-variability in taxonomy (classification), many bacteria are described as *Gram-type-positive* or *Gram-type-negative* according to whether their cell walls are of the Gram-positive or Gram-negative type, respectively (section 2.2.9).

14.9.1.1 Gram staining living cells

The Gram reaction of *living* cells has been determined by differential fluorescence staining: the LIVE *Bac*Light™ bacterial Gram stain kit (Molecular Probes Inc., Eugene, Oregon, USA) uses two fluorescent dyes, one of which (hexidium) selectively stains Gram-positive cells (which fluoresce orange-red); Gram-negative cells fluoresce green.

14.9.2 The Ziehl–Neelsen stain (acid-fast stain)

'Acid-fast' bacteria (AFB) differ from all other bacteria in that once they are stained with hot, concentrated carbolfuchsin they cannot be decolorized by mineral acids or by mixtures of acid and ethanol; such bacteria include e.g. *Mycobacterium tuberculosis*. A heat-fixed smear is flooded with a concentrated solution of carbolfuchsin, and the slide is heated until the solution steams; it should not boil. The slide is kept hot for about 5 minutes, left to cool, and then rinsed in running water. Decolorization is attempted by passing the slide through several changes of acid–alcohol (e.g. 3% v/v concentrated hydrochloric acid in 90% ethanol). After washing in water, the smear is counterstained with a contrasting stain (such as malachite green), washed again, and dried. Acid-fast cells stain red, others green.

14.9.3 Capsule stain

Bacterial capsules may be demonstrated e.g. by *negative staining* (Plate 2.3: *centre, right*). The cells are emulsified with a loopful of e.g. *nigrosin* solution on a clean slide and are then overlaid with a cover-slip; under the oil-immersion or 'high dry' (× 40) objective lens of the microscope the capsule appears as a clear, bright zone between a cell and its dark background.

14.9.4 Endospore stain

See section 16.1.1.4.

14.9.5 Distinguishing live from dead cells by staining

Live and dead bacteria can be distinguished e.g. by a simple one-step staining technique (*Bac*Light™, Molecular Probes Inc., Eugene, Oregon, USA). The cells are treated with a mixture of two fluorescent dyes: SYTO 9 and

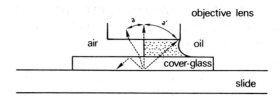

Fig. 14.7 Oil-immersion objective lens: the reason for using oil. In the diagram, the (thin) specimen lies between slide and cover-glass, and rays from one point in the specimen are shown travelling towards the objective lens.

When immersion oil – refractive index (RI) approx. 1.5 – fills the space between lens and cover-glass (right-hand side of diagram), rays of light passing upwards from the cover-glass enter a medium whose RI is similar to that of the glass itself (approx 1.5); consequently, each ray will continue on its original path (unrefracted) as though it were still travelling through glass. Such a ray is shown entering the objective lens; note that rays can enter at the widest angle.

When air (RI = 1) separates lens and cover-glass (left-hand side of diagram), a ray whose angle is similar to that shown on the right-hand side of the diagram cannot leave the glass; instead, it is reflected back into the glass via the upper, inner surface of the cover-glass. Rays which make a smaller angle with the vertical (as shown) can enter the objective lens; however, in this case, less light enters the lens. In this mode, some of the (image-forming) rays from the specimen do not enter the lens.

Numerical aperture (NA) is a characteristic of a lens given by:

$$NA = n \times \sin\theta$$

where n is the refractive index of the material between lens and cover-glass, and θ is *half* the maximum angle at which rays enter the lens. In the diagram, θ is the angle a in air, and a′ with immersion oil.

The maximum resolving power of a lens is related to its NA.

propidium iodide. Under the fluorescence microscope, SYTO 9 fluoresces green when bound to DNA in living cells; propidium iodide, which is generally excluded from living cells, fluoresces red when bound to DNA in dead cells. Living cells fluoresce green, dead cells red.

[Methods for assessing the viability of microorganisms have been discussed by Lloyd & Hayes (1995) FEMSML *133* 1–7.]

14.10 MICROSCOPY

In bacteriology there is often a need to use high magnification (e.g. 1000×), and for this the microscope must have an *oil-immersion* objective lens (magnification about 100×) and a suitable eyepiece (about 10×). When using oil-immersion objectives, a drop of immersion oil fills the space between the lens and the top of the cover-glass (or, as is often the case, between lens and smear).

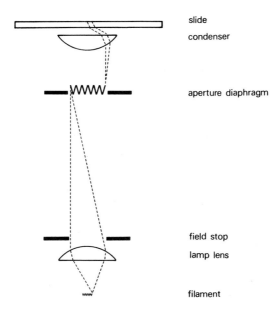

Fig. 14.8 Köhler illumination: ray diagram to show the principle. Rays from the lamp's filament are focussed to form an image (size exaggerated) in the lower focal plane of the substage condenser; coinciding with this plane is an iris diaphragm (*aperture diaphragm*) which forms part of the substage condenser unit. Because the image is in the condenser's focal plane, rays from any given point in the image emerge from the condenser as parallel rays (which illuminate the specimen on the slide). If a microscope has been set up for Köhler illumination, the edge of the field stop should become visible in sharp focus *in the plane of the specimen* if the field stop is gradually closed.

What is the oil for? A powerful objective lens has a very short focal length, and light from the specimen must enter the lens at a *wide* angle; this is possible only if the space between lens and specimen is filled with a material that has a suitable refractive index (Fig. 14.7).

The maximum useful magnification obtainable from a given objective lens is 1000 times its *numerical aperture* (NA) (Fig. 14.7).

14.10.1 Köhler illumination

For the best image, the specimen must be illuminated evenly, regardless of any unevenness in the source of light. Köhler illumination, which is optimal, uses a lamp fitted with a condensing lens and an adjustable diaphragm (=*field stop*) that determines the diameter of the illuminating beam; when set up correctly, an image of the lamp's filament is formed in the lower focal plane of the microscope's substage condenser, and rays from *each* point of this image pass through the condenser to emerge as parallel rays which illuminate the specimen and enter the objective lens (Fig. 14.8).

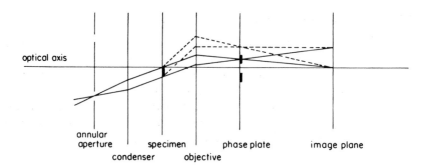

optical axis

annular
aperture   specimen   phase plate   image plane
     condenser   objective

Fig. 14.9 Phase-contrast microscopy: simplified diagrammatic scheme to show the principle.

Within a transparent or translucent specimen, the incident light has illuminated a region which causes diffraction; the first-order waves (*dashed line*) are retarded by about one-quarter of a wavelength ($1/4\lambda$). If these first-order waves and the non-diffracted (zero-order) waves interact in a normal (bright-field) microscope, the resultant wave would have an amplitude similar to that of the background waves so that the feature would not be clearly distinguished; the resultant and background waves would differ slightly in phase, but this cannot be detected by the eye.

In a phase-contrast microscope the condenser has an annular (ring-shaped) aperture in its front focal plane so that a *hollow* cone of light can be focused (as a small bright ring) onto a *phase plate* located in the back focal plane of the objective lens (see diagram). The phase plate is a glass disc on which has been deposited a ring of material (e.g. magnesium fluoride) of such thickness that it retards by $1/4\lambda$ the light that passes through it.

In the absence of a specimen all the light passes through the phase ring. When a specimen (e.g. an unstained cell) is examined, the zero-order and background waves (solid lines) pass through the ring but the first-order waves (dashed lines) pass through the phase plate via regions outside the ring; thus, by retarding the zero-order and background waves by $1/4\lambda$, the phase ring brings all the waves into the same phase. In the image plane, interaction between zero-order and first-order waves gives rise to a visible image because the resultant wave has an amplitude greater than that of the background waves (i.e. the image-forming wave is brighter than the background waves). (This is 'negative' or 'bright' phase-contrast microscopy; the opposite effect (image darker than background) is seen in 'positive' or 'dark' phase-contrast microscopy.)

An image of greater clarity is obtained when the condenser contains a green filter.

Compared with bright-field microscopy, the lamp must be more powerful because only a proportion of the lamp's output is used for image-making.

Care must be taken to ensure that the annular diaphragm in the condenser matches the particular phase-contrast objective lens being used.

The small bright ring on the phase plate must coincide *exactly* with the phase ring (see diagram). Adjustment, if necessary, is made when the condenser is in the correct position and the specimen has been brought into focus to the maximum extent possible. Under these conditions, remove the eyepiece and replace it with the specialized 'telescope' used for this purpose; the telescope is focused on the phase plate, and adjustments to the position of the small bright ring can be made e.g. by adjusting centring screws on the condenser (depending on the particular model of microscope).

Diagram from *Dictionary of Microbiology and Molecular Biology*, Singleton & Sainsbury, 2nd edition, 1987, p. 550, by courtesy of the publisher, John Wiley, Chichester, UK.

14.10.2 Phase-contrast microscopy

Ordinary (*bright-field*) microscopy can reveal fine detail when different parts of the specimen absorb different amounts of light or (perhaps due to staining) absorb different colours. *Un*stained cells can sometimes be seen – perhaps well enough to be counted – but with little detail.

Phase-contrast microscopy can reveal fine detail e.g. in transparent/translucent, unstained living cells by altering the phase difference between diffracted and non-diffracted light from the specimen (Fig. 14.9).

14.10.3 Literature on microscopy

Particularly helpful booklets on microscopy (containing both theoretical and practical information) have been produced by the major manufacturers of microscopes. These include *Microscopy from the Very Beginning*, by Friedrich K. Möllring (Carl Zeiss, Germany), and (a more comprehensive treatment) *The Microscope and its Applications*, by Hans Determann and Friedrich Lepusch (Leitz Wetzlar, Germany).

15 Man against bacteria

Bacteria can be a nuisance, or even dangerous, in many everday situations, and we therefore need methods to eliminate them or to inhibit their activities. Sometimes it is necessary to destroy, completely, all forms of life on a given object – as, for example, when surgical instruments are prepared for use. At other times it may be sufficient merely to eliminate only the potentially harmful organisms. There is also the special problem of inactivating pathogenic bacteria on or within living tissues.

15.1 STERILIZATION

Any procedure guaranteed to kill *all* living organisms – including endospores (section 4.3.1) and viruses – is called a *sterilization* process. (Since sterilization kills *all* organisms, expressions such as 'partial sterilization' are meaningless.) Ideally, sterilization methods should be efficient, quick, simple and cheap, and they should be applicable to a wide range of materials; sterilization is usually carried out by physical methods – commonly by the use of heat.

15.1.1 Sterilization by heat

The cells of different species of bacteria vary in their susceptibility to heat, and endospores are much more resistant than vegetative cells; vegetative cells generally die rapidly in boiling water, while endospores may survive for long periods of time.

The sterilizing power of heat depends not only on temperature but also on factors such as time, the presence of moisture, and the number and condition of the microorganisms present. Note that the larger the initial population of bacteria, the longer will be the time needed to achieve sterility at a given temperature; this can be seen from Fig. 12.2.

15.1.1.1 Fire

Fire is used e.g. for the rapid sterilization of surfaces and loops etc. (sections 14.3, 14.4), while disposable items such as used surgical dressings and one-use syringes may be sterilized – and destroyed – by incineration. However, less destructive methods are used for most other purposes.

15.1.1.2   The hot-air oven

This apparatus is used e.g. for the sterilization of heat-resistant items such as clean glassware. In use, a temperature of 160–170°C is maintained for 60–90 minutes; this denatures proteins, desiccates cytoplasm and oxidizes various components in any organisms present. Within the oven, air should be circulated by a fan to ensure that all parts are kept at the required temperature; items should be well spaced in order not to impede the flow of air.

15.1.1.3   Sterilization by steam: the autoclave

Steam can sterilize at lower temperatures (for shorter times) than those used in a hot-air oven. At normal atmospheric pressure, steam has a temperature of only 100°C – a temperature at which some endospores can survive for long periods – but, when *under pressure*, steam can reach higher temperatures suitable for sterilization; in fact, there is a definite relationship between the pressure and temperature of *pure* steam, i.e. steam containing no air: the higher the pressure the higher the temperature. When sterilizing with steam, items to be sterilized are placed inside a strong, metal, gas-tight chamber (an *autoclave*). Steam is produced within the chamber (Fig. 15.1) or, in larger autoclaves, is piped in from a boiler; air passes out through a valve until the chamber is filled with pure steam – at which time the valve is closed. The pressure and temperature of the steam rise as heating is continued (Fig. 15.1) or as more steam is piped in. At a pre-determined pressure, an (adjustable) valve opens – thus determining the pressure/temperature within the autoclave; steam which escapes (via the valve) is replaced by steam generated in the chamber, or piped in, so that pressure in the autoclave remains constant.

The time allowed for sterilization must be sufficient for all parts of the *load* (i.e. items/materials being sterilized) to reach the sterilizing temperature and to stay at that temperature until any organisms present have been killed.

Typical temperature/time combinations used in autoclaves are (i) 115°C (at a pressure of 0.68 bar [= 69 kPa; 10 lb/inch^2] above atmospheric pressure) for 35 minutes; 121°C (1.02 bar [103 kPa; 15 lb/inch^2]) for 15–20 minutes; 134°C (2.04 bar [207 kPa; 30 lb/inch^2]) for 4 minutes. These times may be varied according to the nature of the load and the nature and degree of contamination (see section 15.1.1). Note that the timing does not start until all parts of the load have reached the sterilizing temperature. Note also that an autoclave's pressure gauge may show pressure either in terms of the pressure *above atmospheric* (as given above) or as the *absolute pressure* (abs), i.e. steam pressure + atmospheric pressure; moreover, the gauge may be calibrated in different units in different autoclaves (approx. equivalents: 1 bar = 101 kPa = 14.7 lb in^2 = 760 mmHg).

For effective sterilization the steam must be saturated, i.e. it must hold as much water in vapour form as is possible for the given temperature and

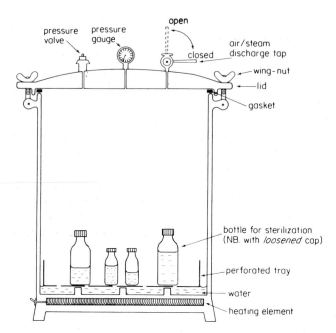

Fig. 15.1 A typical, small laboratory autoclave. Water is placed in the bottom of the chamber. Objects for sterilization are placed on the perforated tray which holds them above the water. The lid, with the air/steam discharge tap *open*, is clamped securely in position; the rubber gasket ensures a gas-tight seal. The heating element is switched on and the water boils. Steam fills the chamber, eventually displacing all the air (which leaves via the discharge tap). Pure steam begins to issue vigorously from the discharge tap, which is then closed. As heating continues, water continues to vaporize so that the pressure (and hence temperature) within the chamber increases. Once the desired pressure/temperature has been reached (see text), a pre-set pressure valve opens; steam escapes, thus maintaining the pressure at a steady level. When the appropriate time has elapsed (see text), the heating element is switched off and the autoclave is allowed to cool until the pressure inside the chamber (indicated by the gauge) does not exceed atmospheric pressure; the lid can then be opened safely and the sterile contents removed.

pressure; no air should be present because air upsets the pressure–temperature relationship: an air–steam mixture at a given pressure has a lower temperature than that of pure steam at the same pressure. Hence, *all* air must be purged from the chamber – and from all items within the chamber – before the valve is closed.

Small portable laboratory autoclaves generally resemble the domestic pressure cooker both in principle and mode of use (Fig. 15.1). In larger autoclaves, such as those in hospitals, steam is usually piped to the autoclave chamber from a boiler, and factors such as timing, pressure and steam quality are often controlled automatically. In some models steam is admitted at the top of the chamber so that air is displaced downwards; this is more effective than upward displacement (used in small autoclaves) since steam is lighter than air under these conditions.

In another type of autoclave, air is removed from the chamber by a vacuum pump before steam is admitted; this allows rapid and thorough penetration by the steam of porous materials such as dressings or bed linen – materials which tend to trap air.

Some materials cannot be sterilized by autoclaving; these include water-repellent substances (e.g. petroleum jelly) and substances which are volatile or are heat-labile (i.e. destroyed by heat). Some of these materials (such as petroleum jelly) can be sterilized in a hot-air oven.

Certain materials, which would be damaged by autoclaving, may be sterilized by a combination of steam at reduced pressure (at e.g. 80°C) and formaldehyde; this method kills endospores within about 2 hours, and is used for sterilizing heat-sensitive surgical instruments, plastic tubing, woollen blankets etc.

15.1.2   Sterilization by ionizing radiation

Ionizing radiation – e.g *beta*-rays (electrons), *gamma*-rays, X-rays – sterilizes by supplying energy for a variety of lethal chemical reactions in the contaminating organisms. *Gamma*-radiation (typically using a cobalt-60 source) is widely used e.g. for the sterilization of pre-packed biological equipment such as plastic Petri dishes and syringes.

15.1.3   Sterilization by filtration

A liquid known to be free of the smallest microorganisms (e.g. viruses and subviral agents) may be sterilized by passing it through a *membrane filter* (of suitable pore size) which can retain all other microorganisms; the liquid may be drawn through the filter by reduced pressure in the (sterile) receiving vessel, or forced through e.g. by a syringe plunger. The filter itself consists of a thin sheet of cellulose acetate, polycarbonate or similar material; the pore size may be e.g. 0.2 μm.

Filtration is used e.g. for the sterilization of serum (for laboratory use), solutions of heat-labile antibiotics and media containing heat-labile sugars.

15.1.4   Sterilization by chemical agents

Chemicals used for sterilization (*sterilants*) are necessarily highly reactive and damaging to living tissues; they therefore require careful handling, and tend to be used only in larger institutions with suitable equipment and personnel.

Ethylene oxide (C_2H_4O) – a water-soluble cyclic ether – is a gas at temperatures above 10.8°C and forms explosive mixtures with air; it is therefore used diluted with another gas such as carbon dioxide or nitrogen.

For sterilization, the gas mixture is used in a special chamber, and the temperature, humidity, time, and concentration of ethylene oxide must be carefully controlled. Ethylene oxide is an alkylating agent which reacts with various groups in proteins and nucleic acids; it is used e.g. for sterilizing clean medical equipment, bed linen, and heat-labile materials such as certain plastics.

Other sterilants include glutaraldehyde and β-propiolactone.

'Gas plasma' refers to a process in which hydrogen peroxide is injected into the sterilizing chamber (a dry atmosphere under reduced pressure) and radio-frequency energy converts the hydrogen peroxide to reactive chemical species (which effect sterilization). Conditions are important; failure to sterilize may be due e.g. to (i) low temperature ($< 42°C$); (ii) the presence of lipids; (iii) the presence of cellulose; or (iv) damp loads. Uses may include e.g. the sterilization of certain medical instruments (such as endoscopes).

15.2 DISINFECTION

Disinfection refers to any procedure which destroys, inactivates or removes *potentially harmful* microbes – without necessarily affecting the other organisms present; it generally has little or no effect on bacterial endospores. 'Disinfection' often refers specifically to the use of certain chemicals (*disinfectants*) for the treatment of non-living objects or surfaces, but the term is sometimes also used to refer to antisepsis (section 15.3). Although chemical disinfection is widely used, physical methods are more suitable for certain purposes.

15.2.1 Disinfection by chemicals

Ideally, disinfectants for general use should be able to kill a wide range of common or potential pathogens. However, any given disinfectant is usually more effective against some organisms than it is against others, and the activity of a disinfectant may be greatly affected by factors such as dilution, temperature, pH, or the presence of organic matter or detergent; to be effective at all, a disinfectant needs appropriate conditions, at a suitable concentration, for an adequate period of time. Some disinfectants (e.g. hypochlorites) tend to be unstable, and some (e.g. 'pine disinfectants') need solubilization in order to be effective. At low concentrations some disinfectants not only cease to be effective, they can actually be metabolized by certain bacteria – e.g. species of *Pseudomonas* can grow in dilute solutions of carbolic acid (phenol).

Disinfectants which *kill* bacteria are said to be *bactericidal*. Others merely halt the growth of bacteria, and if such a disinfectant is inactivated – e.g. by dilution, or by chemical neutralization – the bacteria may be able to resume growth; these disinfectants are said to be *bacteriostatic*. A bactericidal disinfectant may become bacteriostatic when diluted.

Of the many disinfectants in use, only a few of the common ones are mentioned here.

Phenol and its derivatives (e.g. 'phenolics' such as *cresols* and *xylenols*) can be bactericidal at appropriate concentrations; they appear to act mainly by affecting the permeability of the cytoplasmic membrane. *Lysol* is a mixture of methylphenols solubilized by soap; at 0.5% it kills many non-sporing pathogens in 15 minutes, but endospores may survive in 2% Lysol for days.

Chlorine is widely used for the disinfection of water supplies and for the sanitation of water in swimming pools. It acts (directly, and via hypochlorous acid) as an effective disinfectant, though its activity is decreased by the presence of organic matter and by other substances with which it can react.

Quaternary ammonium compounds (QACs) are cationic detergents used e.g. for the disinfection of equipment in the food and dairy industries. They are bacteriostatic at low concentrations, bactericidal at higher concentrations, and are typically more active against Gram-positive than Gram-negative bacteria. QACs appear to disrupt the cytoplasmic membrane; their activity is inhibited e.g. by soaps, some cations (e.g Ca^{2+}, Mg^{2+}), low pH and organic matter.

Hypochlorites are highly effective against a wide range of bacteria (including endospores), the undissociated form of HOCl being strongly bactericidal. Sodium hydroxide is commonly used as a stabilizer in commercial hypochlorite disinfectants.

15.2.2   Disinfection by physical agents

Ultraviolet radiation can damage DNA and can be lethal to bacteria under appropriate conditions. It has poor powers of penetration (being readily absorbed by solids), but ultraviolet lamps (wavelength about 254 nm) are used e.g. for the disinfection of air and exposed surfaces in enclosed areas.

The disinfection of milk by *pasteurization* is described in section 12.2.1.1.

15.3   ANTISEPSIS

Antisepsis is the disinfection of *living* tissues; it may be used prophylactically (i.e. to prevent infection) or therapeutically (i.e. to treat infection).

The comments on disinfectants (section 15.2.1) generally apply also to *antiseptics*, i.e. chemicals used for antisepsis.

Dettol is a general-purpose phenolic antiseptic when used in dilute form, but a domestic disinfectant in more concentrated form; it is based on chloroxylenols.

Hexachlorophene has been used in antiseptic soaps; it is a *bisphenol* (i.e. the molecule contains two phenolic groups) which is much more effective against Gram-positive than Gram-negative bacteria.

A 70:30 *ethanol*:water mixture is used as a general skin antiseptic.

Soaps generally have little or no antibacterial activity unless they contain antiseptics; however, soap can help to remove bacteria from the skin – along with dirt and grease.

QACs (section 15.2.1) include cetyltrimethylammonium bromide (*Cetrimide, Cetavlon* etc.) which is used in antiseptic creams.

Iodine (in alcoholic or aqueous solution) is a potent bactericidal and sporicidal antiseptic.

15.4 ANTIBIOTICS

Originally, 'antibiotic' meant any microbial product which, even at very low concentrations, inhibits or kills certain microorganisms; the term is now generally used in a wider sense to include, in addition, any semi-synthetic or wholly synthetic substance with these properties.

No antibiotic is effective against all bacteria. Some are active against a narrow range of species, while others are active against a *broad spectrum* of organisms – including both Gram-positive and Gram-negative bacteria. In some cases a natural antibiotic (one produced by a microbe) can be chemically modified in the laboratory to form a semi-synthetic antibiotic which may have a significantly different spectrum of activity.

Like disinfectants, antibiotics may be either bactericidal or bacteriostatic, and one which is bactericidal at one concentration may be bacteriostatic at a lower concentration.

An antibiotic characteristically acts at a precise site in the cell; depending on antibiotic, the site of action may be in the cell wall, the cytoplasmic membrane or the protein-synthesizing machinery, or it may be an enzyme involved in nucleic acid synthesis. Since a bacterium differs in many ways from a eukaryotic cell (Table 1.1), the toxic effect of an antibiotic on a bacterium is unlikely to be exerted e.g. on human or animal cells; because of this *selective toxicity*, some antibiotics are useful for treating certain diseases: a bacterial pathogen can be attacked without harming the host. Clearly, any antibiotic used in this way must retain activity within the body for long enough to be effective against the pathogen.

Of the many known antibiotics, relatively few are suitable for treating disease; some of these are mentioned briefly below.

15.4.1 *β*-Lactam antibiotics

These antibiotics (Fig. 15.2) include the penicillins, cephalosporins, carbapenems, clavams and monobactams; in each case the molecule includes a four-membered nitrogen-containing ring, the *β-lactam ring*. *β*-Lactam antibiotics act by disrupting synthesis of the cell envelope in growing cells: they inactivate *penicillin-binding proteins* (PBPs, enzymes involved in peptidoglycan synthesis) and thus inhibit synthesis of the peptidoglycan sacculus (Fig. 2.7). The accompanying cell lysis apparently results from enzymic cleavage of the peptidoglycan; the cell envelope normally contains both hydrolytic and synthetic enzymes which jointly mediate growth of the sacculus (Fig. 3.1), so that selective inhibition of the *synthetic* enzymes (PBPs) may account for lysis.

Note that *only* growing cells are killed by these antibiotics.

Various *β*-lactam antibiotics are mentioned in Plate 15.2.

In many of these antibiotics the *β*-lactam ring is susceptible to cleavage by certain bacterial enzymes (*β-lactamases*: section 15.4.1.2); such cleavage destroys the antibiotic, and organisms which produce these enzymes generally show at least some degree of resistance to particular *β*-lactam antibiotic(s).

Other antibiotics (e.g. *vancomycin*) inhibit the synthesis of peptidoglycan at an earlier stage (see section 6.3.3.1).

15.4.1.1   Penicillins

The original penicillins (e.g. *benzylpenicillin*) have low activity against Gram-negative bacteria owing to poor penetration of the outer membrane (section 2.2.9.2); they also have little or no effect against those Gram-positive bacteria which form *β*-lactamases. Some semi-synthetic penicillins (such as *cloxacillin, methicillin, nafcillin*) are resistant to a number of different *β*-lactamases (including those formed by some staphylococci), but they are still poorly effective against Gram-negative bacteria. *Ampicillin* and its derivatives (e.g. *amoxycillin*) combine resistance to some *β*-lactamases with increased activity against Gram-negative bacteria. [Clinical pharmacology and therapeutic uses of penicillins (review): Nathwani & Wood (1993) Drugs 45 866–894.]

15.4.1.2   β-Lactamases

These enzymes inactivate (susceptible) *β*-lactam antibiotics by hydrolysing the *β*-lactam ring (Fig. 15.2); they may be encoded by chromosome-, plasmid- or transposon-borne genes. Some *β*-lactamases are secreted into the medium; others are retained in the cell envelope.

A given *β*-lactamase inactivates a *particular* range of *β*-lactam antibiotics, being weakly active, or inactive, against other *β*-lactams; antibiotics susceptible to one *β*-lactamase may be unaffected by another. Thus, a bacterium producing a given *β*-lactamase may exhibit resistance to only a limited number of *β*-lactams.

Some *β*-lactamases are inducible, and some *β*-lactam antibiotics (e.g. cefoxitin, imipenem) are particularly good inducers (see e.g. Plate 15.2). Bacterial resistance to *β*-lactams, due to inducible *β*-lactamase(s), may be apparent from a disc diffusion test – see e.g. *Staphylococcus aureus* in section 15.4.11.1, and 'enzyme inactivation' in Plate 15.1.

The production of *β*-lactamases can be readily detected. One method uses a

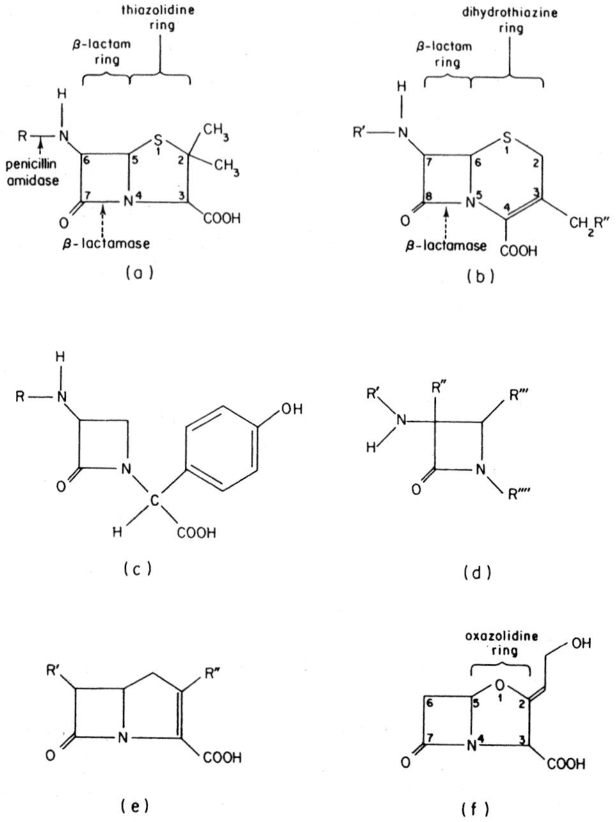

Fig. 15.2 β-Lactam antibiotics (section 15.4.1). Each formula shows the *nucleus* (= essential framework) of the molecule of one member of the family of β-lactam antibiotics; these antibiotics are grouped together because each contains the β-lactam ring.

The nuclei are those of (a) *penicillins*, (b) *cephalosporins*, (c) *nocardicins*, (d) *monobactams* and (e) *carbapenems*.

The structure at (f) is that of clavulanic acid (a *clavam*). Clavulanic acid is a weak antibiotic, but it inactivates many types of β-lactamase (section 15.4.1.2) – apparently by reacting covalently with a specific sequence of amino acids in the enzyme; it has therefore been used in combination with other β-lactam antibiotics – one antibiotic acting on the bacteria while clavulanic acid protects the first antibiotic from β-lactamases. For example, *augmentin* is a mixture of amoxycillin (a penicillin) and clavulanic acid.

The dashed arrow shows the site of action of β-lactamases; these enzymes open the β-lactam ring, thereby inactivating the antibiotic. The site of action of another enzyme, penicillin amidase, is shown at (a).

The formulae are from *Dictionary of Microbiology and Molecular Biology*, 2nd edn, Singleton & Sainsbury, p. 483, by courtesy of the publisher, John Wiley, Chichester, UK.

chromogenic cephalosporin, *nitrocefin* (Unipath, Basingstoke, UK), which, on cleavage by a β-lactamase, changes from yellow to red; a drop of nitrocefin solution added to a colony of a β-lactamase-producing bacterium will give a red coloration either very quickly or after incubation for up to 30 minutes.

Antibiotic-resistance, due to β-lactamases, poses a clinical problem which may be approached e.g. by opting to use antibiotics of a different type, when possible, or using β-lactams which are not susceptible to those β-lactamase(s) produced by the given pathogen. Another possibility is to use a combination of drugs, such as *augmentin* – which contains the *β-lactamase inhibitor* clavulanic acid (Fig. 15.2).

15.4.2   Aminoglycoside antibiotics

These broad-spectrum, typically bactericidal antibiotics include *amikacin*, *gentamicin*, *kanamycin*, *neomycin* and *streptomycin*; they are active against both Gram-positive and Gram-negative bacteria. Aminoglycoside antibiotics act on various bacterial functions, but their main effect results from binding to the 30S ribosomal subunit and interference with protein synthesis. For example, low levels of streptomycin cause misreading of the mRNA (i.e. incorporation of incorrent amino acids), while high levels completely inhibit protein synthesis – apparently by blocking ribosomes specifically at the start of translation (the stage in Fig. 7.9b).

Resistance to streptomycin correlates with mutation(s) in ribosomal proteins of the 30S subunit (affecting the binding of the antibiotic).

Side-effects (particularly with neomycin and streptomycin) include dose-dependent damage to the 8th cranial nerve (causing impairment of hearing). Hypersensitivity reactions also occur.

15.4.3   Tetracyclines

These broad-spectrum, bacteriostatic antibiotics (Fig. 15.3) inhibit protein synthesis by binding to ribosomes (in *E. coli*, preferentially to protein S7 in the 30S subunit) and inhibiting the binding of aminoacyl-tRNAs to the A site (Fig. 7.9); they are used for treating human and animal diseases caused e.g. by *Brucella*, *Chlamydia*, *Mycoplasma* and *Rickettsia* – and, interestingly, for the treatment of certain plant diseases such as coconut lethal yellowing (caused by a *Mycoplasma*-like organism).

Bacteria normally accumulate tetracyclines (in an energy-dependent manner). Some bacteria are resistant to these drugs because they can pump them outwards across the cytoplasmic membrane: see TET protein in section 15.4.11. In another form of resistance ribosome–tRNA binding is less sensitive to tetracyclines: see Tet(M) protein in section 8.4.2.3. A third (apparently less common) mechanism involves the bacterial modification (i.e. detoxification)

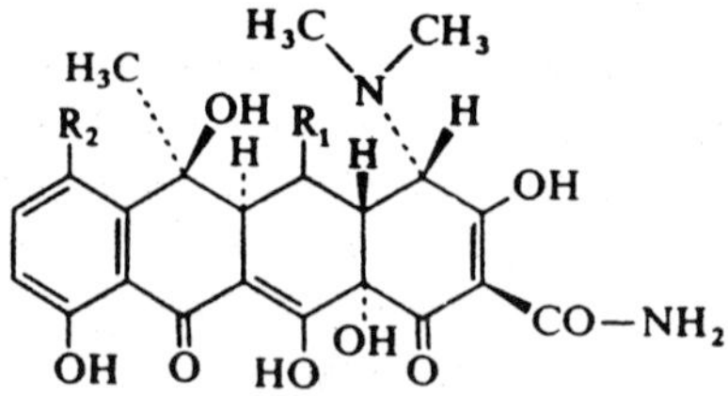

Fig. 15.3 Tetracycline (R_1, R_2 = H); chlortetracycline (R_1 = H, R_2 = Cl); oxytetracycline (R_1 = OH, R_2 = H).

of tetracyclines. Resistance to one tetracycline usually means resistance to all of them.

[Tetracycline resistance (review): Roberts (1996) FEMSMR *19* 1–24.]

15.4.4 Macrolides, chloramphenicol and streptogramins

All of these antibiotics inhibit protein synthesis by binding to the 50S ribosomal subunit (Fig. 7.9).

Macrolides. The molecule consists of a large (> 12-membered) lactone ring substituted with one or more sugars or aminosugars; the important macrolide *erythromycin* has a 14-membered ring linked to cladinose and desosamine. Macrolides, which are typically bacteriostatic, cause premature termination of polypeptide synthesis; they bind to peptidyltransferase (Fig. 7.9d) and may inhibit transpeptidation under certain conditions and/or trigger abortive translocation (Fig. 7.9e) leading to dissociation of an incomplete polypeptide. Macrolides are effective against mainly Gram-positive bacteria. In *E. coli*, mutants resistant to erythromycin may have modified protein(s) in the 50S subunit; in *Staphylococcus aureus*, inducible resistance involves an enzyme which methylates a site in the 23S rRNA of the 50S subunit.

Chloramphenicol. This small molecule (readily synthesized) inhibits the activity of peptidyltransferase, possibly by preventing normal binding of the aminoacyl-tRNA at the ribosomal A site. Chloramphenicol is a broad-spectrum, bacteriostatic agent, but its use is limited by toxicity (e.g. it affects mitochondrial ribosomes in mammalian cells). It is used e.g. against *Salmonella typhi* and in cases where other drugs are unsuitable. *Pseudomonas aeruginosa* is innately resistant. Bacteria with acquired resistance may encode the enzyme *chloramphenicol acetyltransferase* (CAT), which inactivates the antibiotic; CAT is usually inducible in Gram-positive bacteria (see section 7.8.5) but synthesized constitutively in Gram-negative species.

Streptogramins are composite antibiotics, each one consisting of at least two distinct types of molecule – which include a polyunsaturated macrolactone ring (group A component) and a cyclic hexadepsipeptide (group B component); A and B form a synergistic combination which can be bactericidal for certain (primarily Gram-positive) pathogens. Group A molecules inactivate functional sites on peptidyltranferase, thus inhibiting transpeptidation; group B molecules inhibit normal binding of the growing polypeptide chain at the P site during translation of certain amino acids (e.g. proline), thereby promoting early release of an (incomplete) polypeptide chain.

A new streptogramin, quinupristin/dalfopristin (RP 59500; Synercid®) has useful activity against certain pathogens (including MRSA) and has a long post-antibiotic effect (section 15.4.12); it may be valuable e.g. for treating severe infections caused by multi-resistant Gram-positive pathogens [Finch (1996) Drugs 51 (suppl 1) 31–37].

15.4.5 Polymyxins

Polymyxins are peptides which are active against many Gram-negative bacteria; most Gram-positive bacteria are resistant. In Gram-negative bacteria polymyxins act by increasing the permeability of the cytoplasmic membrane and by damaging the outer membrane.

15.4.6 Nalidixic acid and the quinolones; novobiocin

Quinolone antibiotics interact e.g. with the A subunit of *gyrase* (section 7.2.1) in a gyrase–DNA complex, inhibiting enzymic activity and apparently stabilizing a double-stranded gyrase-mediated break in the DNA. Nalidixic acid is active against Gram-negative bacteria. Other quinolones include *oxolinic acid*, *norfloxacin*, *ofloxacin*, *ciprofloxacin* and *fleroxacin* – some (e.g. ciprofloxacin, ofloxacin) being not only more active but also effective against some Gram-positive bacteria. Resistance to these drugs can involve e.g. a mutant gyrase with poor binding ability and/or reduced permeability of the cell envelope.

Novobiocin is a coumarin derivative which inhibits DNA synthesis by binding to the B subunit of gyrase and blocking the enzyme's ATPase activity (which is essential for ongoing gyrase activity). Its use is limited by resistance (which develops readily) and by toxicity.

[Gyrase-targeted antibiotics: Maxwell (1997) TIM 5 102–109.]

15.4.7 Metronidazole

Metronidazole is a nitroimidazole derivative (Fig. 15.4); it is an effective antimicrobial agent only at low (i.e. highly negative) redox potentials, and is therefore used against strictly anaerobic bacteria (and certain pathogenic

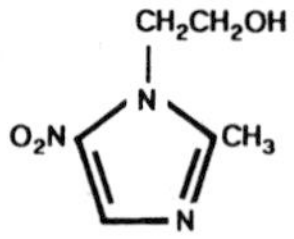

Fig. 15.4 Metronidazole: a nitroimidazole antibiotic used against anaerobic bacteria.

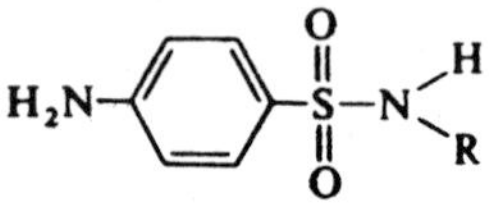

Fig. 15.5 The sulphonamide nucleus; R = H in sulphanilamide (*p*-amino-benzenesulphonamide). The therapeutic sulphonamides may be regarded as derivatives of sulphanilamide: e.g. *sulphadiazine* is 4-amino-*N*-2-pyrimidinyl-benzenesulphonamide, and *sulphamethazine* (= *sulphadimidine*) is 4-amino-*N*-(4,6-dimethyl-2-pyrimidinyl) benzenesulphonamide.

protozoa). In sensitive anaerobes, the drug's nitro group is reduced (by electron transport components such as *ferredoxins* – section 5.1.1.2) to short-lived cytotoxic intermediate(s) which can cleave DNA.

Metronidazole is used e.g. against *Clostridium difficile* in pseudomembranous colitis, and, prophylactically, in bowel surgery to protect against anaerobes such as *Bacteroides* and clostridia.

15.4.8 Rifamycins

These antibiotics are generally active against Gram-positive bacteria (including mycobacteria and staphylococci) and against some Gram-negative bacteria; they inhibit RNA polymerase, thus inhibiting e.g. RNA synthesis (section 7.5).

15.4.9 Sulphonamides, trimethoprim and cotrimoxazole

Sulphonamide antibiotics (Fig. 15.5) are typically bacteriostatic for susceptible Gram-positive and Gram-negative bacteria. Sulphonamides inhibit the synthesis of dihydrofolic acid (DHF), a precursor of the coenzyme tetrahydrofolic acid (THF) – and thus inhibit THF-dependent functions (section 6.3.1); they are believed to act in one or both of two ways: (i) as *competitive inhibitors* of the enzyme dihydropteroate synthetase (DHPS, involved in the synthesis of DHF), (ii) as *substrates* of DHPS – giving rise to functionally defective analogues of DHF. These activities appear to reflect the structural similarity between the sulphonamide molecule and that of *p*-aminobenzoic acid (PABA), the latter being a precursor in the synthesis of DHF.

Certain bacteria use an external source of folate, and are therefore not affected by sulphonamides, but most bacteria synthesize their own folate.

However, resistance to sulphonamides develops readily and may be due e.g. to (i) the production of higher levels of PABA (thus combatting the competitive inhibition) and/or (ii) the development of a mutant form of DHPS whose affinity for sulphonamides is much lower than that for PABA.

In sensitive cells, growth continues for several generations following exposure to sulphonamides; during this time the existing stock of folate is being used up.

Currently, sulphonamides are used e.g. for urinary-tract infections (much of the drug is excreted in the urine). The most frequently used sulphonamides include sulphadiazine and sulphamethazine.

Trimethoprim, a diaminopyrimidine derivative, inhibits the conversion of DHF to THF by inhibiting the enzyme dihydrofolate reductase (DHFR). DHFR occurs in all cells (microbial and human), but trimethoprim inhibits bacterial DHFRs without significantly affecting human DHFR at the concentrations used.

Cotrimoxazole is a (synergistic) combination of trimethoprim and the sulphonamide sulphamethoxazole – two drugs which block different reactions in the same pathway.

15.4.10   Synergism and antagonism between antibiotics

If two antibiotics, acting simultaneously on an organism, produce an effect which is greater than the sum of their individual effects, the antibiotics are said to be acting *synergistically*. Cotrimoxazole (section 15.4.9) is one example of a synergistic combination of antibiotics. Another example is the combination of the A and B components of streptogramins (section 15.4.4).

Antagonism is the converse of synergism. For example, antibiotics which inhibit growth (e.g. chloramphenicol) antagonize those antibiotics (e.g. β-lactams) which act only when cells are growing. Another form of antagonism is exemplified by those antibiotics which induce bacteria to form antibiotic-inactivating enzymes which affect *other* antibiotics; examples include the antagonism of imipenem and cefoxitin to other β-lactam antibiotics shown in Plate 15.2.

15.4.11   Bacterial resistance to antibiotics

Why are some bacteria not affected by some antibiotics? In some cases a bacterium is resistant because it lacks the target structure of a given antibiotic; for example, species of *Mycoplasma* (which lack cell walls) will not be affected by penicillins – whose ultimate target (peptidoglycan) is a cell-wall component. Some bacteria may not carry out the particular process affected by an antibiotic: sulphonamides, for example, will not affect organisms which normally obtain their folic acid, ready-made, from the environment. Resistance can also be due to the ability of a cell to exclude an antibiotic from the target site; in many Gram-negative bacteria the outer membrane is impermeable to certain antibiotics, and in both Gram-positive and Gram-negative species the cytoplasmic membrane may act as a barrier.

Some bacteria produce enzyme(s) which inactivate particular antibiotics. For example, many strains of *Staphylococcus* produce enzymes which inactivate particular penicillins; such 'penicillinases' are examples of β-lactamases (section 15.4.1.2).

Resistance to antibiotics can also be *acquired*. This can happen e.g. through mutation (section 8.1) or by the acquisition of an R plasmid (section 7.1). A mutation may alter the target site so that it is no longer affected by the antibiotic; for example, a mutation conferring resistance to streptomycin may alter the ribosome so that streptomycin no longer binds to it, or, if binding does occur, ribosomal function is not affected. A single mutation usually confers resistance to only one antibiotic, or to closely related antibiotics which have the same target site.

An R plasmid or a transposon may encode resistance to one or many related or unrelated antibiotics. For example, it may encode antibiotic-inactivating enzymes – such as chloramphenicol acetyltransferase (which inactivates chloramphenicol) [O-acetyltransferases for chloramphenicol: Murray & Shaw (1997) AAC *41* 1–6.] or β-lactamases; in some cases the genes encoding these enzymes are inducible (section 7.8), i.e. the presence of a given antibiotic promotes the synthesis of the inactivating enzyme (see e.g. section 7.8.5). In *E. coli*, an inducible cytoplasmic membrane protein, TET (encoded e.g. by transposon Tn*10*), appears to pump tetracycline outwards into the periplasm, away from the ribosomal target site [Thanassi, Suh & Nikaido (1995) JB *177* 998–1007].

Another mode of resistance involves increased production of a particular metabolite in order to overcome competitive inhibition by the antibiotic; this occurs e.g. in one form of resistance to sulphonamides (section 15.4.9).

Exceptionally, *apparent* resistance to an antibiotic may be due to the patient's abnormal metabolism of the drug. For example, treatment of syphilis with a normally adequate dosage of penicillin has failed in a few cases in which the patients were categorized as 'quick penicillin secretors'; in a very few instances, penicillin-sensitive treponemes have been recovered from these patients and have produced typical lesions when inoculated into rabbits.

Although large numbers of new antibiotics have been developed in the last few decades, bacterial resistance to these new agents continues to emerge; this means that there is a continuing need to produce new antibiotics, and a need to prevent or delay the emergence of bacterial resistance by avoiding the inappropriate use of antibiotics [review: Neu (1992) Science *257* 1064–1073]. One example of the problem was the emergence of methicillin-resistant strains of coagulase-positive *Staphylococcus aureus* (so-called MRSA); these strains, which are typically resistant to many other β-lactam antibiotics as well (and sometimes also to e.g. aminoglycosides), are apparently generally susceptible to vancomycin and teicoplanin. Another example was the emergence of so-called 'RISE-resistant tuberculosis'; this involves strains of *Mycobacterium tuberculosis* which are resistant to various common anti-tuberculosis agents, including – at least – rifampin, isoniazid, streptomycin and ethambutol. Other examples include penicillin-resistant *Streptococcus pneumoniae* [see e.g. Appelbaum (1996) Drugs *51* (suppl 1) 1–5], tetracycline-resistant *Neisseria gonorrhoeae* and *Shigella dysenteriae*, vancomycin-resistant enterococci [see e.g. Cormican & Jones (1996) Drugs *51* (suppl 1) 6–12], and multi-drug-resistant strains of *Vibrio cholerae* O1 [Dubon *et al.* (1997) Lancet *349* 924].

A recent Ciba Foundation Symposium (Antibiotic Resistance: Origins, Evolution, Selection and Spread, London, 16–18 July, 1996) [published by John Wiley (1997), ISBN 0471-97105-7] made various recommendations:

- World-wide surveillance of antibiotic resistance
- Better clinical use of antibiotics
- Provision of new antibiotics to give choice for treatment
- Better infection control in hospitals
- Better diagnostic procedures
- Use of new/improved vaccines against common infectious diseases.

15.4.11.1   Antibiotic-sensitivity tests

Tests can be carried out to determine the susceptibility of a pathogen to a range of antibiotics; the pattern of sensitivities of a given strain towards a range of antibiotics is called an *antibiogram*. The results of such tests may enable the clinician to select optimally active antibiotic(s) for chemotherapy (section 11.9).

One common form of test is the *disc diffusion test*. A plate of suitable agar medium is inoculated from a suspension of a pure culture of the pathogen; the entire surface of the medium is inoculated (often with a swab) so that near-confluent growth (section 3.3.1) will develop on incubation. Before incubation, several small absorbent paper discs, each impregnated with a different antibiotic, are placed at different locations on the inoculated medium. On subsequent incubation, antibiotic diffuses from each disc; if the organism if sensitive to a given antibiotic a zone of growth-inhibition develops around that particular disc. Methods must be standardized: the presence or size (diameter) of a no-growth zone can be interpreted correctly only when the whole procedure has been standardized in terms of type of medium, density of inoculum on the plate etc. Reactions which may be seen in a disc diffusion test are shown in Plate 15.1.

The disc diffusion test is not suitable for slow-growing bacteria (e.g.

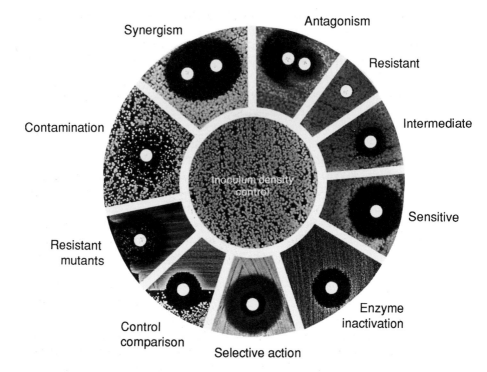

Plate 15.1 Antibiotic-sensitivity testing (disc diffusion test): types of zone which may develop around the discs. (The rationale for the test is given in section 15.4.11.1.)
Resistant. No zone: growth occurs right up to the disc.
Intermediate. A narrow growth-free zone surrounds the disc.
Sensitive. A wide growth-free zone surrounds the disc.
Enzyme inactivation. A narrow growth-free zone surrounds the disc. Unlike the 'intermediate' zone, the edge of the zone is sharply defined and it contains somewhat heavier growth with some normal-sized or relatively large colonies (for explanation see section 15.4.11.1).
Selective action. This result can be obtained e.g. when the inoculum consists of two different strains which differ in their degree of susceptibility to the given antibiotic. Close to the disc, the concentration of antibiotic is high enough to inhibit both strains (narrow growth-free zone). Further from the disc (lower concentration of antibiotic) one strain is still inhibited while the other can grow.
Control comparison. The 'half-zone' obtained with a control strain (one side of the disc) is opposite the 'half-zone' obtained with the test strain (other side of the disc). To make this comparison, the test and control strains are inoculated onto separate halves of the plate and the disc is placed between them.
Resistant mutants. The inoculum contained a small proportion of mutant cells which were able to form colonies under conditions which inhibited non-mutants. The mutants are usually antibiotic-resistant cells. There is, however, another type of mutant which grows *only in the presence* of a given antibiotic, and such mutants would also form colonies in an otherwise growth-free zone; for example, 'streptomycin-dependent' mutants contain non-functional ribosomes which, in the presence of streptomycin, appear to be distorted in such a way that they become functional.

Contamination. The inoculum contained a mixture of organisms, at least one of which is resistant to the antibiotic.

Synergism. The two discs contain different antibiotics. The zone shows an inhibitory effect which is greater than the sum of the effects of each antibiotic acting alone; that is, the antibiotics are acting synergistically.

Antagonism. The two discs contain different antibiotics. Here, the presence of one antibiotic inhibits the activity of the other. Examples of antagonism are seen in Plate 15.2.

Photograph courtesy of Oxoid, Basingstoke, UK.

Mycobacterium tuberculosis) because the antibiotics would become too diffused before visible growth was formed; the test is suitable for those bacteria which produce visible growth after overnight incubation.

In a disc diffusion test, a no-growth zone (simulating sensitivity) may be formed by organisms which encode inducible antibiotic-inactivating enzymes – e.g. strains of *Staphylococcus aureus* which encode an inducible β-lactamase. Cells in the inoculum close to the disc are killed before adequate amounts of enzyme can be synthesized. Cells further from the disc experience gradually increasing concentrations of the outwardly diffusing antibiotic, and, at a certain distance from the disc, some of the cells will have synthesized enough enzyme to permit survival; these cells give rise to normal-sized or relatively large colonies by using the nutrients forfeited by the inactivated cells in their immediate vicinity. An example of a zone characteristic of enzyme inactivation is shown in Plate 15.1.

Some tests use antibiotic-containing tablets or paper/plastic strips (Plate 15.2).

In a *dilution test*, the organism is tested for its ability to grow in the presence of each of a range of concentrations of a given antibiotic (in either solid (agar-based) or liquid media). The lowest concentration of antibiotic which prevents growth is the *minimum inhibitory concentration* (MIC) of that antibiotic under the given conditions for the given strain. For any particular antibiotic, different strains of an organism may have different MICs.

A *breakpoint test* uses the dilution test technique to determine whether the MICs of given strains are above or below certain chosen test concentration(s) of the antibiotic – thus determining which strains are to be regarded as 'resistant' and which 'sensitive'. The actual choice of test concentrations is influenced by clinical, pharmacological and microbiological factors. The concentrations used in breakpoint tests are not the same in all countries.

The *E test* (Plate 15.2) is a diffusion test which can give an MIC. One side of a plastic strip (placed in contact with the solid medium) carries the antibiotic in a concentration gradient, while the other (uppermost) side is graduated with the MIC scale; after incubation, the test is read by noting the lowest concentration of antibiotic which prevents growth.

[Developments in antimicrobial susceptibility testing: Brown (1994) RMM 5 65–75.]

Plate 15.2 Antibiotic-sensitivity testing without paper discs.

Top. Testing five strains of *Staphylococcus aureus* for resistance to methicillin (a penicillin which is resistant to many β-lactamases). The plate is first inoculated with each of the strains, inoculation being carried out in five parallel lines across the plate. These lines are then overlaid, at right-angles, with a paper strip impregnated with methicillin. After incubation (at e.g. 30 °C for 18 hours), the result is as shown: the first and fourth strains (from the left) are resistant, the others are sensitive.

Centre. The E test (see section 15.4.11.1 for rationale). Here, a strain of *Pseudomonas aeruginosa* has been tested for its reaction to four different antibiotics; for each antibiotic, the MIC is indicated where the edge of the no-growth zone intersects the scale. When compared with other forms of antibiotic-sensitivity testing (including dilution methods), the E test was found to be as reliable [Baker *et al.* (1991) JCM 29 533–538]. Other assessments of the E test include a study involving *Streptococcus pneumoniae* [Macias *et al.* (1994) JCM 32 430–432].

Bottom. Testing with tablets. Neo-Sensitabs are antibiotic-containing tablets, rather than impregnated paper discs; they are 9 mm in diameter, are colour-coded for identification, and – at room temperature – most are stable for a minimum of 4 years. Neo-Sensitabs are used in the standard procedure for bacterial sensitivity testing in Denmark, Holland, Belgium, Norway and Finland, and are also used in other countries.

A strain of *Pseudomonas aeruginosa* has been tested (on a plate of Mueller-Hinton agar) against a range of β-lactam antibiotics: the penicillins piperacillin (P) and ticarcillin (T); the cephalosporins cefsulodin (Cs), cefotaxime (Ct), cefoxitin (Cf), ceftriaxone (Cx), cephalothin (Cp) and cefamandole (Cl); imipenem (Ip – a carbapenem); and aztreonam (A – a monobactam). Note the antagonism of imipenem towards e.g. piperacillin and aztreonam; imipenem has induced the synthesis of β-lactamases in cells of the inoculum – enzymes to which imipenem itself is resistant but to which piperacillin and aztreonam are sensitive. Note also the resistance of this strain of *P. aeruginosa* to several of the antibiotics.

Photograph of the methicillin test courtesy of Mast Diagnostics Ltd, Bootle, Merseyside L20 1EA, UK.
Photograph of the E test courtesy of Dr Carolyn Baker, Centers for Disease Control, Atlanta, Georgia, USA.
Photograph of Neo-Sensitabs courtesy of A/S ROSCO, Taastrup, Denmark.

15.4.12   The post-antibiotic effect

'Post-antibiotic effect' (PAE) refers to the suppression of bacterial growth (typically for a number of hours) following a short exposure to certain antimicrobial agents – including aminoglycosides, β-lactams, 4-quinolones, macrolides and streptogramins; PAE is a factor when considering the dose/dosage frequency of these antibiotics. Clinically, PAE can be advantageous when it coincides with low concentrations of an intermittently administered antibiotic. A further advantage is that, during the period of suppressed growth, bacteria may be more readily eliminated by the immune system.

The pattern of DNA synthesis during PAE has been monitored in various organism–antibiotic combinations. The absence of a common pattern has been taken to indicate that PAE may involve different mechanisms in different combinations [Gottfredsson *et al.* (1995) AAC 39 1314–1319].

Table 15.1 Ongoing activity of antibiotics after uptake by phagocytes

Antibiotic	Activity within phagocytes
Aminoglycosides	Low-level, apparently due to poor penetration of the phagocyte; activity found to be increased against some bacteria in long-term experiments
Fluoroquinolones	High-level; the antibiotics (e.g. ciprofloxacin) accumulate within phagocytes
β-Lactams	Generally considered to be none, but activity reported in some cases
Macrolides	Conflicting results on activity
Rifampin	High-level

15.4.13   Activity of antibiotics within phagocytes

Some pathogenic bacteria (e.g. *Listeria monocytogenes* [Cossart (1994) BCID *1* 285–304], *Mycobacterium* spp and even *Staphylococcus aureus*) can survive, and sometimes grow, inside phagocytes. For this reason it is useful to know whether antibiotics, administered to patients, can (i) penetrate phagocytes, and (ii) act against bacteria *within* phagocytes. This aspect of antibiotic activity has been discussed by Pascual [(1995) RMM *6* 228–235], and some of the findings are summarized in Table 15.1.

16 The identification and classification of bacteria

16.1 IDENTIFICATION

How do we identify a bacterium? Traditionally, the organism is first obtained in pure culture (section 14.6) and then – after certain observations and tests – checked against known, named species until a match is found. This can be quite time-consuming, and rapid methods of identification are now being developed for medical and veterinary work (where prompt diagnosis can be crucial) and also e.g. for use in the food industry; rapid, nucleic-acid-based indentification is referred to in section 16.1.6.

Traditional methods are included here because (i) they are still widely used, and (ii) they are useful for illustrating the types of characteristic which distinguish one bacterium from another.

For traditional tests we need to start with a pure culture; with an impure culture (mixture of organisms) the reaction of one organism may differ from that of another so that, generally, a meaningful result would not be obtained. However, even with a pure culture, it sometimes happens that the characteristics of the unknown organism do not exactly match those of any species in a manual of identification. This can occur, for example, if a mutation (section 8.1) has altered one or more of the organism's typical characteristics. Similarly, a plasmid (section 7.1) may confer characteristics which are not typical of the species to which the unknown organism belongs. Nevertheless, the organism is not always to blame: inexplicable results can sometimes be due to slight variations in the methods and/or materials used in the tests themselves.

Essentially, in the traditional approach, the characteristics of the unknown organism are compared with those of each of a number of known, named species until a match is found. The principle is simple enough, but, in practice, which criteria are used in a given case – and must the unknown organism be compared with each of the thousands of known species of bacteria? Fortunately, the source of an organism often gives clues which, together with a few simple observations and tests, may indicate the possible identity of the organism, or, at least, may serve to narrow the search to one of the major groups of bacteria. For example, if the bacterium comes from sewage, and is found to be a Gram-negative, motile, facultatively fermentative bacillus, the bacteriologist would immediately think of the family Enterobacteriaceae – since this family contains many bacteria of that type, some of which are common in sewage. Identification to species level may then be possible by comparing the characteristics of the unknown organism with those of genera and species of the Enterobacteriaceae.

Clearly, the practice of identification is made easier if the bacteriologist (i) has a knowledge of the types of organism likely to be present in a given environment, and (ii) is familiar with the main distinguishing features of the families, genera and species of common bacteria – information of the type given in the Appendix.

The following characteristics are often the first to be determined because they have the greatest *differential* value, each helping to exclude one or more of the major groups of bacteria: (i) reaction to certain stains, particularly the Gram stain (section 14.9.1); (ii) morphology (coccus, bacillus etc.); (iii) motility; (iv) the ability to form endospores; (v) the ability to grow under aerobic and/or anaerobic conditions; (vi) the ability to produce the enzyme *catalase*. Fortunately, these characteristics are among the easiest to determine.

16.1.1 Preliminary observations and tests

16.1.1.1 *Reaction to stains*

A smear (section 14.9) from a pure culture is generally Gram-stained (section 14.9.1). A capsule stain (section 14.9.3) is used when capsulation is an important differential feature.

16.1.1.2 *Morphology*

Morphology is generally determined by examining a stained smear under the microscope. The smear may be stained either by Gram's method or by a simpler procedure – for example, by treating a heat-fixed smear for one minute with methlyene blue or carbolfuchsin. Usually the stained smear is examined under the oil-immersion objective of the microscope (total magnification about 1000×), although the cells of some species (e.g. *Bacillus megaterium*) are clearly visible under the 'high-dry' lens (total magnification about 400×).

16.1.1.3 *Motility*

Motility (section 2.2.15.1) can often be determined by examining a 'hanging drop' preparation (Fig. 16.1) under the microscope. Even with unstained cells and ordinary (bright-field) microscopy, it is often possible to see whether or not the cells are motile; however, cells can be seen more clearly with dark-field or phase-contrast microscopy. Motility should be distinguished from *Brownian motion*: small, random movements exhibited by any small

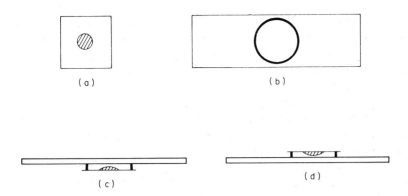

Fig. 16.1 The hanging-drop method for determining motility. (a) A drop of culture containing live, unstained bacteria is placed on a clean cover-slip. (b) A ring of plasticine is pressed onto a microscope slide. (c) The slide is inverted and pressed onto the cover-slip. (d) The whole is inverted for examination under the microscope.

particle of about 1 μm or less when freely suspended in a liquid medium; these movements (seen e.g. in particles of colloidal clay) are due to bombardment of the particles by molecules of the liquid.

Motility can sometimes be inferred from the way an organism grows on solid media: motile species may tend to spread outwards from the inoculated area as organisms swim in the thin layer of surface moisture.

16.1.1.4 Endospore formation

Relatively few bacteria can form endospores (section 4.3.1). If an endospore-former is grown for several days or a week on a solid medium, endospores can usually be detected by treating a heat-fixed smear of the growth with a 'spore stain'. This resembles the Ziehl–Neelsen stain (section 14.9.2) but uses e.g. ethanol for decolorization; vegetative cells are decolorized but endospores retain the (red) dye. Vegetative cells may be counterstained.

Endospores may be detected indirectly by heating a culture to 80°C for 10 minutes – a procedure which endospores usually survive but which kills most types of vegetative cell. Any growth following subculture to a fresh medium suggests the presence of endospores.

The shape of an endospore and its position within the cell are features sometimes used for identification. These observations are best made with the phase-contrast microscope (section 14.10.2, Fig. 14.9) on a thin wet film of unstained cells overlaid with a cover-slip; in such a preparation, the spore-containing mother cells have not undergone shrinkage – as have cells in a heat-fixed smear – so that the spore's relation to the mother cell can be more easily determined. A spore is seen as a bright body (oval or round, depending on species) with a dark margin.

Spores can be distinguished from other intracellular inclusions by differential staining. For example, granules of PHB (section 2.2.4.1), but not endospores, can be stained by dyes such as Sudan black B.

16.1.1.5   Aerobic/anaerobic growth

Whether an organism is an aerobe, anaerobe or facultative anaerobe (section 3.1.6) is easily determined by attempting culture anaerobically (section 14.7) and aerobically.

16.1.1.6   Catalase production

Catalase is an iron-containing enzyme which catalyses the decomposition of hydrogen peroxide (H_2O_2) to water and oxygen; it is formed by most aerobic bacteria, and it de-toxifies hydrogen peroxide produced during aerobic metabolism. The catalase test is used to detect the presence of catalase in a given strain of bacterium. Essentially, bacteria are exposed to hydrogen peroxide, the presence of catalase being indicated by bubbles of gas (oxygen). In the traditional form of the test, a speck of bacterial growth is transferred, with a loop, to a drop of hydrogen peroxide on a slide; however, in a positive test the bursting bubbles will give rise to an aerosol (section 14.1). The author's method (Fig. 16.2) avoids this problem.

Some bacteria (e.g. certain strains of *Lactobacillus* and *Enterococcus* (*Streptococcus*) *faecalis*) produce *pseudocatalase*, a non-iron-containing enzyme which behaves like catalase.

If growth used for the catalase test is obtained from a colony on blood agar (section 14.2.1), care should be taken to exclude erythrocytes (red blood cells) from the sample since they contain catalase – and may therefore give a false-positive result.

16.1.2   Secondary observations and metabolic ('biochemical') tests

Once the search for identity has been narrowed to one or a few families, the bacteriologist uses some simple 'biochemical' tests; these tests distinguish between bacteria of different genera and species by detecting differences in their metabolism. For example, a test may distinguish between species which can and cannot ferment a particular carbohydrate, or which produce different products when metabolizing a particular substrate. The following are a few (of many) tests which are frequently carried out in bacteriological laboratories – sometimes as *micromethods* (section 16.1.3).

16.1.2.1   The oxidase test

This test detects a particular type of respiratory chain (section 5.1.1.2): one containing a terminal cytochrome *c* and its associated *oxidase*. Bacteria which

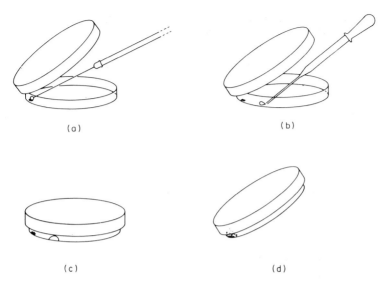

(a)

(b)

(c)

(d)

Fig. 16.2   The catalase test: a method which avoids contamination of the environment by aerosols. (a) A small quantity of bacterial growth is placed in a clean, empty, non-vented Petri dish. (b) Two drops of hydrogen peroxide are placed in the Petri dish a short distance from the growth. (c) The Petri dish is closed. (d) The closed Petri dish is tilted so that the hydrogen peroxide runs onto the bacterial growth. A positive reaction is indicated by the appearance of bubbles. The lid can be taped to the base before disposal in the proper container.

contain such a chain can oxidize chemicals such as *Kovács' oxidase reagent* (1% tetramethyl-*p*-phenylenediamine dihydrochloride); electrons are transferred from this reagent to cytochrome *c* and thence, via the oxidase, to oxygen. When oxidized in this way, the reagent develops an intense violet colour. In the test, a small area of filter paper is moistened with a few drops of Kovács' oxidase reagent, and a small amount of bacterial growth is smeared onto the moist filter paper with a glass spatula or a platinum loop (but *not* a nichrome loop); oxidase-positive species give a violet coloration immediately or within 10 seconds. Oxidase-positive bacteria include e.g. species of *Neisseria*, *Pseudomonas* and *Vibrio*; a negative reaction is given e.g. by members of the family Enterobacteriaceae. (The used test paper should be disposed of safely.)

16.1.2.2   *The coagulase test (for staphylococci)*

Some ('coagulase-positive') strains of *Staphylococcus* produce one or (usually) both of two different proteins called *coagulases*; this test is used to detect coagulase production.

Free coagulase (= true coagulase, or staphylocoagulase) is released into the medium and is detected (in a *tube test*) by its ability to coagulate (i.e. clot) plasma containing an anticoagulant such as citrate, oxalate or heparin; the anticoagulant is necessary since, without it, the plasma would clot spontaneously. In one form of tube test, 0.5 ml of an 18–24 hour broth culture of the strain under test is added to 1 ml of plasma in a test-tube; the tube is kept at 37°C and examined for the presence of a clot after 1, 2, 3 and 4 hours, and at 24 hours. Free coagulase triggers conversion of the plasma protein *fibrinogen* to *fibrin* – which forms the clot. Those coagulase-positive strains which also produce a *fibrinolysin* (an enzyme which lyses fibrin) may not form a clot, or may lyse a clot soon after its formation – hence the need for frequent examination of the tube.

Bound coagulase (= clumping factor) is a protein component of the cell surface; it binds to fibrinogen, resulting in the clumping of cells (*paracoagulation*). (Contrast this with the clotting of plasma.) Bound coagulase is detected by a *slide test*: a loopful of citrated or oxalated plasma is stirred into a drop of thick bacterial suspension on a microscope slide; in a positive test, cells clump within 5 seconds.

Known coagulase-positive and coagulase-negative strains should be used as controls in each form of the test.

Some other bacteria (e.g. strains of *Yersinia pestis*) also form coagulases.

16.1.2.3 The oxidation–fermentation (O–F) test

This is an early form of test (= Hugh & Leifson's test) for determining whether an organism uses oxidative (respiratory) or fermentative metabolism for the utilization of a given carbohydrate (usually glucose). Two test-tubes are filled to a depth of about 8 cm with a peptone–agar medium containing the given carbohydrate and a pH indicator, bromthymol blue, which makes the medium green (pH 7.1). One of the tubes is steamed (to remove dissolved oxygen) and is quickly cooled just before use. Each tube is then stab-inoculated (section 14.5.2) with the test organism to a depth of about 5 cm; in the 'steamed' tube the medium is immediately covered with a layer of sterile liquid paraffin about 1 cm deep (to exclude oxygen). Both tubes are then incubated and later (1–14 days) examined for evidence of carbohydrate utilization, namely, acid production – indicated by yellowing of the pH indicator.

Respiratory organisms (such as *Pseudomonas* species) cause yellowing only in the uncovered ('aerobic') medium. *E. coli* causes yellowing in both media; glucose is fermented in the covered medium, and is attacked first by respiration and then by fermentation in the uncovered medium.

16.1.2.4 Acid/gas from carbohydrates ('sugars')

In some genera the species can be distinguished from one another by differences in the types of carbohydrate ('sugar') which they can metabolize. The range of sugars utilized by any particular organism can be determined simply by growing the organism in a series of media, each containing a different sugar (as a source of carbon) and a system for detecting sugar

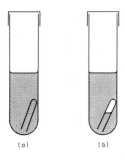

Fig. 16.3 The detection of gas produced during growth in a liquid medium. (a) An uninoculated tube of liquid medium containing an inverted Durham tube. (b) A gas-producing organism has been grown in the medium; some of the gas has collected in the Durham tube.

utilization. The medium may be based on peptone-water or nutrient broth; it contains, in addition to the sugar, a pH indicator to detect acidification due to metabolism of the sugar. Such 'peptone-water sugars' or 'broth sugars' are used e.g. in tests on bacteria of the family Enterobacteriaceae. If a particular sugar is metabolized, acid products will be formed, and the acidity is detected by the pH indicator.

Certain bacteria cannot be tested in media containing peptone-water or broth – for example, species of *Bacillus* may form excess alkaline products (from the peptone or broth) so that any acid formed from sugar metabolism would not be detected. For some species the tests may be carried out in media containing inorganic salts and a given sugar. For other types of bacteria the medium must be enriched with e.g. serum, otherwise growth will not occur.

Test media are generally used in test-tubes, or in bijoux (section 14.2), and the tube or bijou may contain an inverted Durham tube (Fig. 16.3) to collect gas that may be formed during the metabolism of the sugar. Gas may be formed e.g. in the mixed acid and butanediol fermentations (Figs 5.5 and 5.6), formic acid being split into carbon dioxide and hydrogen by the *formate hydrogen lyase* enzyme system.

16.1.2.5 IMViC tests

The IMViC tests are a group of tests used particularly for identifying bacteria of the family Enterobacteriaceae. IMViC derives from: *i*ndole test, *m*ethyl red test, *V*oges–Proskauer test and *c*itrate test. (See e.g. *Escherichia* in the Appendix.)

The indole test. This test detects the ability of an organism to produce indole from the amino acid tryptophan. The organism is grown in peptone-water or tryptone-water for 48 hours; to the culture is then added Kovács' indole reagent (0.5 ml per 5.0 ml culture) and the (closed) container is gently shaken. In a positive test, indole (present in the culture) dissolves in the reagent – which then becomes pink, or red, and forms a layer at the surface of the medium.

The methyl red test (MR test). The MR test detects the ability of an organism, growing in a phosphate-buffered glucose–peptone medium, to produce sufficient acid (from the metabolism of glucose) to reduce the pH of the medium from 7.5 to about 4.4 or below. The medium is inoculated and is then incubated for at least 48 hours at 37°C, following which the pH of the culture is tested by adding a few drops of 0.04% methyl red (yellow at pH 6.2, red at pH 4.4); with an MR-positive organism the culture becomes red.

The Voges–Proskauer test (VP test). This test detects the ability of an organism to form acetoin (acetylmethylcarbinol) – see butanediol fermentation, Fig. 5.6. A phosphate-buffered glucose–peptone medium is inoculated with the test strain and incubated at 37°C for 2 days, or at 30°C for at least 5 days. In one form of the test (*Barritt's method*), 0.6 ml of an ethanolic solution of 5% α-naphthol, and 0.2 ml of 40% potassium hydroxide solution, are added sequentially to 1 ml of culture; the (stoppered) tube or bottle is then shaken vigorously, placed in a sloping position (for maximum exposure of the culture to air), and examined after 30 and 60 minutes. Acetoin (if present) is apparently oxidized to diacetyl ($CH_3.CO.CO.CH_3$) which, under test conditions, gives a red coloration (a positive VP test).

The citrate test. This test detects the ability of an organism to use citrate as the sole source of carbon. Media used for the test – e.g. Koser's citrate medium (a liquid), and Simmons' citrate agar – include citric acid or citrate, ammonium dihydrogen phosphate (as a source of nitrogen and phosphorus), sodium chloride and magnesium sulphate. A saline suspension of the test organism is made from growth on a solid medium; using a *straight wire* (section 14.4), Koser's medium is inoculated from the suspension and is then incubated and examined for signs of growth (turbidity) after one or two days. Organisms which grow in the medium are designated 'citrate-positive'. A straight wire is used so that little or no nutrient is carried over in the inoculum; any nutrient carried over from the original medium may permit a small amount of growth in the citrate medium, thus giving a false-positive result. In an alternative method of inoculation, a straight wire is used to transfer a small quantity of growth from the *top* of a colony direct to the test medium.

16.1.2.6   Hydrogen sulphide production

Many species of bacteria produce hydrogen sulphide, e.g. by the reduction of sulphate (section 5.1.1.2) or from the metabolism of sulphur-containing amino acids. A sensitive test for sulphide is likely to be positive even for those species which form very small amounts of sulphide. A test of *low* sensitivity can distinguish between those species which form negligible or small amounts of sulphide and those which form large amounts. In one form of test, the organism is stab-inoculated into a tube of solid, gelatin-based medium containing peptone and a low concentration of ferrous chloride; organisms which form a lot of sulphide form visible amounts of black ferrous sulphide. In a more sensitive test, a strip of lead acetate paper is placed above the medium on/in which the test organism is growing; hydrogen sulphide production is indicated by the formation of lead sulphide, which causes blackening of the strip.

16.1.2.7 The urease test

Ureases are enzymes which hydrolyse urea, $(NH_2)_2.CO$, to carbon dioxide and ammonia. Urease production in enterobacteria can be detected by culture on e.g. *Christensen's urea agar*: a phosphate-buffered medium containing glucose, peptone, urea, and the pH indicator phenol red (yellow at pH 6.8, red at pH 8.4); when grown on this medium, 'urease-positive' strains liberate ammonia which raises the pH and causes the pH indicator to turn red. Urease-positive bacteria include e.g. *Klebsiella pneumoniae* and *Helicobacter pylori*.

16.1.2.8 Decarboxylase tests

These tests detect the ability of an organism to form specific *decarboxylases* – enzymes which decarboxylate the amino acids arginine, lysine and ornithine to agmatine, cadaverine and putrescine, respectively. Three tubes of *Møller's decarboxylase broth*, each including glucose, peptone, one of the amino acids, and the pH indicators bromcresol purple and cresol red, are inoculated with the test organism; each broth is covered with a layer of sterile paraffin (to exclude air), incubated at 37°C, and examined daily for 4 days. Initially the medium becomes acidic (yellow) due to glucose metabolism; if a decarboxylase is *not* formed the medium remains yellow. Decarboxylation of the amino acid produces an alkaline product which subsequently raises the pH, causing the medium to become purple. A *control* medium resembles the test medium but lacks an amino acid; it should become, and remain, yellow.

16.1.2.9 The phenylalanine deaminase test (PPA test)

This test detects the ability of an organism to deaminate phenylalanine to phenylpyruvic acid (PPA). The organism is grown overnight on phenylalanine agar (containing yeast extract, Na_2HPO_4, sodium chloride and DL-phenylalanine); 0.2 ml of a 10% solution of ferric chloride is then added to the growth. PPA, if present, gives a green coloration with the ferric chloride. PPA-positive bacteria include e.g. *Proteus vulgaris*.

16.1.2.10 The ONPG test

The utilization of lactose often involves two enzymes: (i) a galactoside 'permease' (which facilitates the uptake of lactose), and (ii) β-D-galactosidase (which splits lactose into glucose and galactose); species such as *E. coli* can usually synthesize both of these enzymes. Certain bacteria which do not utilize lactose, or which metabolize it very slowly, may nevertheless form the enzyme β-D-galactosidase; the inability of such organisms to carry out normal lactose metabolism may be due, for example, to an inability to synthesize galactoside permease. To detect the presence of β-D-galactosidase in such organisms, use is made of a substance, o-nitrophenyl-β-galactopyranoside (ONPG), which can enter the cell without a specific permease; once inside the cell, ONPG is cleaved by the galactosidase to galactose and the yellow-coloured o-nitrophenol. In the ONPG test, the organism is grown for 18–24 hours in broth containing ONPG; a positive test (β-D-galactosidase present) is indicated by the appearance of the (yellow) o-nitrophenol in the medium.

16.1.2.11   The phosphatase test

Phosphatases, enzymes which hydrolyse organic phosphates, are produced by a number of bacteria and can be detected by the phosphatase test. The organism is grown for 18–24 hours on a solid medium which includes sodium phenolphthalein diphosphate; this substance is hydrolysed by phosphatases with the liberation of phenolphthalein – a pH indicator which is colourless at pH 8.3 and red at pH 10.0. To detect phenolphthalein (a positive test), the culture is exposed to gaseous ammonia, which causes phosphatase-containing colonies to turn red.

16.1.2.12   The nitrate reduction test

This test detects the ability of an organism to reduce nitrate (see e.g. anaerobic respiration, section 5.1.1.2). The following test can be used e.g. for enterobacteria and pseudomonads. The organism is grown for one or more days in nitrate broth (e.g. peptone water containing 0.1–0.2% w/v potassium nitrate), and the medium is then examined for evidence of nitrate reduction. To test for *nitrite*, 0.5 ml of 'nitrite reagent A' and 0.5 ml of 'nitrite reagent B' are added to the culture; these reagents combine with any nitrite present to form a soluble red azo dye. The *absence* of red coloration could mean that either (i) nitrate had not been reduced, or (ii) nitrite was formed but had been subsequently reduced e.g. to nitrogen or ammonia. To distinguish between these two possibilities, the medium is tested for the presence of nitrate by adding a trace of zinc dust – which reduces nitrate to nitrite; if nitrate is present (i.e. it has not been reduced by the test organism), the addition of zinc will bring about a red coloration since the newly-formed nitrite will react with the reagents present in the medium.

16.1.2.13   Reactions in litmus milk

Many species of bacteria give characteristic reactions when they grow in *litmus milk* (skim-milk containing the pH indicator litmus). A given strain of bacterium may produce one or more of the following effects: (i) no visible change; (ii) acid production from the milk sugar (lactose) indicated by the litmus; (iii) alkali production, usually due to hydrolysis of the milk protein (casein); (iv) reduction (decolorization) of the litmus; (v) the production of an acid clot, which is soluble in alkali; (vi) the formation of a clot at or near pH 7 due to the action of rennin-like enzymes produced by the bacteria; (vii) the production of acid *and* gas which may give rise to a *stormy clot*: a clot which has been disturbed and is permeated by bubbles of gas.

16.1.2.14   *Aesculin hydrolysis*

Some bacteria can hydrolyse aesculin (the 6-β-D-glucosyl derivative of 6,7-dihydroxycoumarin); hydrolysis releases 6,7-dihydroxycoumarin – detected by the formation of a brown coloration with soluble ferric salts. In one form of the test, the organism is grown on an agar-based medium which includes aesculin (0.1%) and ferric chloride (0.05%); a brown coloration indicates a positive test. Organisms which hydrolyse aesculin include e.g. strains of *Bacteroides*, *Enterococcus* and *Streptococcus*, and *Listeria monocytogenes*.

16.1.2.15   *The MUG test for* Escherichia coli

The MUG test helps to detect *E. coli* e.g. in cultures. A reagent, 4-methyl-umbelliferyl-β-D-glucuronide (MUG), is added to the culture medium prior to inoculation. Most strains of *E. coli* contain the enzyme β-glucuronidase which cleaves MUG to a fluorescent compound, 4-methylumbelliferone; this compound (and, hence, the presumptive presence of *E. coli*) is detected as green-blue fluorescence when the culture is exposed to ultraviolet radiation of wavelength 366 nm. (Note that external sources of glucuronidase – found e.g. in shellfish samples – can lead to false-positive indications.) An alternative method is to smear a colony onto a filter paper which has been impregnated with MUG and examine it under ultraviolet radiation.

16.1.3   Micromethods (= multitest systems)

In clinical bacteriology, *micromethods* are miniaturized test procedures which are used to carry out, simultaneously, a range of routine biochemical identification tests on a given organism. These procedures involve the use of commercial 'kits' which save time, space and materials; a few (of many types) are mentioned below.

The *API* system consists of a plastic strip holding a number of microtubes, each containing a different dehydrated medium. Each microtube is inoculated from a suspension of the test organism; where necessary, mineral oil is added to particular microtubes (to exclude air) and the strip is then incubated. Later, reagents are added, where appropriate, to detect particular metabolic products. Separate API test systems are used for enterobacteria, for streptococci, and for anaerobic bacteria (as different tests are appropriate to each group of organisms).

Staph-Ident resembles the *API* system but contains tests that are particularly appropriate for staphylococci.

Enterotube II consists of a tube divided into a sequence of 12 compartments, each containing a different agar-based gel medium. The media are inoculated by passing an inoculum-bearing straight wire axially through the tube.

The *PathoTec* system consists of various test strips, each impregnated with a dehydrated medium appropriate to a given test. Each strip is incubated in a suspension of the test organism (or inoculated directly from a colony) and the test is read after a specified time.

The *Rapidec Staph* test can detect coagulase-positive staphylococci by detecting an enzyme, aurease, which is specific to these bacteria. Apparently, aurease and a test reagent, prothrombin, form a complex that can lyse a substrate used in the test; such lysis produces a compound which is detected by its fluorescence under ultraviolet radiation.

16.1.4 Other observations and tests

16.1.4.1 Haemolysis

When certain bacteria grow on *blood agar* (section 14.2.1), each colony is surrounded by a 'halo' of differentiated medium in which the erythrocytes have been lysed or in which the blood has been discoloured; the lysis of erythrocytes (*haemolysis*) is due to the action of substances (*haemolysins*) released by the bacteria. Some species produce glass-clear, colourless haemolysis which contrasts sharply with the opaque red medium; this is formed e.g. by *Streptococcus pyogenes* and by some strains of *Staphylococcus aureus*. (When formed by *Streptococcus*, glass-clear haemolysis is often called β-haemolysis, but when formed by *S. aureus* it may be referred to as α-haemolysis; it is probably best to refer to this type of haemolysis simply as 'clear haemolysis'.)

The so-called *viridans streptococci*, and some other bacteria, form zones of greenish-brown discoloration (*greening*) around their colonies.

For any given haemolytic bacterium, haemolysis – or a particular form of it – may develop only if the organism has been grown on media containing the blood of specific type(s) of animal (e.g. horse, rabbit, man etc.)

16.1.4.2 Elek plate

This method detects the production of a diffusible toxin by an organism growing on an agar medium. A plate is inoculated in a single, straight line with the organism being tested. The inoculated line is then overlaid, at right angles, with a strip of paper impregnated with the relevant *antitoxin* (i.e. antibody to the toxin); the inoculated line and paper strip thus form a cross. After incubation, and growth of the organism, toxin production is indicated by the formation of a line of whitish precipitate which approximately bisects each of the four right-angles bounded by the paper strip and the line of microbial growth; the precipitate results from the combination of toxin and antitoxin following their diffusion from the line of growth and the paper strip, respectively. Known positive and negative organisms (i.e. toxin- and non-toxin-producers) are used as controls.

This method has been used e.g. to detect toxin formation by strains of *Corynebacterium diphtheriae*.

16.1.5   Typing

'Typing' means (i) matching an unknown strain with a specific known strain of a given species (a form of identification), *and* (ii) distinguishing between different strains of a given species (a form of classification); *these two procedures use similar criteria (and methods)*. Clearly, (ii) must precede (i).

In sense (i), it should be appreciated that no system of typing can prove that a given unknown strain is *totally* identical to a particular known strain – although it can indicate *non*-identity; this is because any given typing system is designed to reveal similarity, or otherwise, in respect of only one (or a few) characteristics – so that even if an unknown strain is identical to a known strain in *these* characteristics it may well differ in others. Hence, for a given unknown strain, different typing systems may give different results – they frequently do.

Typing is particularly useful e.g. in epidemiology. The essential premise here is that various isolates of a given pathogen from within the same *chain of infection*, or *outbreak of disease*, will be progeny derived from the same ancestral cell; such a clonal relationship will be detectable by the high degree of similarity of the genotypes and/or phenotypes of the isolates – as compared with other, randomly acquired isolates of the same species. Thus, if different strains of a pathogen have been stably distinguished (typed), we can often match a fresh isolate of the pathogen with one of the known strains; this may enable a particular case or outbreak of the relevant disease to be linked with a particular source of infection (e.g. by noting the prevalence of particular strains in particular geographical areas). [Evaluation and use of epidemiological typing systems: Struelens *et al.* (1996) CMI 2 2–11.]

Typing may be based on e.g. serology (section 16.1.5.1); susceptibility to phages (16.1.5.2); electrophoresis of enzymes (16.1.5.3); nucleic acid sequences (16.1.6); and (in *Pseudomonas aeruginosa*) variability in *pyoverdin* (section 11.5.5) [Meyer *et al.* (1997) Microbiology *143* 35–43].

16.1.5.1 Serological tests

These tests can distinguish between closely related bacteria by detecting differences in their cell-surface *antigens* (section 11.4.2.1). For example, using serological tests, thousands of different strains of *Salmonella* can be distinguished primarily by slight chemical differences in their O antigens (lipopolysaccharide–protein antigens) and H antigens (flagellar antigens); differences in the O antigens, for example, occur in the O-specific chains (section 2.2.9.2). Strains which are distinguished mainly on the basis of their antigens are called *serotypes* (see e.g. *Salmonella* in the Appendix).

In practice, we can detect (and identify) a given serotype by using specific *antibodies* which are known to combine only with the antigens of known serotype(s). Antibodies can be obtained by injecting an experimental animal with antigens from a known serotype; after an interval of time, the animal's serum will contain antibodies to those antigens. Such a serum is called a specific *antiserum*. On mixing this antiserum with cells of the given serotype, combination will occur between cell-surface antigens and their corresponding antibodies in the antiserum; when this happens, the bacterium–antibody complexes commonly form a visible whitish suspended mass, or a sediment, in the test-tube (an *agglutination reaction*). If this same antiserum can agglutinate an unidentified strain, it may be concluded that the unknown strain has antigen(s) in common with the original serotype. Hence, an unknown serotype can be identified by testing it with each of a range of antisera, each antiserum containing antibodies of particular, known serotype(s).

Rapid results can be obtained with *latex slide-agglutination tests* in which the reagent consists of minute (microscopic) particles of latex coated with specific antibodies (whose antigen-binding sites face outwards). Bacteria with the matching cell-surface antigens will bind these particles to each other, causing them to agglutinate into visible clumps; hence, clumping indicates a positive test. One example is the *E. coli* O157 latex test (product DR620, Oxoid, Basingstoke, UK).

16.1.5.2 Bacteriophage typing ('phage typing')

This procedure distinguishes between different strains of closely related bacteria by exploiting differences in their susceptibility to a range of bacteriophages (Chapter 9). A *flood plate* (section 14.5.2) is prepared from a culture of a given strain; the plate is 'dried' (section 14.2.2), and a grid is drawn on the base of the Petri dish. Next, the agar over each square of the grid is inoculated with one drop of a suspension of phage – each square being inoculated with a different phage; the drops are allowed to dry and the plate is incubated. Usually, one, two or more of the phages will be found to be lytic for a given strain; lysis (susceptibility to a given phage) is indicated on the lawn plate by the formation of a *plaque* (section 9.1.3) at the point of inoculation of each phage particle. In this way, strains can be defined (and identified) by the range of phages to which they are susceptible. Phage typing is used e.g. for *Staphylococcus aureus* and *Yersinia enterocolitica*.

16.1.5.3   *Multilocus enzyme electrophoresis (MEE)*

In this method, strains are distinguished by comparing the electrophoretic mobilities of a range of enzymes from one organism with the electrophoretic mobilities of equivalent enzymes from one or more closely related organisms. Enzymes must be isolated and tested under conditions in which their activity is retained. The enzymes can be identified in gels by the use of specific colour-generating substrates.

16.1.6   Rapid detection/identification and typing with nucleic-acid-based methods

16.1.6.1   *Rapid detection/identification methods*

Given very specific primers, PCR (section 8.5.4) may permit simultaneous detection and identification of a given organism in clinical or other material – rapidly, and without the need for culture or for any traditional tests; this approach is particularly useful for the early diagnosis of certain diseases (see e.g. section 8.5.4.1, item 4). Theoretically, identification is possible if the sample contains only a single accessible copy of the given chromosomal template.

Species- or strain-specific *probes* (section 8.5.3) are an alternative to PCR-based investigation. (See also section 11.7.5.)

Mycobacterium spp can be detected in clinical samples by a (non-PCR) procedure based on NASBA (section 8.5.9.2); this procedure (Gen-Probe) amplifies a specific sequence within the rRNA and detects the amplified product with a (chemiluminescent) probe. [Evaluation of the Gen-Probe Amplified *Mycobacterium tuberculosis* Direct Test: Miller, Hernandez & Cleary (1994) JCM 32 393–397.] The Gen-Probe test has recently been approved by the US Food and Drug Administration for use with *smear-positive* specimens, i.e. specimens from which a smear containing AFB (section 14.9.2) can be prepared [see e.g. Catanzaro (1996) Lancet 347 1500–1501].

16.1.6.2   *Typing with nucleic-acid-based methods*

Typing (section 16.1.5) may involve e.g. DNA fingerprinting/ribotyping, PCR-based methods or RFLP analysis (sections 16.2.2.3–16.2.2.6).

Several PCR-based methods can be used without prior knowledge of the genome; some (of many) examples from the literature include AP-PCR (section 16.2.2.4) used for distinguishing strains of *Borrelia burgdorferi* [Welsh *et al.* (1992) IJSB *42* 370–377], and RAPD (section 16.2.2.4) used for typing *Staphylococcus aureus* [Young *et al.* (1994) LAM *18* 86–89] and *Campylobacter* spp [Madden, Moran & Scates (1996) LAM *23* 167–170].

REAP (restriction endonuclease analysis of plasmid DNA) types bacteria on the basis of their plasmids; it involves restriction and electrophoresis of the isolated plasmids. Simple electrophoresis (without restriction) can be unreliable owing e.g. to the different electrophoretic behaviour of supercoiled and other forms of the same plasmid, and the possibility of polymeric forms.

Traditional and molecular methods for typing *Staphylococcus aureus* have been compared by Tenover and colleagues [JCM (1994) *32* 407–415].

The use of nucleic-acid-based typing in epidemiological studies has been called *molecular epidemiology*.

16.2   THE CLASSIFICATION (TAXONOMY) OF PROKARYOTES

Ideally, biological (taxonomic) classification should be *phyletic*, i.e. based on the natural (evolutionary) relationships between organisms. This is traditional in higher organisms – in which evolutionary relationships can be deduced from structural and other features. In prokaryotes, however, the 'simple' structure offers too few clues for phyletic classification, and these organisms have been classified traditionally on the basis of observable (phenotypic) characteristics of the type described in section 16.1.

The modern, phyletic classification of prokaryotes is based on *molecular* criteria: organisms are classified primarily according to differences (and similarities) in their nucleic acids; this approach is discussed in the following pages. (It's appropriate to mention here that there is also a 'consensus' form of taxonomy – *polyphasic taxonomy* – which aims to base classification on the *maximum* amount of information available, including both phenotypic and molecular data [polyphasic taxonomy: Vandamme *et al.* (1996) MR *60* 407–438].) Before considering the 'new taxonomy' we look briefly at an earlier molecular criterion: the GC ratio.

16.2.1   The GC ratio (GC%) of DNA

In DNA, the GC ratio (= GC%) is the amount of guanine and cytosine as a percentage of total nitrogen bases:

(guanine + cytosine)/(guanine + cytosine + adenine + thymine) × 100%

The GC% of a sample of DNA can be estimated by various physical methods, e.g. by measuring its density in a *density gradient column* (section 8.5.1.4). GC% has been used as a crude yardstick for comparing DNA from different organisms; however, similar values do not necessarily mean a close taxonomic relationship, though widely differing values suggest the absence of such a relationship. Among bacteria, GC% values range from 24 to 76. Some GC% values are given in the Appendix.

16.2.2   Sequences of nucleotides: memory in molecules

In heredity (section 7.1), the features of an organism are maintained from generation to generation because the chromosome is faithfully copied at each cell division. However, during long (evolutionary) periods of time, new sequences of nucleotides – and hence, new organisms – emerge from pre-existing ones. Thus, a chromosome contains more than just the blueprint of an organism: it contains details of history and development which may be revealed when sequences from different organisms are compared. Organisms have therefore been classified according to the unique sequences of nucleotides in their chromosomes.

If we examine a specific nucleotide sequence – of either DNA or RNA – then differences in the corresponding sequence in different organisms may indicate major evolutionary divergence between the organisms (e.g. at kingdom level) or perhaps minor variation (e.g. at species level); whether a major or minor taxonomic divergence is indicated depends e.g. on the particular sequence examined and on the apparent relative stability of that sequence during evolution – differences in a stable sequence seeming to be more significant than those in a less stable sequence.

How do we compare the nucleotide sequences of different cells? Commonly, the cells are initially *lysed* (broken open) and their nucleic acids isolated by suitable techniques (section 8.5.1.4). In some cases we isolate DNA, in others RNA. Sequences are then compared by methods such as those discussed in the following sections and outlined in Figs 16.4 and 16.6.

Since classification necessarily involves characterization and comparison of organisms, certain of these methods are also useful in identification; methods which distinguish between related strains are particularly useful for typing (section 16.1.5).

16.2.2.1   *DNA–DNA hybridization*

In this method, DNA from two different organisms is compared by measuring the extent to which the two samples can hybridize (*hybridization* referring to base-pairing between strands from different organisms); the greater the degree of hybridization the closer the relationship between the two organisms (Fig. 16.4a). The method is useful e.g. for investigating relationships at the species level; organisms whose DNA shows >70% stable hybridization would be classified in the same species.

Results from DNA–DNA hybridization (and from other sequence-based methods) can differ markedly from earlier, traditional classifications. For example, in a traditional classification of *Listeria*, the two species *L. grayi* and

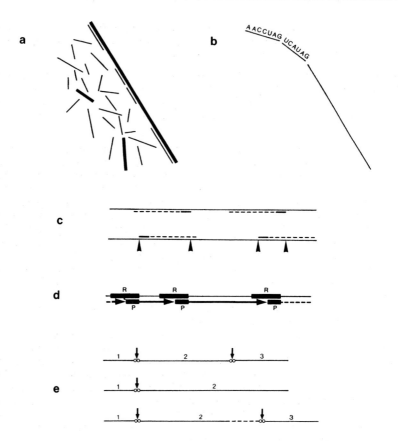

Fig. 16.4 Some methods for comparing sequences of nucleotides from different organisms (diagrammatic, to show principles).

(a) DNA–DNA hybridization (section 16.2.2.1). In one form of this method, single-stranded (heat-denatured) chromosomal DNA from strain A is bound to a sheet of nitrocellulose (or other material); in the diagram this DNA is shown as a long thick line. Chromosomal DNA from strain B is broken into fragments, by suitable methods, and the fragments are denatured by heating; similar, single-stranded fragments are also made from *labelled* DNA of strain A. Under suitable conditions, the *un*labelled DNA of strain A (bound to nitrocellulose) is then exposed to the fragments of strain B (short thin lines) and (simultaneously) to the (labelled) fragments of strain A (short thick lines); this allows fragments to *hybridize* (by base-pairing) with complementary sequences in the bound DNA.

The fragments from strains A and B compete for sequences on the bound DNA. However, the concentration of B fragments is *very* much higher than that of A fragments; consequently, if strains A and B are very similar, few (if any) A fragments will hybridize. Conversely, many A fragments will hybridize if strains A and B are very different. Thus, the similarity of strains A and B is indicated by the number of A fragments which hybridize – this being determined by measuring the label after all unbound fragments have been washed away. Clearly, 'controls' are necessary; in one control, only labelled A fragments are used (no B fragments) – this giving a measure of the maximum amount of binding by A fragments.

Results are meaningful only when the fragments hybridize *stably*; stability is subsequently assessed by monitoring the dissociation of fragments from the bound DNA as the temperature is gradually increased.

(b) 16S rRNA oligonucleotide cataloguing (section 16.2.2.2). In this method, 16S rRNA is cleaved by particular enzymes – e.g. RNase T1, which cleaves RNA specifically at Gp↓N (where Gp is guanosine 3'-monophosphate, and N is the next nucleotide); the resulting fragments form the 'catalogue'. Strains are compared by comparing their catalogues.

(c) Arbitrarily primed PCR (AP-PCR) (section 16.2.2.4). Copies of an arbitrarily chosen primer bind, at various locations, to each strand of (heat-denatured) chromosomal DNA under *low*-stringency conditions (section 8.5.4); the primers bind at 'best-fit' sequences, albeit with mismatches. In some cases, two primers will bind relatively efficiently, on opposite strands, at locations a few hundred bases apart; if strand elongation can occur efficiently from these primers, and if elongation is time-limited, the resulting PCR products will be two short single-stranded fragments of DNA.

In the diagram, the two strands of chromosomal DNA are shown as long parallel lines. On the right, two primers (short lines) have bound, close together, on opposite strands; on the left, another two primers have bound, further apart, on opposite strands. Strand elongation from each primer (dashed line) has produced the four fragments shown. (Note that a fragment synthesized on one strand contains a copy of the best-fit sequence of the *other* strand.) Another cycle of low-stringency PCR is used to produce more copies of the fragments.

Subsequently, many cycles of PCR are carried out under higher stringency, using copies of the same primer. Under these conditions, primers bind to best-fit sequences (rather than elsewhere) on the fragments formed by low-stringency PCR, though (due to the higher stringency) primers may not bind to best-fit sequences on *all* of the fragments – so that only a proportion of the fragments formed under low stringency may be amplified under higher stringency. The distance between each pair of arrowheads shows the length of the fragments amplified under higher stringency. On electrophoresis, the fragments from a given strain form a characteristic pattern of bands (the 'fingerprint') (see e.g. Plate 16.1).

(d) Repetitive sequence-based PCR (REP-PCR) (section 16.2.2.5). In this method, PCR (section 8.5.4) involves primers that bind to REP sequences. In the diagram, three REP sequences (R) are shown in one strand of chromosomal DNA. A primer (P) has bound to each REP sequence, and elongation has proceeded from left to right; an arrowhead marks the end of a newly formed fragment. Note that elongation from a given primer cannot continue beyond the next primer. (The fragments do not join together because the reaction mixture does not contain the type of enzyme (a ligase) which could make such a join.) After a number of cycles of PCR the different-sized molecules are separated by electrophoresis, yielding a fingerprint characteristic of the given chromosome.

(e) Restriction fragment length polymorphism (RFLP) (section 16.2.2.6). Horizontal lines represent related DNA duplexes. The top duplex has two sites for a given restriction endonuclease; enzymic cleavage (arrow) at each of these two sites produces three restriction fragments. In the centre duplex, the second cleavage site has been lost through mutation; enzymic cleavage produces only two fragments, fragment 1 being the same as before. In the lower duplex, a new short sequence of nucleotides (dashed line) has been inserted; enzymic cleavage of this duplex produces three fragments, but fragment 2 is longer than that in the top duplex. Electrophoresis of the fragments from each duplex will give different fingerprints.

L. murrayi were grouped close to the type species, *L. monocytogenes*; however, according to DNA–DNA hybridization, these species are rather distant from *L. monocytogenes* – hybridization between *L. grayi* and *L. monocytogenes*, for example, being <25% [Hartford & Sneath (1993) IJSB 43 26–31]. Why do such results differ? In traditional classification organisms are compared mainly on the basis of *phenotypic* (observable, measurable) features such as enzymes, subcellular structure, motility and metabolic products. By contrast, DNA–DNA hybridization (which examines the *whole* chromosome) compares (i) sequences that encode the phenotypic features, *and* (ii) various sequences in the chromosome which are *not* expressed phenotypically by the cell (e.g. REP sequences – section 16.2.2.5). Thus, compared with traditional methods, hybridization compares organisms on a broader basis.

16.2.2.2 16S rRNA: a record of species and kingdoms

16S rRNA is useful for phyletic classification because (i) it occurs in all bacteria, and (ii) it contains highly conserved (stable) sequences of nucleotides as well as more variable sequences. Feature (ii) permits classification at both ends of the taxonomic spectrum; thus, divergence at the level of higher taxonomic ranks (e.g. kingdom, domain) may be evident when comparing the highly conserved sequences of different organisms, while divergence at the strain or species level may be seen when comparisons are made of the more variable regions.

In *16S rRNA oligonucleotide cataloguing*, 16S rRNA from a given organism is cleaved, enzymatically, into small pieces (oligonucleotides) in each of which the actual sequence is determined (Fig. 16.4b); the resulting 'catalogue' is characteristic of the organism, and different organisms can be compared, and classified, on the basis of their catalogues. By this means, bacteria were originally divided into two kingdoms: the Eubacteria and the Archaebacteria (see Appendix). Subsequently, 16S rRNA was 'sequenced' (i.e. the nucleotide sequence of the whole molecule was determined), and this improved method revealed two main groups of archaebacteria: (i) the sulphur-dependent archaebacteria (e.g. *Sulfolobus*), and (ii) methanogens, extreme halophiles, and the wall-less species *Thermoplasma acidophilum* [Yang, Kaine & Woese (1985) SAM 6 251–256]. Subsequently, Eubacteria and Archaebacteria were each elevated to the taxonomic rank *domain* and were re-named Bacteria and Archaea, respectively. Some groupings and species within these two domains are shown in Fig. 16.5.

16S rRNA analysis is also used at lower taxonomic levels. For example, in a given genus, the 16S rRNAs of different species have been found generally to differ by at least 1.5%. However, certain species of *Bacillus* – which are clearly distinguishable by DNA–DNA hybridization – were shown to have virtually identical 16S rRNAs [Fox, Wisotzkey & Jurtshuk (1992) IJSB 42 166–170]; a possible explanation is that these species have diverged recently (on an

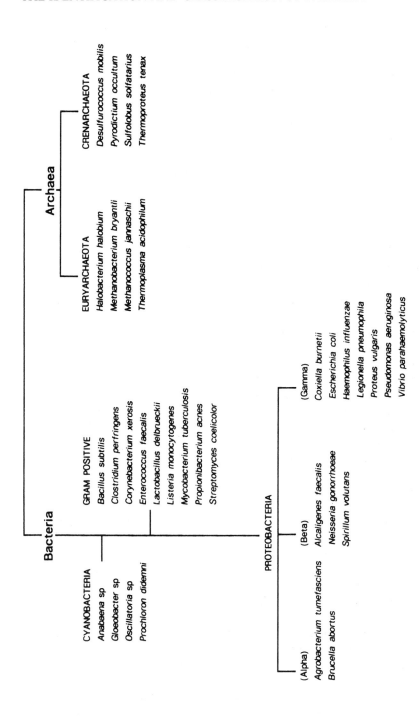

Fig. 16.5 The domains Bacteria and Archaea: examples of species and simplified groupings in each domain. Major bacterial groups not shown include the spirochaetes, the large *Cytophaga/Flexibacter/Bacteroides* group, and the 'delta and epsilon' subdivision of the Proteobacteria (which includes *Helicobacter pylori* and species of *Bdellovibrio* and *Desulfuromonas*); moreover, the Gram-positive bacteria are divided into 'low GC%' and 'high GC%' groups.

evolutionary time-scale) so that their 16S rRNAs have had insufficient time to change.

Many taxonomic studies now use the 16S rRNA *gene* (rather than 16S rRNA itself). Many copies of the gene are needed for analysis, and in some cases they have been obtained by PCR (section 8.5.4). PCR-amplified genes are particularly useful when bacteria cannot be cultured by standard techniques [see e.g. Haygood & Distel (1993) Nature *363* 154–156].

[Bacterial phylogeny based on 16S and 23S rRNA sequence analysis (review): Ludwig & Schleifer (1994) FEMSMR *15* 155–173.]

16.2.2.3 *DNA fingerprinting (chromosomal fingerprinting) and ribotyping*

DNA fingerprinting (= *restriction enzyme analysis*, REA) is used e.g. for identifying/classifying organisms at the species/strain level. In the original method, the chromosomal DNA from a given strain is cleaved by certain restriction endonuclease(s) (section 7.4; Table 8.1) and the fragments of different lengths are separated by gel electrophoresis (section 8.5.1.4); the fragments are denatured (made single-stranded) *in situ* (i.e. within the gel) and then blotted onto a sheet of nitrocellulose or other medium (Fig. 8.13). After blotting, the bands of fragments are stained; for a given organism, the pattern of stained fragments (i.e. the 'fingerprint') reflects the number and location of restriction sites (of the given enzyme(s)) in the chromosome – and is thus characteristic of the organism. Different organisms can therefore be classified/identified by their fingerprints. (Currently, 'fingerprinting' also refers to newer methods, such as AP-PCR (see later), which generate fingerprints in other ways.)

One problem with the original method is that it can yield too many fragments – thus giving a complex fingerprint which is difficult to interpret. One solution is to use a 'rare-cutting' enzyme (Table 8.1); the fewer, larger fragments thus formed can be separated for analysis e.g. by PFGE (section 8.5.1.4).

Another solution is to use a labelled *probe* (section 8.5.3) that binds only to those (few) blotted fragments of the chromosome which contain the sequence complementary to the probe; since only the bands of *labelled* fragments will be made visible, the fingerprint will consist of only a small number of bands. One such probe is labelled rRNA; this binds to those fragments of the chromosome containing the genes for rRNA. Many bacteria (though not e.g. *Mycobacterium tuberculosis*) contain multiple copies of the rRNA genes, so that probe-binding sequences will occur on a small number of chromosomal fragments – thus giving a fingerprint consisting of a small number of bands. This method (using a labelled rRNA probe, or a labelled complementary DNA probe) is called *ribotyping* (Fig. 16.6a). Ribotyping has been used e.g. for molecular epidemiology; in one study it was able to reveal differences among strains of *Legionella pneumophila* serogroup 1, i.e. strains which are identical by routine serotyping [Matsiota-Bernard *et al.* (1994) FEMSIMM *9* 23–28].

16.2.2.4   *Methods using PCR with an arbitrary primer*

Several very similar methods – commonly used e.g. for typing (section 16.1.5) – involve PCR (section 8.5.4) with primers of random (arbitrary) sequence; collectively, these methods have been called *amplification fragment length polymorphism (AFLP) analysis* – but cf. AFLP fingerprinting on page 364.

In these methods, the primer is used to amplify discrete but random sequences of chromosomal DNA; only one type of primer is needed (compare basic PCR, in which two primers are used). Copies of the arbitrary primer bind at various 'best-fit' sequences under appropriate conditions of stringency (section 8.5.4), and the various PCR products are then examined by gel electrophoresis and staining (thus revealing the 'fingerprint' or 'profile' of the organism). Since the primer is arbitrary, we cannot predict which sequences in the chromosome will be copied; however, the method can be used for typing because, under given conditions, the same sequences are copied from the same chromosome when the same primer (and polymerase) are used.

These methods have several advantages for typing: (i) the entire chromosome is potentially available for examination, and (ii) there is no requirement for prior knowledge of the genome – so that, potentially, any isolate can be typed.

AP-PCR (arbitrarily primed PCR), as a representative of these methods, is described in Fig. 16.4c and is compared with the other methods in Table 16.1.

RAPD (random amplified polymorphic DNA) analysis closely resembles AP-PCR, and is considered to be the same method by some authors; it differs e.g. in that it tends to use shorter primers (Table 16.1). When a group of strains is compared by RAPD analysis, it is usual for the comparison to be repeated with each of a range of different primers; different primers give different PCR products (as they bind to different 'best-fit' sequences) so that this procedure increases the chance of detecting differences between strains. RAPD has been used e.g. for molecular epidemiology.

DAF (direct amplification fingerprinting) differs from the other two methods e.g. in that it uses very short primers (Table 16.1).

One problem with these methods is that, although results are reproducible when methods are standardized in a given laboratory, comparable results will not necessarily be obtained by other laboratories unless *identical* procedures are used; comparability of fingerprints depends not only on the primer but also e.g. on the use of a specific polymerase [Schierwater & Ender (1993) NAR

21 4647–4648] and on the procedure used for preparing the sample DNA [Micheli *et al.* (1994) NAR 22 1921–1922].

16.2.2.5 *Repetitive sequence-based PCR*

In this method, PCR is used to generate fingerprints by copying *particular* sequences in the chromosome (rather than random sequences, as in the methods just discussed). The method can be used only with those bacteria (e.g. *Escherichia coli*) whose chromosome contains multiple copies of a specific sequence of nucleotides – such as a *REP sequence*; each REP sequence (repetitive extragenic palindromic sequence) includes a *palindromic sequence* (i.e. a region of nucleic acid containing a pair of inverted repeats – see section 8.3). REP sequences occur in non-coding (phenotypically non-expressed) parts of the chromosome.

In REP-PCR the primers are designed so that they can bind to various forms of REP sequence that occur in different strains. REP-PCR yields DNA molecules of various sizes (Fig. 16.4d) – the differing lengths apparently reflecting differences in the distance between consecutive REP sequences in a given chromosome; when separated by gel electrophoresis, these molecules give a characteristic fingerprint – so that different organisms can be compared/classified by their fingerprints.

16.2.2.6 *Restriction fragment length polymorphism (RFLP) and PCR-RFLP*

Related sequences of nucleotides (e.g. variant forms of a given part of a chromosome) can be compared by exposing them to the same restriction endonuclease(s). Clearly, with given enzyme(s), identical sequences of nucleotides will yield identical fragments. However, different fragments

Table 16.1 Typing by PCR with arbitrary primers: comparison of methods

Method	Primers (typical length, in nucleotides)	Temperature[1] (°C)	Electrophoresis (type of gel used)	Stain/bands[2]
AP-PCR[3]	20–50	40	Agarose	Ethidium bromide/few
RAPD[4]	10–20	36	Agarose	Ethidium bromide/few
DAF[5]	5–8	30	Polyacrylamide	Silver/many

[1] These are examples, only, of the types of temperature used for primer binding, though higher temperatures generally correlate with longer primers.
[2] Ethidium bromide fluoresces under ultraviolet radiation, and this effect is used to detect the bands of PCR product (which are stained with the dye). The bands formed in DAF tend to be very close together; fluorescence from such bands would tend to merge (obscuring the presence of some individual bands) so that silver staining is used instead.
[3] Arbitrarily primed PCR.
[4] Random amplified polymorphic DNA analysis.
[5] Direct amplification fingerprinting.

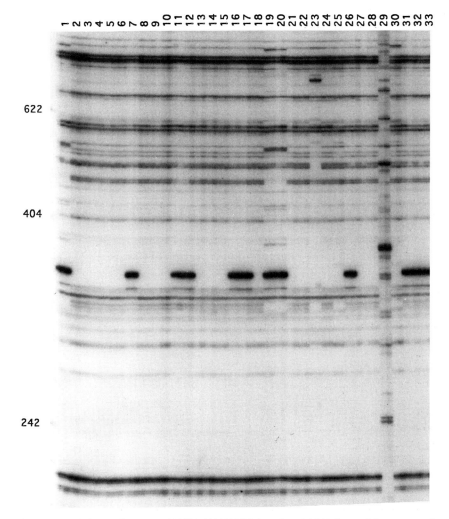

Plate 16.1   Arbitrarily primed PCR (AP-PCR) (section 16.2.2.4) used for comparing and classifying 33 strains of methicillin-resistant *Staphylococcus aureus* from patients in San Diego hospitals. For each strain, the short pieces of DNA (copied from the chromosome) have been separated by gel electrophoresis (moving from top to bottom in the photograph); the scale on the left gives the size of the pieces in terms of the number of bases they contain. (Note that, during electrophoresis, the smaller pieces of DNA have moved further.) One of the strains (29) is obviously different from the rest. Other strains can be grouped together according to shared sequences.

Photograph courtesy of Dr Michael McClelland, California Institute of Biological Research, La Jolla, California, USA.

would be formed if, for example, mutation in one sequence had destroyed (or created) a restriction (cutting) site, or if nucleotide(s) had been inserted or deleted (Fig. 16.4e). Electrophoresis, and staining, of fragments from a given sequence will yield a characteristic fingerprint, so that different sequences can be compared by comparing their fingerprints. (The fragments formed in this method are themselves often called RFLPs.)

RFLP analysis, as described above, is used e.g. for identification/classification at the species/strain level.

In conventional RFLP analysis one of the problems is to prepare a sufficient quantity of the particular sequence. Another potential problem is the effect of methylation (section 7.4) in the DNA isolated from cells: this may affect the activity of the restriction enzymes chosen for the analysis. PCR-RFLP analysis can overcome these problems. A specific target sequence (typically 1–2 kb long) is initially chosen and amplified by PCR; the amplified product (non-methylated, as synthesized *in vitro*) is then subjected to RFLP analysis as described above. PCR-RFLP has been used e.g. for analysing the 23S rRNA of strains of *Campylobacter jejuni* [Iriarte & Owen (1996) LAM *23* 163–166].

16.2.2.7 PCR ribotyping

Conventional ribotyping (section 16.2.2.3; Fig. 16.6a) is rather time-consuming. As an alternative, PCR (section 8.5.4) can be used to look for differences between strains by detecting differences in a specific part of the rRNA-encoding region of the chromosome; this method, PCR ribotyping, and its rationale, are described in Fig. 16.6b. The method has been used e.g. for distinguishing between strains of *Burkholderia* (*Pseudomonas*) *cepacia* [Kostman *et al.* (1992) JCM *30* 2084–2087].

16.2.2.8 AFLP DNA fingerprinting

This method (Fig. 16.7) involves (i) digesting the chromosome with *two* types of restriction enzyme, (ii) adding a short DNA 'adaptor' sequence to both ends of each fragment, and (iii) amplifying, by PCR, *certain* of the fragments. (This is *not* the 'AFLP' mentioned in section 16.2.2.4.)

Restriction fragments are mixed with two types of adaptor molecule (types 'A' and 'B'), each adaptor molecule having *one* sticky end corresponding to one or other of the REs; ligation therefore produces three kinds of sequence: A-fragment-A, A-fragment-B and B-fragment-B.

PCR primers are designed to be complementary to the adaptor molecules (including the restriction site); importantly, the 3' end of each primer extends one (or a few) nucleotide(s) *beyond* the restriction site – so that a given fragment will be amplified only if these terminal 'selective' nucleotide(s) are

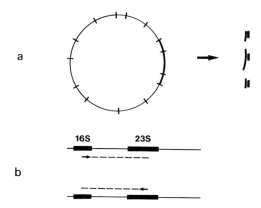

Fig. 16.6 Ribotyping and PCR ribotyping (diagrammatic).
(a) Ribotyping (section 16.2.2.3). Chromosomes from bacteria of a given strain are cut into fragments by a restriction endonuclease; one such chromosome is shown with bars representing the locations of the cuts. The genes which encode rRNAs are shown as a thickened line on the right-hand side (relative length exaggerated for clarity); these genes have been cut into three fragments. All the fragments from the chromosomes are subjected to electrophoresis and blotting and are then exposed to labelled probes; three bands of DNA will bind the probe and will be detected by the probe's label. (The diagram identifies those (three) fragments to which probes (short, thick lines) will bind following blotting and probing.) A radioactive label can be detected by *autoradiography* (for the basic idea see Fig. 8.10). Ribotyping thus reflects the distribution of cutting sites (of the given enzyme) among the rRNA genes of the strain examined. Bacteria of a different strain, with different cutting sites in their rRNA genes, would yield fragments of correspondingly different lengths – detected as bands in different relative positions on the autoradiograph.

(b) PCR ribotyping (section 16.2.2.7). In most or all bacteria the three types of rRNA (section 2.2.3) are encoded by genes which form an operon (section 7.8.1); in e.g. *E. coli*, the order of transcription is: 16S rRNA, 23S rRNA and 5S rRNA. In many bacteria (though not e.g. in *Mycobacterium tuberculosis*) the chromosome contains multiple copies of the rRNA operon. The diagram shows that part of an rRNA operon which encodes the 16S and 23S molecules; the two strands of the duplex are shown, separated, as thin parallel lines, and the *coding* regions of the two genes are shown as thick lines (━━━). The nucleotide sequence between the coding regions is called the *spacer* region. Primers (small arrows) are designed such that they bind to certain highly conserved sequences within the 16S and 23S coding regions; the PCR products formed by elongation (dashed lines) will therefore include the spacer region. The length of the spacer can vary in different copies of the rRNA operon within the same chromosome; hence, after electrophoresis, more than one band may be obtained – though, for a given strain, this will be reproducible. In general, variation in the length of spacer DNA among different isolates can be exploited for typing purposes.

aligned with *complementary* bases in the fragment. To ensure specificity, high-stringency conditions (section 8.5.4) are used in the earlier cycles of PCR. Either primer can be labelled, allowing detection of the bands of PCR-amplified products following electrophoresis.

(a)   ----NNNNG-3'   5'-AATTNNNN-3'
                       ----NNNNCTTAA-5'   NNNN-5'

(b)   ----NNNNGAATTNNNN-3'
                       TCTTAANNNN-5'

Fig. 16.7 AFLP fingerprinting (section 16.2.2.8) (diagrammatic).
(a) On the left is shown one end of a fragment which has been cut by *Eco*RI (Table 8.1).
('N' is a nucleotide, i.e. A, T, C or G.) On the right is one of the two types of adaptor
molecule; this molecule has a sticky end (5'-AATT) corresponding to the *Eco*RI
restriction site.

(b) The fragment-end and the adaptor molecule, shown at (a), have been ligated, and
strand separation has subsequently occurred during PCR. The upper strand has
bound a primer (5'-NNNNAATTCT). The 3'-end of the primer (T) can be extended by
PCR only if it pairs with the complementary base – in this case adenine (A) – in the
fragment strand. The same argument applies to each primer-binding site so that only
certain primers will be extended. PCR products are examined by gel electrophoresis,
the labelled primer forming bands which constitute the fingerprint.

AFLP can differentiate closely related strains and is useful for typing and
epidemiology [Janssen *et al.* (1996) Microbiology *142* 1881–1893]. The method
has been used e.g. for carrying out a taxonomic evaluation of *Bacillus anthracis*
and some related species [Keim *et al.* (1997) JB *179* 818–824].

16.2.3   Epilogue

The classification of prokaryotes is only part of a larger aim: an integrated,
phyletic classification of *all* organisms, both prokaryotic and eukaryotic. Such
'global' taxonomy requires e.g. that we understand the ultimate link between
prokaryotes and eukaryotes – that is, the point at which eukaryotes emerged
from prokaryotic ancestors. Taxonomy in this area has followed two main
approaches: one based e.g. on the sequence and structure of rRNA [Woese,
Kandler & Wheelis (1990) PNAS *87* 4576–4579; Wheelis, Kandler & Woese
(1992) PNAS *89* 2930–2934], and one based e.g. on the structure of certain
proteins [Rivera & Lake (1992) Science *257* 74–76]; these two approaches have
reached different conclusions (Fig. 16.8).

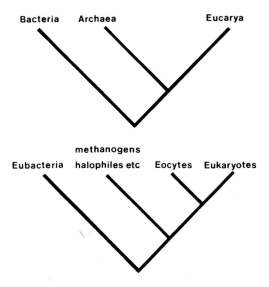

Fig. 16.8 Simplified evolutionary trees illustrating different views on the relationships between major groups of prokaryotes and the origin of eukaryotes; evolutionary distances are not drawn to scale.

The upper tree shows relationships deduced from rRNA sequences and structure [Woese, Kandler & Wheelis (1990) PNAS *87* 4576–4579]; each of the three named groups is a newly created taxonomic category, a *domain*, which is ranked immediately above kingdom. The domain 'Bacteria' contains prokaryotes previously classified in the kingdom Eubacteria (see Appendix). Archaea contains prokaryotes previously classified in the kingdom Archaebacteria. Eucarya includes all eukaryotic organisms. Note that, according to this view, members of the Bacteria are more distantly related to eukaryotes than are members of the Archaea.

The lower tree is based e.g. on a specific sequence of amino acids in a particular protein involved in the process of protein synthesis; this sequence of 11 amino acids occurs in eukaryotes and in some prokaryotes (the eocytes) but not in other prokaryotes [Rivera & Lake (1992) Science *257* 74–76]. Eocytes are thermophilic, sulphur-dependent organisms which include e.g. species of *Desulfurococcus*, *Sulfolobus* and *Thermoproteus* (see Appendix); in the alternative scheme (upper tree) these organisms are included in a separate kingdom (Crenarchaeota – not shown) within the Archaea. One of the arguments supporting the lower tree is that the required amino acid sequence would have had to develop only once (in the common ancestor of the Eocytes and Eukaryotes) whereas two identical events would have had to occur independently in the upper tree. As in the upper tree, members of the Eubacteria (= 'Bacteria') are more distantly related to eukaryotes than are other prokaryotes.

Interestingly, evidence from an entirely different source relates to the period of time *pre-dating* that covered in the diagram. Thus, recent work has shown that the bacterial cell-cycle-associated FtsZ ring (section 3.2.1) also occurs in members of the domain Archaea – indicating that this component of the cell division apparatus occurred in a common prokaryotic ancestor and was retained by both major groups after the evolutionary split [Wang & Lutkenhaus (1996) MM *21* 313–319].

Appendix Minidescriptions of some genera, families, orders and other categories of bacteria

The following minidescriptions give the essential features of many of the bacteria mentioned in the text; they are intended simply for the rapid orientation of the reader. (They are *not* intended to be read like text and memorized!) Further details of these and many other bacteria (and other microorganisms) can be found in *Dictionary of Microbiology and Molecular Biology* [Singleton & Sainsbury (2nd edition, 1987) published by John Wiley, Chichester, UK; ISBN 0–471–94052–6]; entries in the dictionary also include terms, tests, techniques, biochemical pathways, and topics in genetics, molecular biology, medicine and immunology.

The difference between 'Gram-positive' and 'Gram-type-positive' etc. is explained in section 14.9.1.

The GC% (section 16.2.1) gives the range for the genus or other category.

The *type species* is the species which is regarded as the 'permanent representative' of a given genus.

Acetobacter Genus. Gram-type-negative. Ovoid cells/rods, 0.6–0.8 × 1–4 μm. Non-motile or flagellate. Strictly aerobic. Opt. 25–30°C. Chemoorganoheterotrophic. Respiratory. Many strains can oxidize ethanol to acetic acid/CO_2; used in vinegar production. Sugars probably metabolized mainly via the HMP and TCA pathways. GC% 51–65. Type species: *A. aceti.*

Acinetobacter Genus. Gram-type-negative. Rods, 0.9–1.6 × 1.5–2.5 μm. Non-motile. Strictly aerobic. Opt. usually 33–35°C. Chemoorganoheterotrophic. Respiratory. Most strains can grow on minimal salts together with acetate, ethanol or lactate; few strains use glucose. Oxidase −ve. Found in soil, water; opportunist pathogens in man. [Taxonomy and epidemiology: Gerner-Smidt (1995) RMM 6 186–195.] GC% ca. 38–47. Type species: *A. calcoaceticus.*

Actinobacillus Genus. Gram negative. Rods/cocci, ca. 0.3–0.5 × 0.6–1.4μm; non-motile. Facultatively anaerobic. Opt. ca. 37°C. Chemoorganoheterotrophic. Respiratory and fermentative. Complex nutrients needed. No gas from the fermentation of glucose, lactose etc. Found in man, animals; can be pathogenic. GC% 40–43. Type species: *A. lignieresii.*

Actinomyces Genus. Gram-type-positive. Rods or filaments, often branched; non-motile. No spores. Typically anaerobic/microaerophilic. Opt. ca. 37°C. Chemoorganoheterotrophic. Primarily fermentative. Carbohydrates fermented anaerogenically. Found in warm-blooded animals, e.g. in the mouth; can be pathogenic. GC% ca. 57–73. Type species: *A. bovis*.

Actinomycetales Order. Gram-type-positive. Most genera aerobic. Cocci, rods, mycelium (depending on genus); typically non-motile. Spores in many genera. Found in soil, composts, water etc.; some species symbiotic in plants, some pathogenic in man, other animals or plants. Many genera, including *Actinomyces, Arthrobacter, Corynebacterium, Mycobacterium* and *Streptomyces*.

Aeromonas Genus. Gram negative. Rods or coccobacilli, 0.3–1.0 × 1–3.5 μm; singly, pairs, chains, filaments. Some species motile (usually monotrichous); psychrotrophic species non-motile. Facultatively anaerobic. Chemoorganoheterotrophic. Respiratory and fermentative. Sugars, organic acids used as carbon sources. Oxidase+ve. Found in marine and fresh waters; *A. salmonicida* (opt. 22–25°C) is parasitic/pathogenic in fish. There is evidence for a pathogenic role (e.g. diarrhoeal disease, wound infections) in man [Thornley *et al.* (1997) RMM *8* 61–72]. GC% 57–63. Type species: *A. hydrophila*.

Agrobacterium Genus. Gram negative. Rods, 0.6–1 × 1.5–3 μm, capsulated; motile. Aerobic. Opt. 25–28°C. Chemoorganoheterotrophic. Respiratory. Glucose metabolized mainly via Entner–Doudoroff and HMP pathways (section 6.2). Found in soil; most strains can induce tumours in plants, pathogenicity being plasmid-encoded. GC% 57–63. Type species: *A. tumefaciens*.

Alcaligenes Genus (taxonomically unsettled). Gram negative. Rods, coccobacilli, cocci; motile. Aerobic. Chemoorganoheterotrophic, some strains chemolithotrophic. Respiratory. Acetate, lactate, amino acids etc. used as carbon sources. Oxidase +ve. Found in soil, water, vertebrates etc. GC% ca. 56–70. Type species: *A. faecalis*.

Alteromonas Genus. Gram negative. Rods, 0.7–1.5 × 1.8–3 μm, some pigmented (yellow, orange, violet etc.); monotrichously flagellate. Aerobic. Chemoorganoheterotrophic. Respiratory. Carbon sources include acetate, alcohols, amino acids, sugars. Found in marine waters. GC% ca. 38–50. Type species: *A. macleodii*.

Anabaena Genus. Gram-type-negative. Filamentous cyanobacteria (q.v.); trichome: spherical, ovoid or cylindrical cells. Gas vacuoles. Heterocysts. *A. flos-aquae* can form 'blooms' (section 10.1.1), and produce toxins (*anatoxins*).

Aphanizomenon Genus. Gram-type-negative. Filamentous cyanobacteria (q.v.); trichomes: individual cells cylindrical, end cells are often tapering colourless 'hair cells'. Gas vacuoles. Heterocysts. May form 'blooms' (section 10.1.1) in fresh and brackish waters; some strains produce toxins.

Aquaspirillum Genus. Gram negative. Cells: typically helical, rigid, 0.2–1.4 × 2–30 μm (longer in some species). Motile. Aerobic/facultatively anaerobic. Chemoorganoheterotrophic/chemolithoautotrophic. Respiratory. Carbon sources: e.g. amino acids, not usually carbohydrates. Some species (e.g. *A. peregrinum*) can fix nitrogen anaerobically. Typically oxidase +ve. Found in various freshwater habitats. GC% 49–66. Type species: *A. serpens*.

Archaea Domain. See section 1.1.1 and Figs 16.5 and 16.8.

Archaebacteria Kingdom. Differ from Eubacteria (q.v.) (and from eukaryotic organisms) e.g. in nucleotide sequences in 16S rRNA (section 16.2.2.2), and in the chemistry of the cytoplasmic membrane and cell wall (which lacks peptidoglycan). Archaebacteria are often found in 'harsh' environments – many being e.g. extreme thermophiles or halophiles (sections 3.1.4 and 3.1.7). Genera include e.g *Desulfurococcus, Halobacterium, Thermoproteus*. Now re-classified as the domain Archaea.

Azotobacter Genus. Gram negative. Rods/coccobacilli/filaments; motile or non-motile. Cysts (section 4.3.3). Aerobic. Chemoorganoheterotrophic. Carbon sources: e.g. sugars, ethanol. Can fix nitrogen (section 10.3.2.1). Most strains are oxidase +ve. Found e.g. in fertile soils of near-neutral pH. GC% 63–68. Type species: *A. chroococcum*.

Bacillus Genus. Gram-type-positive. Rods, often $0.5–1.5 \times 2–6$ μm, typically motile. Endospores (section 4.3.1). Aerobic or facultatively anaerobic, depending on species. Respiratory or facultatively fermentative. Most species chemoorganoheterotrophic; many can grow on nutrient agar. Some species (e.g. *B. polymyxa*) can fix nitrogen (section 10.3.2.1). *B. schlegelii* can grow chemolithoautotrophically. Found e.g. as saprotrophs in soil and water; some species cause disease in man and other animals (including some insects). GC% 30–70. Type species: *B. subtilis*.

Bacteroides Genus. Gram negative. Rods or filaments (some pigmented), non-motile or motile. Anaerobic. Characteristically fermentative; some strains can carry out anaerobic respiration. Most species use sugars; others use peptones. Found e.g. in the alimentary tract in warm-blooded animals; some species are opportunist pathogens. GC% 28–61. Type species: *B. fragilis*. (As indicated e.g. by the GC% range, the genus is heterogeneous; some believe that it should contain only *B. fragilis* and closely-related organisms [Shah & Collins (1989) IJSB *39* 85–87].)

Bartonella Genus. Gram-negative, oxidase-negative. Rods. *B. bacilliformis* is the causal agent of Oroya fever (section 11.3.3.3). The genus now includes bacteria formerly classified in the genus *Rochalimaea*. The organisms can be grown e.g. on agar media enriched with sheep blood in the presence of 5% CO_2. [*Bartonella* (review): Adal (1995) RMM *6* 155–164.]

Bdellovibrio Genus. Gram negative. Cells: vibrioid, $0.2–0.5 \times 0.5–1.4$ μm, each with one sheathed flagellum. Aerobic. Respiratory. Chemoorganoheterotrophic. Predatory: grow within the periplasmic space, and digest, other bacteria (e.g. *Aquaspirillum serpens*, *Escherichia coli*, *Pseudomonas* spp). Found e.g. in soil and sewage. GC% 33–51. Type species: *B. bacteriovorus*.

Beggiatoa Genus. Gram negative. Trichomes. Aerobic/microaerophilic/anaerobic. Respiratory. Typically chemoorganoheterotrophic; carbon sources: e.g. acetate, but hexoses (e.g. glucose) are not used. Found in various aquatic habitats.

Beijerinckia Genus. Gram negative. Rods, typically $0.5–1.5 \times 1.7–4.5$ μm, motile or non-motile. Aerobic. Respiratory. Chemoorganoheterotrophic; sugars (e.g. glucose) are used as carbon sources. Can fix nitrogen (section 10.3.2.1). Opt. 20–30°C; no growth at 37°C. Found e.g. in soil and on leaf surfaces. GC% 55–61. Type species: *B. indica*.

Bordetella Genus. Gram negative. Coccobacilli, approx. $0.2–0.5 \times 0.5–2$ μm, non-motile or motile. Aerobic. Respiratory. Carbon sources: e.g. amino acids; sugars not used. Enriched media are needed for culture. Found e.g. as parasites/pathogens of the mammalian respiratory tract. GC% 66–70. Type species: *B. pertussis*.

Borrelia Genus (order Spirochaetales – q.v. for basic details). Cells: about $0.2–0.5 \times 3–20$ μm. Anaerobic/microaerophilic. Some species can be grown in complex media. Found as parasites/pathogens in man and other animals. Type species: *B. anserina*.

Brucella Genus. Gram negative. Rods, coccobacilli or coccoid cells, about $0.5–0.7 \times 0.6–1.5$ μm; non-motile. Aerobic. Respiratory. Chemoorganoheterotrophic. Complex, enriched media are needed for culture. Most strains are oxidase +ve; typically urease +ve. Found, typically, as intracellular parasites/pathogens of animals, including man. GC% 55–58. Type species: *B. melitensis*.

Burkholderia Genus. Gram negative. Rods. Aerobic. *B. cepacia* (formerly *Pseudomonas cepacia*) can cause disease in plants and also in man (see section 11.3.2.1). Type species: *B. cepacia*. [*B. cepacia* (medical, taxonomic and ecological issues): Govan, Hughes & Vandamme (1996) JMM *45* 395–407.]

Campylobacter Genus. Gram negative. Cells: spiral, typically $0.2–0.5 \times 0.5–5$ μm (see Plate 12.1); motile, with a single unsheathed flagellum at one or both poles. Microaerophilic, needing 3–5% CO_2 for growth. Respiratory. Chemoorganoheterotrophic; carbon sources: amino acids or TCA cycle intermediates (Fig. 5.10) but not carbohydrates. Oxidase +ve. Found e.g. in the reproductive and intestinal tracts in man and other animals. GC% 30–38. Type species: *C. fetus*. (*C. pylori* has been transferred to a new genus, *Helicobacter* [Goodwin *et al.* (1989) IJSB *39* 397–405].)

Caulobacter Genus. Gram negative. Complex cell cycle (section 4.1). Some strains are pigmented. Aerobic. Respiratory. Chemoorganoheterotrophic. Found in certain soils and waters. GC% 64–67.

Cellulomonas Genus. Gram-type-positive. Rods/filaments/coccoid cells, non-motile or motile. Aerobic/facultatively anaerobic. Respiratory and fermentative. Chemoorganoheterotrophic; starch and cellulose are attacked. Found e.g. in soil. GC% 71–77. Type species: *C. flavigena*.

Chlamydia Genus. Gram negative. Cells: vary according to stage in the developmental cycle, but are non-motile, coccoid, 0.2–1.5 μm in diameter, and pleomorphic. Obligate intracellular parasites/pathogens in man and other animals; have been cultured in laboratory animals, in the yolk sac of chick embryos, and in cell cultures. GC% 41–44. Type species: *C. trachomatis*.

Chlorobium Genus. Gram negative. Rods or vibrios, about 1–2 μm long, non-motile. Obligately anaerobic. Primarily photolithoautotrophic (electron donors: e.g. sulphide). Chlorophyll occurs in *chlorosomes* (section 2.2.7). Found e.g. in sulphide-rich mud.

Clostridium Genus. Gram-type-positive. Cells: typically rods, about $0.3–1.9 \times 2–10$ μm, motile or non-motile. Endospores (section 4.3.1). Obligately anaerobic (or, in a few cases, aerotolerant). Chemoorganoheterotrophic. Typically fermentative, though some strains (of e.g. *C. perfringens*) can carry out nitrate respiration (section 5.1.1.2). Growth often poor in/on basal media. Found e.g. in soil and in the intestines of man and other animals; some species pathogenic. GC% 22–55. Type species: *C. butyricum*. (According to a recent study of 16S rRNA gene sequences, the genus *Clostridium* is very heterogeneous, and it has been proposed that some of the species be placed in five newly created genera: *Caloramator, Filifactor, Moorella, Oxobacter* and *Oxalophagus* [Collins *et al.* (1994) IJSB *44* 812–826].

coliform In general: any Gram-negative, non-sporing, facultatively anaerobic bacillus which can ferment lactose within 48 hours with the formation of acid and gas at 37°C. For water bacteriologists (in the UK): any member of the Enterobacteriaceae which grows at 37°C and which normally possesses the enzyme β-galactosidase [see Report 71 (1994) HMSO, London; ISBN 0 11 753010 7]. *Escherichia coli* is a typical coliform. (See also Table 13.2.)

Corynebacterium Genus. Gram positive. Rods, often curved/pleomorphic, non-motile. Facultatively anaerobic. Chemoorganoheterotrophic. Respiratory and fermentative. Found e.g. in soil and vegetable matter; some species parasitic/pathogenic in man and other animals. GC% 51–59. Type species: *C. diphtheriae*.

Coxiella Genus. Gram negative. Rods (highly pleomorphic), $0.2–0.4 \times 0.4–1$ μm, non-motile. Endospores (section 4.3.1). The sole species, *C. burnetii*, is an obligate intracellular parasite/pathogen in vertebrates and arthropods; it undergoes a developmental cycle. GC% about 43.

cyanobacteria ('blue-green algae') Non-taxonomic category. Photosynthetic bacteria which differ from most other phototrophic prokaryotes (i) in having chlorophyll *a*, and (ii) in carrying out *oxygenic* photosynthesis (section 5.2.1.1) (cf. *Prochloron*). Cells: Gram-type-negative, single or e.g. in trichomes (according to species). No flagella; some exhibit gliding motility (section 2.2.15.1). Some have gas vacuoles (section 2.2.5). Depending on e.g. pigments, cells may appear blue-green, yellowish, red, purple or almost black etc. Some species form akinetes, heterocysts and/or hormogonia (section 4.4).

Typically photolithoautotrophic, fixing CO_2 via the Calvin cycle (section 6.1.1), and respiratory, using oxygen as terminal electron acceptor. Some can grow as chemoorganoheterotrophs, carrying out e.g. anaerobic respiration or even fermentation. Some can carry out facultative *anoxygenic* photosynthesis, using photosystem I (Fig. 5.11) with e.g. sulphide as electron donor.

Found in a wide range of aquatic and terrestrial habitats; some form 'blooms' (section 10.1.1). (See also section 13.5.)

Desulfomonas Genus. Gram negative. Rods, non-motile. Anaerobic. Respiratory: sulphate respiration (section 5.1.1.2) using e.g. pyruvate as electron donor. Found e.g. in the human intestine. GC% 66–67. Type species: *D. pigra*.

Desulfurococcus Genus. Archaea. Cocci, about 1 μm diameter, motile or non-motile. Anaerobic, thermophilic, chemolithoautotrophic and/or chemolithoheterotrophic. Respiratory (sulphur respiration: section 5.1.1.2). Found e.g. in Icelandic solfataras. GC% 51.

Enterobacteriaceae Family. Gram-negative, non-sporing, facultatively anaerobic bacilli, typically 0.3–1 × 1–6 μm; motile (most peritrichously flagellate) or non-motile. Cells occur singly or in pairs. Chemoorganoheterotrophic; typically grow well in/on basal media (section 14.2.1). Carbon sources include sugars. Respiratory *and* fermentative. Oxidase −ve. All except a few strains are catalase +ve. Found e.g. as parasites, pathogens or commensals in man and other animals, and as saprotrophs in soil and water.

Genera (and species) are differentiated e.g. by biochemical tests – particularly IMViC tests (section 16.1.2.5), the urease test (section 16.1.2.7), and the decarboxylase tests (section 16.1.2.8). Typically, lactose is fermented by e.g. *E. coli*, *Klebsiella pneumoniae* and some strains of *Citrobacter*, but not by *Salmonella*, *Shigella*, *Proteus* or *Yersinia*; a strain which normally does not ferment lactose may do so if it acquires a Lac plasmid (which encodes the uptake and metabolism of lactose). Genera include e.g. *Citrobacter*, *Enterobacter*, *Erwinia*, *Escherichia*, *Klebsiella*, *Proteus*, *Salmonella*, *Serratia*, *Shigella* and *Yersinia*.

Enterococcus Genus. See notes under *Streptococcus*.

Eocytes See Fig. 16.8.

Erwinia Genus (family Enterobacteriaceae – q.v. for basic details). Saprotrophic, or pathogenic in plants and animals. Typically motile. Acid (little/no gas) from sugars. Opt. 27–30°C. GC% 50–58. Type species: *E. amylovora*.

Escherichia Genus (family Enterobacteriaceae – q.v. for basic details). The following refers to *E. coli*. Cells: single or in pairs, typically motile (peritrichously flagellate) and fimbriate (section 2.2.14.2); see also items on Plates 2.1, 2.3 and 8.1. Opt. 37°C. Respiratory under aerobic conditions; fermentation (section 5.1.1.1) or e.g. nitrate respiration (section 5.1.1.2) carried out anaerobically. Glucose is fermented (usually with gas) via the mixed acid fermentation (Fig. 5.5). *Typical* reactions as follows. IMViC tests (section 16.1.2.5): +, +, −, −; citrate +ve in strains containing the Cit plasmid (section 7.1); urease −ve; H_2S −ve; lactose +ve (acid and gas). Found e.g. as part of the normal microflora of the intestine in man and other animals; some strains can be pathogenic (Table 11.2). GC% 48–52. Type species: *E. coli*.

Eubacteria Kingdom. Includes prokaryotes not classified in the Archaebacteria (q.v.) – e.g. all the cyanobacteria and the anoxygenic photosynthetic bacteria, all enterobacteria and pseudomonads, and the mycoplasmas. Eubacteria differ from members of the Archaebacteria e.g. in their 16S rRNA (section 16.2.2.2) and in the chemistry of their cytoplasmic membrane and cell wall. The medically important prokaryotes – and those species most likely to be encountered in an introductory course in bacteriology – are eubacteria. Now re-classified as the domain Bacteria (see section 1.1.1).

Francisella Genus. Gram negative. Cocci, coccobacilli or rods (depending on species and conditions), non-motile. Aerobic. Chemoorganoheterotrophic; carbohydrates metabolized slowly, without gas. Opt. 37°C. Oxidase −ve. Catalase weakly +ve.

Found as parasites/pathogens of man and other animals. GC% 33–36. Type species: *F. tularensis* (formerly *Pasteurella tularensis*).

Gardnerella Genus. Gram-type-negative (?). Rods (pleomorphic), about 0.5 × 1.5–2.5 μm. Obligately anaerobic and facultatively anaerobic strains. Chemoorganoheterotrophic; growth occurs only on enriched media. Oxidase –ve. Catalase –ve. Opt. 35–37°C. Found in the human genital/urinary tract; (see vaginosis, section 11.10). GC% about 42–44. Type species: *G. vaginalis* (formerly *Haemophilus vaginalis*).

Haemophilus Genus. Gram negative. Rods/coccobacilli (pleomorphic), often about 0.4 × 1–2 μm, or filaments; non-motile. Facultatively anaerobic. Respiratory and fermentative. Chemoorganoheterotrophic; growth occurs on enriched media, e.g. chocolate agar (section 14.2.1). Typically, glucose, but not lactose, is fermented. Aerobic growth in *H. influenzae* needs X factor (haemin) and V factor (NAD) – both found in lysed RBCs. Opt. 35–37°C. Found as parasites/pathogens in man and other animals. GC% 37–44. Type species: *H. influenzae*.

Halobacterium Genus. Domain Archaea. Rods or filaments, motile or non-motile. Gas vacuoles (section 2.2.5) common. Extremely halophilic. Facultatively anaerobic. Aerobic metabolism: chemoorganoheterotrophic and respiratory, with e.g. amino acids or carbohydrates as carbon sources. Oxidase +ve. Some strains obtain energy from a purple membrane (section 5.2.2). Found e.g. in evapourated brines, salted fish etc. GC% 66–68. Type species: *H. salinarium* (formerly *H. halobium*).

Helicobacter Genus. Gram negative. Cells: helical, motile with several sheathed flagella (section 2.2.14.1, Plate 2.1, *top, left*). Microaerophilic. Chemoorganoheterotrophic. Urease +ve *H. pylori* [see Goodwin *et al.* (1989) IJSB *39* 397–405] is associated with disease of the human gastrointestinal tract (gastritis/peptic ulcer disease). [Prospects for the development of a (therapeutic) vaccine: Telford & Ghiara (1996) Drugs *52* 799–804. Diagnosis and treatment of *H. pylori* infection: Goodwin, Mendall & Northfield (1997) Lancet *349* 265–269.]

Klebsiella Genus (family Enterobacteriaceae – q.v. for basic details). Cells: single, pairs, short chains; capsulated. Non-motile. Often MR –ve, VP +ve. Found e.g. in soil, water, and as parasites/pathogens in man and other animals. GC% 53–58. Type species: *K. pneumoniae*.

Kurthia Genus. Gram positive. Rods or filaments; rods are peritrichously flagellate. Aerobic. Respiratory. Chemoorganoheterotrophic; amino acids, alcohols, fatty acids used as carbon sources. Found e.g. on meat and meat products. GC% about 36–38. Type species: *K. zopfii*.

Lactobacillus Genus. Gram positive. Rods or coccobacilli, singly or in chains; typically non-motile. Anaerobic, microaerophilic or facultatively aerobic; usually catalase –ve. Chemoorganoheterotrophic, using e.g. sugars as carbon sources. Characteristically fermentative, lactic acid being formed from glucose by homolactic fermentation (section 5.1.1.1) or by a heterolactic fermentation in which mixed products, including lactic acid, are formed. Found e.g. on vegetation, as part of the natural microflora in man, and in various fermented food products. GC% about 32–53. Type species: *L. delbrueckii*.

Lactococcus Genus. See notes under *Streptococcus*.

Legionella Genus. Gram negative. Rods/filaments, 0.3–0.9 × 2–>20 μm; motile. Aerobic. Chemoorganoheterotrophic, using amino acids (non-fermentatively) for carbon and energy; growth occurs e.g. on buffered charcoal yeast-extract agar supplemented with L-cysteine. Catalase +ve. Oxidase –ve. Opt. 35–37°C. Found e.g. in various aquatic habitats (such as thermally polluted streams); most/all species can be pathogenic for man. [Taxonomy and typing of legionellae: Harrison & Saunders (1994) RMM *5* 79–90.] Type species: *L. pneumophila*.

Leuconostoc Genus. Gram positive. Cells: coccoid, about 1 μm in diameter, in pairs or chains; non-motile. Facultatively anaerobic. Fermentative and respiratory; anaerobically, glucose is fermented mainly to lactic acid, ethanol and CO_2. Found e.g. in various dairy products and fermented drinks. GC% about 38–44. Type species: *L mesenteroides*.

Listeria Genus. Gram positive. Rods or coccobacilli, about 0.5×0.5–2 μm; usually motile at 25°C, apparently always non-motile at 37°C. Aerobic, facultatively anaerobic. Chemoorganoheterotrophic. Catalase +ve. Oxidase –ve. Urease –ve. Sugars are fermented (acid, no gas). Aesculin is hydrolysed. Growth occurs in up to 10–12% sodium chloride. Found in soil, decaying vegetation, certain foods, and as pathogens in man and other animals. Type species: *L. monocytogenes*.

methanogens Non-taxonomic category; includes all those organisms which can produce methane. All methanogens are obligately anaerobic members of the domain Archaea; they occur e.g. in river mud and in the rumen of cows and other ruminants. The names of some genera are mentioned in section 5.1.2.2.

Methylococcus Genus. Gram negative. Cocci, about 1 μm in diameter; non-motile. Aerobic/microaerophilic. Obligately methylotrophic (section 6.4); methane can be used as sole source of carbon and energy. Found e.g. in mud, soil. GC% about 63. Type species: *M. capsulatus*.

Mycobacterium Genus. Gram positive. Rods, 0.2–0.8×1–10 μm, coccoid forms, branched rods or fragile filaments; some strains pigmented. Non-motile. Acid-fast (section 14.9.2) during at least some stage of growth. Microaerophilic. Respiratory. Typically chemoorganoheterotrophic, though some strains may be chemolithotrophic; typically not nutritionally fastidious, though growth in at least some can be stimulated e.g. by serum or egg-yolk. Found e.g. as free-living saprotrophs in soil and water, or on plants, and as parasites/pathogens of man and other animals. GC% about 62–70. Type species: *M. tuberculosis*.

Mycoplasma Genus. Cells: pleomorphic, ranging from coccoid (about 0.3–0.8 μm in diameter) to branched filamentous forms; some capable of gliding motility (section 2.2.15.1). No cell wall. Facultatively or obligately anaerobic. Chemoorganoheterotrophic; growth occurs on complex media, and all species need cholesterol or related sterols. Catalase –ve. Found as parasites/pathogens e.g. in the respiratory and urogenital tracts in man and other animals. GC% about 23–40. Type species: *M. mycoides*.

Neisseria Genus. Gram-type-negative. Typically cocci, 0.6–1 μm in diameter; non-motile. Aerobic. Chemoorganoheterotrophic; some species need enriched media (e.g. chocolate agar). Oxidase +ve. Found e.g. as parasites/pathogens of man and other animals. GC% about 46–54. Type species: *N. gonorrhoeae*.

Nitrobacter Genus. Gram negative. Rods, about 0.6–0.8×1–2 μm; usually non-motile. Reproduce by budding (section 3.2.2). Obligately aerobic. Some strains obligately chemolithoautotrophic (nitrifying bacteria: section 5.1.2; Fig. 10.2), others facultatively chemoorganoheterotrophic. Opt. 25–30°C. Found e.g. in soil. GC% about 61. Type species: *N. winogradskyi*.

Nitrosococcus Genus. Gram negative. Cocci, about 1.5 μm in diameter; motile or non-motile. Obligately aerobic. Obligate chemolithoautotrophs, oxidizing ammonia to nitrite (section 5.1.2; Fig. 10.2). Found e.g. in soil.

Oscillatoria Genus. Gram negative. Filamentous cyanobacteria (q.v.); trichomes: motile, composed of flattened, disc-shaped cells. Gas vacuoles. Hormogonia. Found in various aquatic and terrestrial habitats. GC% 40–50.

Pasteurella Genus. Gram negative. Cells coccoid, rod-shaped/pleomorphic, about 0.3–1×1–2 μm, occurring singly, in pairs or short chains. Non-motile. Facultatively anaerobic. Chemoorganoheterotrophic. Opt. growth temperature: 37°C. Catalase +ve. Generally oxidase +ve. Found as parasites/pathogens in man and other animals. GC% 40–45. Type species: *P. multocida*.

Pelodictyon Genus. Gram negative. Rods/coccoid forms which may form chains/three-dimensional networks; non-motile. Gas vacuoles. Anaerobic. Phototrophic. Found e.g. in sulphide-rich mud.

Prochloron Genus. Gram negative. Cells contain chlorophylls *a* and *b* and carry out oxygenic photosynthesis. Taxonomy uncertain: prochlorophytes, which include e.g. *P. didemni* (found on warm-water sea-squirts), resemble cyanobacteria in having chlorophyll *a*, but differ in also having chlorophyll *b* and in lacking certain typical cyanobacterial pigments. Possibly related to ancestral chloroplasts. [Photosynthetic machinery in Prochlorophytes: Post & Bullerjahn (1994) FEMSMR *13* 393–414.]

Propionibacterium Genus. Gram positive. Pleomorphic branched/unbranched rods or coccoid forms; non-motile. Anaerobic. Chemoorganoheterotrophic. Fermentative: hexoses (e.g. glucose) or lactate fermented mainly to propionic acid. Growth occurs e.g. on yeast extract–lactate–peptone media. Found e.g. in dairy products. [*P. acnes* (pathogenic potential in man): Eady & Ingham (1994) RMM *5* 163–173.] GC% about 57–67. Type species: *P. freudenreichii*.

Proteus Genus (family Enterobacteriaceae – q.v. for basic details). Motile, often swarming (section 4.2). Typically H_2S +ve, urease +ve. Growth requires nicotinic acid. Found e.g. in soil, polluted waters, and the mammalian intestine; some species (e.g. *P. mirabilis*) can be pathogenic. GC% 38–41. Type species: *P. vulgaris*.

Pseudomonas Genus. Gram negative. Rods, $0.5–1 \times 1.5–5$ μm; most species have one/several unsheathed, typically polar flagella per cell, though *P. mallei* is non-motile (i.e. it lacks flagella), and some species have sheathed flagella (section 2.2.14.1). Aerobic or facultatively anaerobic. Respiratory; many species can carry out nitrate respiration (section 5.1.1.2). Typically chemoorganoheterotrophic and nutritionally highly versatile; many strains will grow on inorganic salts with an organic carbon source, while some can grow chemolithoautotrophically. Catalase +ve. Commonly oxidase +ve. Found e.g. in soil and water, and as pathogens in man, other animals, and plants. GC% 58–70. Type species: *P. aeruginosa*.

Pyrodictium Genus. Domain Archaea. The organisms grow as a network of filaments associated with 'discs', each disc being 0.3–2.5 μm in diameter. Anaerobic. Energy obtained by sulphur-dependent metabolism. Chemolithoautotrophic. Thermophilic (section 3.1.4). Halotolerant. Found in an underwater volcanic region.

Rhizobium Genus. Gram negative. Rods, $0.5–0.9 \times 1.2–3$ μm; motile. Aerobic. Chemoorganoheterotrophic; carbon sources include sugars. Found e.g. in soil and in root nodules (section 10.2.4.1). GC% 59–64. Type species: *R. leguminosarum*.

Rickettsia Genus. Gram negative. Rods, $0.3–0.6 \times 0.8–2$ μm; non-motile. Obligate intracellular parasites/pathogens in vertebrates (including man) and arthropods (ticks, mites etc.). Apparently respiratory, with glutamate as the main energy subtrate; glucose is not used. Opt. 32–35°C. GC% about 29–33. Type species: *R. prowazekii*.

Ruminococcus Genus. Gram-type-positive. Cocci, about 1 μm in diameter, in pairs or chains. Anaerobic. Chemoorganoheterotrophic; typically heterofermentative, forming e.g. acetic and formic acids from carbohydrates. Many strains can use cellulose. Found in the rumen. GC% about 40–45. Type species: *R. flavefaciens*.

Salmonella Genus (family Enterobacteriaceae – q.v. for basic details). Typically motile. *Typical* reactions as follows. IMViC tests (section 16.1.2.5): −, +, −, +; glucose (acid and gas at 37°C) +ve; lactose usually −ve (but the ability to ferment lactose can be plasmid-encoded); H_2S +ve; urease −ve; lysine and ornithine decarboxylases (section 16.1.2.8) +ve. Salmonellae can grow on basal media and e.g. on MacConkey's agar and DCA (Table 14.1); enrichment media include e.g. selenite broth (Table

14.1). Found e.g. as pathogens in man and other animals. GC% 50–52. Type species: *S. choleraesuis*.

Unlike most bacteria, the salmonellae are commonly identified and named as *serotypes* (section 16.1.5.1) rather than as species. In the *Kauffmann–White classification scheme* there are about 2000 named serotypes; each serotype is defined by its O and H antigens (section 16.1.5.1) and, in some serotypes, by the *Vi antigen*: a polysaccharide antigen in a *microcapsule* (section 2.2.11) associated with virulence for particular host(s). Each serotype is given an antigenic formula which lists, in order, the organism's O, Vi (if present) and H antigens; in many serotypes the H antigens can switch, owing to *phase variation* (see Fig. 8.3c), so that the formula of such a serotype includes two alternative H antigens (or two alternative *sets* of H antigens). For example, the antigenic formula of *S. typhimurium* is 1,4,[5],12:i:1,2. This means: O antigens 1, 4, 5 ([ ] indicates variable presence) and $\overline{12}$, phase 1 H antigen 'i', and phase 2 H antigens 1 and 2; O antigen $\underline{1}$ is underlined to show that that antigen is present as a result of *phage conversion* (section 9.4).

Special note. The genus *Salmonella* was named after the American bacteriologist D. E. Salmon. The correct pronunciation of the genus is accordingly 'Salmon-ella'.

Serratia Genus (family Enterobacteriaceae – q.v. for basic details). Usually motile. Some strains form a red pigment, *prodigiosin*. *Typical* reactions: MR –ve; VP +ve (at 30°C, but may be –ve at 37°C); citrate +ve; lactose +ve or –ve, according to species. Glucose is fermented by the Entner–Doudoroff pathway (Fig. 6.2). Found e.g. in soil and water, on plants, and in man and other animals. GC% 52–60. Type species: *S. marcescens*.

Shigella Genus (family Enterobacteriaceae – q.v. for basic details). Non-motile. Sugars fermented usually without gas. MR +ve; VP –ve; citrate –ve; H_2S –ve; lysine decarboxylase –ve. Found e.g. as intestinal pathogens of man and other primates. GC% 49–53. Type species: *S. dysenteriae*.

Simonsiella Genus. Gram negative. Flat, multicellular filaments, the outer face of each terminal cell being rounded (Plate 2.1: *top, right*). Gliding motility. Chemoorganoheterotrophic. Found e.g. in the mouth (human and animal).

Spirochaetales Order. Motile, Gram-negative, typical helical cells which have a characteristic structure (see section 2.2.14.1); 0.1–3 × 5–250 μm, according to species. While most species are assumed to be helical, *Borrelia burgdorferi* swims as a *flattened waveform* – 'axial twists' along the length of the cell causing different sections to lie in different planes. [Structure/motility of spirochaetes: Goldstein, Buttle & Charon (1996) JB *178* 6539–6545.] The spirochaetes include free-living and pathogenic species, anaerobic and aerobic species. Chemoorganoheterotrophic. Respiratory and/or fermentative. Genera include *Borrelia, Leptospira, Spirochaeta* and *Treponema*.

Staphylococcus Genus. Gram positive. Cocci, about 1 μm in diameter, often in clusters, some containing orange or yellow carotenoid pigments; non-motile. Facultatively anaerobic. Chemoorganoheterotrophic. Carbon sources include various sugars. Commonly halotolerant (section 3.1.7). Catalase +ve. The staphylococci are divided into coagulase +ve and coagulase –ve strains (section 16.1.2.2), the former including *S. aureus* and *S. intermedius*, the latter including *S. epidermidis* (formerly *S. albus*). Found e.g. as commensals and pathogens of man and other animals. GC% about 30–39. Type species: *S. aureus*.

Streptococcus Genus. Gram positive. Cocci, typically about 1 μm in diameter, often in pairs or chains. Non-sporing. Capsulation common. Facultatively or strictly anaerobic. Catalase –ve. Chemoorganoheterotrophic. Typically fermentative, sugars being metabolized usually without gas. Found e.g. as commensals and pathogens of man and other animals. GC% 34–46. Type species: *S. pyogenes*.

Streptococci can be classified and identified e.g. by *Lancefield's grouping test*. This involves extraction and identification of certain cell-envelope-associated carbohydrates

('C substances'); typically, only one type of C substance occurs in a given strain. To extract the C substance, a cell suspension in 0.5 ml saline can be autoclaved for 15 minutes at 121°C and the supernatant used for the test. In one form of the test, the extract is layered onto an antiserum (containing antibodies to a given C substance) in a small tube; the test is repeated – with antisera to different C substances – until a positive result is obtained, i.e. a whitish precipitate at the extract–antiserum interface. All strains whose C substance reacts with a given antiserum are placed in the same Lancefield group. Lancefield groups are designated A, B, C . . . etc.; strains of *S. pyogenes* belong to group A.

A number of species, formerly in the genus *Streptococcus*, have been transferred to other genera. For example, *S. faecalis* and *S. faecium* have been transferred to *Enterococcus* as *E. faecalis* and *E. faecium*, respectively [Schleifer & Kilpper-Bälz (1984) IJSB *34* 31–34]. Both species usually grow at 10°C and 45°C, survive 60°C/30 minutes, and can grow in 6% salt (NaCl) and at pH 9.6; they differ e.g. in that *E. faecalis*, but not *E. faecium*, can obtain energy from pyruvate, citrate and malate.

S. lactis has been transferred to *Lactococcus* as *L. lactis* [Validation of the genus *Lactococcus*: (1986) IJSB *36* 354–356]. Lactococci are cocci/coccoid forms which occur singly, in pairs or in chains. Growth occurs at 10°C but not at 45°C. The organisms are fermentative, L(+) lactic acid being the main product from glucose metabolism. Lactococci are found e.g. in dairy products. Type species: *L. lactis*.

Some recent developments in the taxonomy (classification) of streptococci have been reviewed by Hardie & Whiley [(1994) RMM *5* 151–162].

Streptomyces Genus (order Actinomycetales – q.v.). Gram positive. Mycelium (section 2.1.1), part of which fragments to form chains of spores (section 4.3.2; Fig. 4.3). Aerobic. Respiratory. Chemoorganoheterotrophic; carbon sources include glucose, lactate and starch. Antibiotics formed by *Streptomyces* species include chloramphenicol (section 15.4.4) and streptomycin (section 15.4.2). Found e.g. in soil and as pathogens of plants. GC% 69–78. Type species: *S. albus*.

Sulfolobus Genus. Domain Archaea. Cocci, coccoid or irregularly-shaped cells in which the cell wall consists of only an S layer (section 2.2.12). Thermophilic (growth occurs between 50 and 90°C). Acidophilic (section 3.1.5). Aerobic and facultatively anaerobic. Energy is obtained by the (respiratory, aerobic) oxidation of sulphur (or Fe^{2+}) and/or by sulphur respiration (section 5.1.1.2) in which elemental sulphur is used as terminal electron acceptor. Obligately heterotrophic or facultatively autotrophic. Found e.g. in certain hot springs.

Thermoproteus Genus. Domain Archaea. Rods, filaments, about 0.5×1–80 μm. Anaerobic. Energy obtained by sulphur respiration (see also *Sulfolobus*). Thermophilic. Autotrophic and/or heterotrophic (carbon sources include glucose, ethanol, formate). Found e.g. in Icelandic solfataras.

Thiobacillus Genus. Gram negative. Rods, about 0.5×1–3 μm; typically motile. Obligately aerobic or (some) facultatively anaerobic. Respiratory; energy commonly obtained by the oxidation of sulphur and/or reduced sulphur compounds. Obligately or facultatively chemolithoautotrophic. Found e.g. in soil, mud, hot springs. GC% about 50–68. Type species: *T. thioparus*.

Treponema Genus (order Spirochaetales – q.v. for basic details). Cells: 0.1–0.4 × 5–20 μm. Anaerobic or microaerophilic. Some species can be grown in complex media; others (including *T. pallidum*) cannot, and are grown e.g. intratesticularly in rabbits. Found e.g. as parasites/pathogens in man and other animals. GC% 25–54. Type species: *T. pallidum*.

Vibrio Genus. Gram negative. Rods, curved (vibrios) or straight, 0.5–0.8×1.4–2.6 μm; motile, flagella typically sheathed (section 2.2.14.1). Facultatively anaerobic. Typically oxidase +ve. Chemoorganoheterotrophic; glucose is fermented by the mixed acid fermentation (Fig. 5.5), usually without gas. All species can grow at 20°C, most at

30–35°C, and some at 40°C. Some species tolerate high pH (e.g. *V. cholerae* can grow at pH 10). Found e.g. in various aquatic habitats (freshwater, estuarine and marine) and as pathogens in man, fish and shellfish. GC% 38–51. Type species: *V. cholerae*.

Xanthomonas Genus. Gram negative. Rods, 0.4–0.7 × 0.7–1.8 μm, typically containing yellow pigment(s); some strains form extracellular slime. Motile. Aerobic. Chemoorganoheterotrophic. In strains of *X. campestris*, glucose is metabolized e.g. via the Entner–Duodoroff pathway (Fig. 6.2). Oxidase –ve (or weakly +ve). Catalase +ve. Found e.g. as pathogens in plants. GC% 63–71. Type species: *X. campestris*.

Yersinia Genus (family Enterobacteriaceae – q.v. for basic details). Cells: 0.5–0.8 × 1–3 μm. Most species are motile 30°C, non-motile at 37°C; *Y. pestis* is non-motile. Growth occurs on basal media. VP –ve at 37°C (+ve in some species at 25°C); MR +ve; acid (little/no gas) from glucose; lactose typically –ve. Urease –ve (e.g. *Y. pestis*) or +ve (e.g. *Y. enterocolitica*). Opt. growth temperature: 28–29°C; *Y. enterocolitica* is psychrotrophic and can grow at 4°C. Found e.g. as parasites/pathogens in man and other animals. GC% 46–50. Type species: *Y. pestis*.

Zoogloea Genus. Gram negative. Rods, monotrichously flagellated, 1–1.3 × 2.1–3.6 μm, typically in masses in a polysaccharide matrix. Aerobic. Respiratory. Chemoorganoheterotrophic. Oxidase +ve. Catalase +ve. Opt. growth temperature: 28–37°C. Found e.g. in organically polluted freshwater habitats and in aerobic sewage-treatment plants. GC% about 65. Type species: *Z. ramigera*.

Index